LIPIDS

NUTRITION and HEALTH

LIPIDS

NUTRITION and HEALTH

CLAUDE LERAY

CRC Press
Taylor & Francis Group
Boca Raton London New York

CRC Press is an imprint of the
Taylor & Francis Group, an **informa** business

Translated from Les Lipides – Nutrition et Santé

CRC Press
Taylor & Francis Group
6000 Broken Sound Parkway NW, Suite 300
Boca Raton, FL 33487-2742

© 2015 by Taylor & Francis Group, LLC
CRC Press is an imprint of Taylor & Francis Group, an Informa business

No claim to original U.S. Government works

Printed on acid-free paper
Version Date: 20140916

International Standard Book Number-13: 978-1-4822-4231-7 (Paperback)

Visit the Taylor & Francis Web site at
http://www.taylorandfrancis.com

and the CRC Press Web site at
http://www.crcpress.com

Contents

Foreword

The complexity of lipids is one of the most exciting areas in human nutrition. Most neophytes have actually a simplistic view, reducing them to their energy intake, enhanced by prevention messages that have demonized them for a long time: fats make you fat; they "give" cholesterol (even it is ugly and vulgar!), flee! Ideas die hard, and when they are wrong, they can last for several generations. This shows the importance of careful scientific discourse, any simplification or overgeneralization is harmful, as the data changes. Today we know, for example, that the saturated fatty acids are not a homogenous group, as are the polyunsaturated fatty acids. We know that the saturated fatty acids are nonessential but useful and that the polyunsaturated fatty acids exert considerable structural and functional features. We know that omega-3 fatty acids are molecules essential for brain and retina functions very early in life. No fats, no life . . .

Claude Leray accomplished a feat with the data he has collected, encompassing extremely large physiological, nutritional, metabolic, and epidemiological data. One of the advantages of this book is to be able to offer in summary a very broad overview of all aspects of lipids, from the fatty acids to the other forms of fats and vice versa. One of the most exciting and original parts is the historical approach of the great names of "fats," from Chevreul to Claude Bernard, through Mege-Mouries, in the nineteenth century, and then Burr and Hollman and Lynen, through the twentieth century. We must wait for the end of the twenty-first century to choose our best writers!

Today, it is in the field of epidemiology that data is accelerating because, beyond the few true deficiency diseases and neonatal consequences of deficits, research is developing to prevent chronic diseases. Thus, the equilibrium in particular fatty acids through the famous ratio of omega-6/omega-3 not only plays a significant role in the occurrence of coronary heart disease, but also obesity, depression, disorders of the attention deficit and hyperactivity disorder, age-associated cognitive decline, and age-related macular degeneration. But new works now involve fatty acids that have long been regarded as minor, as conjugated fatty acids and *trans* fatty acids.

Finally, the author was not merely limited to the fatty acids; he has expanded his review about the complex lipids, sterols, and fat-soluble vitamins.

With an extensive bibliography, the present work can be considered as one of the best syntheses of both scientific and practical syntheses of lipids and human health. A book that will long be a reference, full of nuances and prudence of true science.

Dr. Jean M. Lecerf
Endocrinology, Nutritionist, and Lipidologist
Institut Pasteur de Lille

Acknowledgment

I sincerely thank Louis Sarliève, INSERM research director, for his attentive and friendly second reading of the French manuscript.

Author

Claude Leray, PhD, earned his PhD in biology from Marseille University, France, in 1968. During and after this first research period, he specialized in fish biology. In 1971, he joined the faculty at the Department of Biology, University of Montréal, Québec, Canada, where he held the rank of associate professor for three years.

In 1974, Dr. Leray joined the National Center for Scientific Research as research director, at the University of Strasbourg, France, where he developed works on fish, focusing on the influence of dietary lipids on the gill and intestine membrane composition, mainly in the context of the animal salinity changes.

He collaborated on several projects including the importance of n-3 fatty acids in the physiology of rat adipose tissue (Ecology and Energetic Physiology Center, National Center for Scientific Research, Strasbourg) and human platelets (Blood Transfusion Center, National Institute of Health and Medical Research, Strasbourg). His main fields of interest are fatty acids and phospholipids.

Dr. Leray has published extensively in the field of lipid and membrane biochemistry. He is the founder and creator of the website: www.cyberlipid.org. He published two books on lipidomics: one in French (*Les lipides dans le monde vivant—Introduction à la lipidomique*, Lavoisier, 2010) and one in English (*Introduction to Lipidomics—From Bacteria to Man*, CRC Press, 2013). Recently, he published a book in French on the relationships between lipids and health (*Les Lipides—Nutrition et Santé*, Lavoisier, 2013).

1 Introduction, History, and Evolution

1.1 GENERAL INFORMATION

Since antiquity until the beginning of the eighteenth century, the notion of lipids was limited to olive oil, at least in the Western world. Toward the end of that period, this concept was extended to the distinction among oils, greases, and waxes. It is only starting from the works of the great French chemist Michel Eugene Chevreul (Figure 1.1), founder of the science of lipids in 1823, when appears a classification of the "fats." They were distributed in six groups defined by their physical property (mainly distillation), an important chemical property (saponification), and also according to the nature of the components of fatty substances. Broadly, this first classification remains the model of those accepted still today.

Let us recall that Chevreul has established for the first time the general structure of fats (oils and greases) as a combination of glycerol and fatty acids, new compounds of which he isolated and characterized several species. Conversely, the hydrolysis of these fats, under the effect of the potash (saponification), enabled him to specify the mechanism of soap formation (Chevreul 1815). This chemical research was completed by the experiments of Claude Bernard (Section 3.2, Figure 3.1) who found in 1849 this same property, but under the effect of the pancreatic juice of dog (Bernard 1849): "pancreatic juice has the property to emulsify simultaneously and in a complete way the neutral fat content, and then to duplicate them into fatty acids and glycerin." This proposal may be considered as the foundation of the physiology of lipids.

That discovery was the first great contribution of Claude Bernard to animal physiology and has shown clearly that the pancreatic juice played "a special and crucial role in digestion." At the same time, he had just discovered the reality of an enzymatic action.

FIGURE 1.1 Michel Eugene Chevreul (1786–1889).

FIGURE 1.2   Marcelin Berthelot (1827–1907).

These fundamental discoveries were confirmed definitively by the works of Marcellin Berthelot (Figure 1.2) in 1854, who carried out for the first time the synthesis of acylglycerides (mono-, di-, and triglycerides) starting from glycerol and fatty acids. That research was first devoted to the synthesis of complex natural molecules in the laboratory.

During nearly one century, the disciples of Chevreul considered the true fats as molecules that can contain only fatty acids. In 1920, the discovery of many phosphorylated or glycosylated lipids encouraged Johann L. Thudichum to propose a classification of "lipoids" into three groups: the simple lipoids (greases and waxes), the complex lipoids (phospholipids and glycolipids), and the derived lipoids (fatty acids, alcohols, and sterols). That classification, slightly modified by more modern authors, still remains of actuality.

1.2   HISTORY OF THE PRODUCTION AND THE USE OF LIPIDS

Apart from their natural ingestion from plants or of animal preys, fats were certainly collected very early by humans in the form of oily paste as soon as they controlled fire (at least 500,000 years ago) (Pick 2005). The sheep fats, lard, butter, and even fish oil were known since prehistoric times. Precise research have shown the presence of lipids originating from milk in potteries dating from the Bronze Age (Copley et al. 2005). Concurrently to these food uses, lipids were also used by our remote ancestors for artistic applications, because fatty acids of vegetable origin have been detected in the binder of the cave paintings of the Magdalenian period (about 12,000 years ago) (Pepe et al. 1991).

Among the first sources of vegetal oils used by man, one has been able to quote sesame, known in the Middle East (Mesopotamia) for more than 60 centuries; soya, mentioned in China 48 centuries ago; rapeseed, known in India for about 40 centuries; olive, located in the Mediterranean zone for more than 37 centuries; and groundnut of Incas, for more than 30 centuries. Mentioned several times in the Bible, quoted in Babylon in 1597 BC, olive oil is the most known model of a crop production, considered at the same time as a source of myths, light, and food. From Minor Asia, it accompanies the Greek, Phoenician, and Roman civilizations typically in their expansions, considering the culture of the olive tree, the extraction of oil, its conservation, or its use. Many years later, during the nineteenth century, the colonial policies of the European countries allowed a diversified use of vegetal oil sources by exploiting already known species but more generous in lipids, such as groundnut, coconut, cottonseed, or soybean. The economic interest of the various sources of food lipids was fluctuating during the previous century, some being stagnant (olive and sesame), some being increasing (soya and palm tree), and others disappearing (whale and fish).

Concurrently to these food uses, it is necessary to stress the importance of lipids in the manufacture of soap. If that production starting from vegetal oils seemed known from the Sumerian time, about 3000 BC, the use of soap was without any doubt developed and spread with the Roman civilization.

1.3 LIPIDS AND HUMAN NUTRITION: HISTORY AND EVOLUTION

Since the most ancient times, man quite naturally consumed lipids with his food, in the form of plants, game, and fish. The progressive improvements of the chemical analysis allowed a more rational knowledge of the main components of food, this leading to an awakening of the importance of the lipids in human consumption.

Thus, it was not until 1827 that the English chemist and doctor William Prout (Figure 1.3) recognized the importance of fats in the animal feeds, besides carbohydrates and proteins.

That fundamental discovery led this famous scientist to receive the Copley medal, the oldest distinction in the field of sciences awarded by the Royal Society of London. The evolution of physiological sciences led in 1912 the chemist and English doctor Frederick G. Hopkins (Figure 1.4) to prove that purified lipids could not ensure the normal growth of young rats because of the lack of some liposoluble factors, actually known as vitamins. For that work he was awarded the Nobel Prize in 1929.

The purely calorigenic aspect of the lipids (heat of combustion) has been established in 1866 by the English chemist Edward Frankland, who demonstrated that lipids were approximately twice as rich in energy than carbohydrates. It is only in 1907 that experiments of human calorimetry carried out by the famous American

FIGURE 1.3 William Prout (1785–1850).

FIGURE 1.4   Frederick G. Hopkins (1861–1947).

FIGURE 1.5   Francis G. Benedict (1870–1957).

physiologist Francis G. Benedict (Figure 1.5) made it possible to establish that the food fats are used to produce muscular energy as effectively as carbohydrates. The basis of nutritional knowledge had just been founded.

The human evolution was marked durably since the lower Paleolithic by the increase in the volume of the skull. Indeed, the cerebral volume, which was of 450 cm³ in *Australopithecus* (about 4 million years ago), rose to 600 cm³ in *Homo habilis* (about 2.5 million years ago), and then to 1100 cm³ in *H. erectus* (about 1.7 million years ago), to reach approximately 1350 cm³ in *H. sapiens*, the current man. If the selection and the ecological pressures undoubtedly supported that slow transformation, the most recent assumptions agreed to stress the importance of a food dietary change that was characterized by an increased intake of *n*-3 polyunsaturated fatty acids of aquatic origin (molluscs and fish) in several areas of the planet (Broadhurst et al. 1998) (Section 3.2.1.3.2).

It has been established that human adapted himself genetically to his diet throughout the Paleolithic Age (from 40,000 to 10,000 years ago). If that mode varied according to the areas and the seasons, one can ensure that it was dictated by the gathering of crop products (fruits, nuts, and roots), game hunting (mammals and birds), and fish catch. From archaeological artifacts, it was thus estimated that the "Paleolithic diet," still praised by certain nutritionists nowadays, did not bring to Cro-Magnons not more than 25% of his energy intake in the form of lipids rich in polyunsaturated fatty acids (Eaton 1992). That situation was modified only at the Neolithic era when humans became farmer and stockbreeder and consumed dairy products and larger quantities of plants in the form of cereals. During that time, the food intake became richer in lipids compensated by a lessening of the protein part, a change probably connected to a reduction in the body size of the individuals. That evolution was further developed, later, at the time of the industrial revolution during which the diet became gradually richer in lipids. These lipids originated from increasingly fatty meat from breeding stocks and from lipids purified from vegetals (various oils), both sources contributing sometimes for more than 40% of the dietary energy. Moreover, the increase in meat consumption is a general law in almost all the industrialized countries and for a certain time even in less-developed countries.

On the qualitative level, the increased consumption of vegetal oils and fatty meat produced from animals fed cereals seriously increased the importance of the n-6 fatty acids compared to the n-3 fatty acids in human tissues. That increase is currently denounced by the most qualified medical authorities (Section 4.2.3.3).

These changes initiated during the modern time, especially after the World War II, are perhaps connected to the new manners of culture and production and to the increased supply of cities with food substances. To that, it is certainly necessary to add the participation of the lipids in the gustatory perception of foods. Indeed, besides the nutritional aspect, one cannot be any more unaware of the importance of lipids in the perception of savors and the improvement of food palatability. This physiological incentive to fat consumption, reinforced by advertisements for increasingly rich products, may only contribute to the increase in the energy overload leading to overweight in modern man.

The modern man having for lipids, as for other nutrients, practically the same metabolic capacities as his Paleolithic ancestors, the evolution of his diet has been certainly at the origin of many pathologies, such as the metabolic syndrome observed nowadays. Let us remember Hippocrates (Figure 1.6) who said more than 25 centuries ago that "To eat more than nature requests, it certainly exposes to several diseases."

FIGURE 1.6 Hippocrate de Cos (460–377 BC).

1.4 LIPIDS AND HEALTH

Hippocrates, considered the father of dietetics, wished that "your food be your first medicine." This proverb was transmitted through generations of doctors having at their disposal only little means of intervention, some powders resulting from the mineral world and extracts of animal tissues and mainly of plants. Finally, over the nineteenth century, the doctors, the chemists, and the physiologists established the bases of the science of nutrition after having explored the labyrinth of the metabolisms. In the twentieth century, the clinicians and the researchers noticed more precisely that nutrition was one of the major factors contributing to the onset of various pathologies, thus modern dietetics was born. If nutrition is not the single cause of these pathologies, it constitutes a supporting factor, essential among other factors belonging to the environment or the genetics. As Hippocrates thought it, doctors must from now on integrate the concept of nutrition in their therapeutic panoply. More especially, if diet represents a risk factor, paradoxically, it can also be a factor of protection and even of cure.

As regards lipids, progress in the knowledge of their beneficial or noxious effects for human health, slowly acquired during the last century, constitutes now an important field leading to recommendations appearing frequently in the media.

Although the concept of optimal proportion of lipids in our diet remains discussed, the main health organizations agreed around a value of 30% of the energy intake, to not endanger our health capital and longevity. This higher limit naturally takes account of a reduced energy expenditure in consequence of the mechanical helps we are currently profiting (car, elevator, etc.). In 2003, the World Health Organization published the nature of the various pathological risks linked to a too fatty diet. These pathologies rise directly from the importance of the lipids in the constitution of cellular membranes, in the energy storage, and in the transport of cholesterol and liposoluble vitamins. Any excess or any imbalance within the lipid ration may target several systems, inducing disorders that lead rapidly to obesity, atherosclerosis, hypertension, immunizing disorders, and even to cancer processes.

Targets of many enzymes, phospholipids are the source of smaller molecules, derived from fatty acids, involved in cellular and intercellular communications. These fatty acids are also essential in the functional modification of proteins, enabling them to become active while passing from the cytoplasmic compartment toward the plasma membrane. They are also essential compounds for the regulation of the functions exerted by other molecules as various peptide hormones, inflammatory neurotransmitters, growth factors, and cytokines. Fatty acids were suspected for a long time to act through their influence on the structure of cellular membranes. Now, they are more commonly considered able to control the expression of specific genes at the nucleus level, thus influencing not only their own metabolism but also other cellular mechanisms.

In several models, even for the brain that remains the more "preserved" organ, it became obvious that the lipid composition of food is able to influence membrane structure, and consequently cellular machinery. It is remarkable, and even worrying, to note that the nature of the ingested lipids appears likely to influence the performance of the nervous system and possibly to alter the behavior and the cognitive functions, this being able to lead to disorders concerning psychiatry. Even if an unquestionable proof is not provided yet, it is not without significance to note that large international health organizations regularly deliver their recommendations on the abundance and the lipid composition of our plate contents. These detailed recommendations relate mainly to the proportion of saturated fatty acids, the value of the n-6 to n-3 fatty acid ratio; the ingested amounts of cholesterol; and vitamins A, D, E, and K. In contrast, the other nutrients, carbohydrates and proteins, are not the objects of so precise recommendations, their excess being in general poorly considered by nutritionists. Nobody worries, at least in our modern societies, of the balance of the protein amino acids present in meats and fish, or of the type of carbohydrate present in the ingested vegetables. None of these food components is suspected to influence health as directly as lipid compounds such as fatty acids, carotenoids, cholesterol, or certain vitamins, to take only well-established examples.

Twenty-four centuries after Hippocrates, the manner is with the "nutraceuticals," but apart from any industrial exploitation guided by the profit, this magic belief that "the diet is our first medicine" curiously is becoming justified in the field of the lipid components. It is remarkable that the most recent research confirm clearly that the human behavior can be influenced by lipids, thus giving credit to that primitive notion of the "principle of incorporation" according to which we are made of the food we eat (Fischler 2001).

After having examined the vegetable and animal sources of the lipids ingested with the human foods (Chapter 2), it will be proceeded to a review of the proven or possible needs of the human being in various fatty acids, sterols, and vitamins belonging to the group of lipids (Chapter 3). The emphasis will be placed on their lower and higher limits of consumption, even if they are already specified in official recommendations or little known for certain lipids. A great place will be reserved for the metabolic and pathological implications linked with a nutritional imbalance concerning the major lipids of the diet, such as certain fatty acids or cholesterol, but also some more rarefied lipids such as vegetable sterols, liposoluble vitamins, some phospholipids, and glycolipids (Chapter 4). Some lipids used in particular therapeutic situations, such as synthetic glycolipids or diacylglycerols, will also be exposed.

REFERENCES

Bernard, C., 1849. Recherches sur les usages du suc pancréatique dans la digestion. *C. R. Acad. Sci. Paris* 28:249–53.

Broadhurst, C.L., Cunnane, S.C. et al., 1998. Rift valley lake fish and shellfish provided brain-specific nutrition for early homo. *Brit. J. Nutr.* 79:3–21.

Chevreul, M.E., 1815. De la saponification de la graisse de porc, et de sa composition. *Ann. Chim.* 94:113–44.

Copley, M.S., Berstan, R. et al., 2005. Dairying in antiquity I. Evidence from absorbed lipid residues dating to the British iron age. *J. Archaeol. Sci.* 32:485–503.

Eaton, S.B., 1992. Evolution, diet and health. *Lipids* 27:814–20.

Fischler, C., 2001. *L'homnivore*, Odile Jacob, Paris.

Pepe, C., Clottes, J. et al., 1991. Le liant des peintures paléolithiques ariégeoises. *C. R. Acad. Sci. Paris* 312:929–34.

Pick, P., 2005. *Les origines de l'homme*, Tallandier, Paris.

2 Nature and Sources of the Main Lipids

2.1 INTRODUCTION

Vegetal oils and animal fats are products coming from agriculture and livestock, and they have been used as food since prehistoric times until modern times when they also became an industrial raw material as well as an energy source. These lipid sources are spread in all countries, each one being integrated within a historical scope linked to geography, sociology, culture, and political economy. Whereas the production of some oils originate from oleaginous plants, especially cultivated for that purpose (oil palm, coconut tree, and rapeseed), other oils are generally only products deriving from textile industry (cottonseed oil) or animal feeds (soybean oil).

Human, from very long time, has selected advantageous natural oleaginous plants for his diet, but now he tries to modify plants by genetic engineering to improve the fatty acid composition of various oils in order to answer specific requests (pharmacy, dietetics, and oleochemistry). Thus, erucic acid has been eliminated from rapeseed oil, the amount of oleic acid has been increased in sunflower oil, and the amount of linolenic acid has been lowered in other oils (linseed and soybean). Recent works even tried to induce the production by plants of fatty acids usually isolated from fish oils, an unforeseen consequence of an important drop of the harvesting of fish or the fisheries production and an increasing marine pollution.

The registered annual worldwide production of oils and fats was approximately 153 million tons in 2006/2007. The production of vegetal oils constitutes approximately 84% of this total (128 million tons), the remainder (16% or 24 million tons) consisting of animal fats. Although the production of these fats remains relatively stable, that of oils is in continuous increase, the growth rate being closely correlated with the gross national product of the consumer countries. The production of vegetal oils has been estimated to reach 168 million tons in 2015. What could be the upper limit? That observed growth is even faster than that of the world population. Thus, since the end of the World War II (1945), the total production of oils and fats increased more than seven times, whereas the world population grew only of approximately three times. That expansion of the world lipid production has been interpreted as the reflection of a general rise in the standard of living, with a parallel increased demand for consumption of meat, fried dishes, and industrial preparations of ready meals, dairy products, and eggs. The population growth in Asia and the concomitant rise in the standard of living of the populations make it possible to envision an important increase in the worldwide market of vegetable oils. These observations establishing a close relationship between wealth and consumption of fatty food let foresee a rapid development of intensive agriculture and a great difficulty to fight against obesity in the majority of the developed and developing countries. Only a specific education should be able to slow down the human tendency to prefer very tasty meals but with a strong fat content. But, is this possible? A partial answer is given still by the industry, which proposed fat substitutes with pleasant flavor but with lower energetic content.

Some oil productions (sunflower) are stable, while others are in constant rise for the last 10 years (7.5% a year for soybean oil and 13% for palm oil). The culture of oleaginous plants gradually was extended from the tropical zones toward the temperate zones, sometimes at the cost of a deforestation denounced in many countries of the intertropical zone. It is now recognized that soya and palm tree are the object of cultures having the most social and environmental negative impact in the world, but a general awakening seems to come up at the horizon.

It must be noticed that approximately 80% of the worldwide production of oils and fats is consumed by humans, whereas 10% is consumed by livestock, and the remaining 10% is processed by chemical industry (oleochemistry) and as energy source (biofuel). Considering the total intake of lipids and the number of humans on our planet, one can estimate the individual consumption at approximately 19 kg/year, for example, approximately 53 g of lipids per day (Metzger 2009). These figures do not obviously take account of the wasting and especially of the hidden (or invisible) lipids contained in food (various oleaginous seeds, ready-made meals, meats, and dairy products). It is appreciated that on a world level the visible lipids (oil, butter, and lard) constitute less than half (47%) of the total consumed lipids, the remainder (53%) being represented by the invisible lipids included in food themselves (source: Food and Agriculture Organization [FAO]). Moreover, that consumption varies according to the countries (generally in a range of ~1–5) and according to the social classes in the same country.

The world annual production of oils and fats (~153 million tons) is comparable with that of other refined products being the object of an international trade as sugar (~135 million tons), but lower than that of meat (~250 million tons). Considering human energetics, oils and fats have a predominant position because these nutrients are approximately 1.4–3 times more important than each other source.

For the whole production of oils and fats, six countries have a total of 64% of the worldwide production: the United States (14%), the European Union (14%), China (12%), Malaysia (11%), Indonesia (7%), and India (6%). Brazil and Argentina produce less than 5% each.

Almost the totality of the lipids consumed by humans, whatever are their sources, consists of fatty acids (Section 2.2) esterifying glycerol, thus forming the triacylglycerols (Section 2.3). These lipid components constitute the essential part of vegetable oils and animal fats. It should be stressed that the fatty acids constitute on average 96% of the weight of the triacylglycerols (71% of the weight of phospholipids). Although present in much less amounts, other lipid molecules appeared, during development of the science of nutrition, more important for their biological properties than for their energetic value. Among these lipids, this work will explore also phospholipids (Section 2.4), glycolipids (Section 2.5), and sterols (Section 2.6), all three groups being components of the cellular membranes ingested with food, of vegetable or animal origin. More rarefied lipids, such as liposoluble pigments (carotenoids) and vitamins (vitamins A, D, E, and K) (Section 2.7), are also part of our dietary lipids. A review of the lipid substitutes (Section 2.8) used in modern human nutrition to reduce the energy intake will finally be carried out.

2.2 FATTY ACIDS

The fatty acid group is made of a great number of molecular species, but only some of them are important in human nutrition. Among the traditional structures with a linear carbon chain, one can distinguish the saturated fatty acids and the unsaturated fatty acids. The former may be divided in several categories according to the length of the chain, the later according to the number, the position, and the configuration of their double bonds.

2.2.1 SATURATED FATTY ACIDS

As all natural fatty acids, the saturated ones have almost always an even number of carbon atoms and have as a general formula: $CH_3(CH_2)_nCOOH$, "n" generally with a value between 2 and 22, they are generally represented by a notation such as 16:0 for saturated fatty acids with 16 carbons (palmitic acid). Their physical and physiological properties depend on the length of the carbon chain. Thus, the saturated fatty acids with less than six carbons, called short-chain fatty acids, are more or less water soluble and do not have the same nutritional behavior as the others because they are quickly absorbed in the intestine. They can even control intestinal sodium and water movements. It is the case of the butyric acid (4:0) present at the level of approximately 3% in weight in butter. Fatty acids having from 6 to 12 carbons (caproic acid, 6:0; caprylic acid, 8:0; capric acid, 10:0; lauric acid, 12:0) are called medium-chain fatty acids. They are especially present in palm kernel oils (Section 2.3.1.6), in coconut oil (Section 2.3.1.9), and to a lesser extent in milk fats (Section 2.3.2.3).

The fatty acids with short and medium chain have a metabolism different from that of fatty acids with a long chain (from 14 to 24 carbons). The main representatives of the latter are the lauric acid (14:0), abundant in oils of palm kernel and coconut, as in milk, but especially the palmitic acid (16:0, Figure 2.1) and the stearic acid (18:0), present in all the vegetal and animal lipids. The nutritional importance of all these fatty acids is discussed in Section 3.2.1.3.2.

Besides saturated fatty acids with linear chain, species with branched chain are present in the lipids of milk and ruminant meat. Considering their low concentrations, these branched-chain fatty acids do not play any energetic role but have a well-known physiological importance (Section 4.2.1.3).

They are generally monomethylated, the methyl group being on the penultimate carbon atom (one from the methyl end) (*iso* position), or on the ante-penultimate

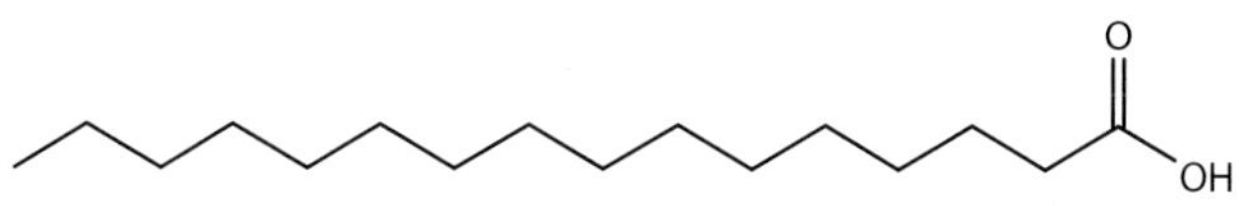

FIGURE 2.1 Palmitic acid (hexadecanoic acid).

carbon atom (two from the methyl end) (*anteiso* position). The carbon chain generally has from 14 to 17 carbons. Cow's milk contains approximately 1.88 g of branched-chain fatty acids for 100 g of total fatty acids. The *iso*-17:0 (Figure 2.2) is the most abundant among the fatty acids of the *iso* series (0.27 g/100 g) and the *anteiso*-17:0 (Figure 2.3) is the most abundant of the *anteiso* series (0.50 g/100 g). The amount of branched-chain fatty acids contained in dairy products depends on the type of food consumed by the animals (grass or corn).

Concurrently to these monomethylated fatty acids, multibranched fatty acids are also present in the cow's milk but with weaker concentrations. The two most abundant are the phytanic acid (3,7,11,15-tetramethylhexadecanoic acid; Figure 2.4) and the pristanic acid (2,6,10,14-tetramethylpentadecanoic acid; Figure 2.5). The concentration of the former is from 0.16 to 0.59 g/100 g of lipids and of the later is from 0.03 to 0.09 g/100 g of lipids. Phytanic acid originates from the degradation of vegetal chlorophyll in the rumen, and may undergo a transformation into pristanic acid in the peroxysomes, a metabolism well studied within the field of genetic diseases.

Few data exist on the ingested amounts and the metabolism of branched-chain fatty acids in man. Analyses have shown that the fetus is very early in contact with these fatty acids, because they are present in the still nonfunctional digestive tract and constitute until a third of the lipids of the vernix, a film covering the fetus (Ran-Ressler et al. 2008). A recent study has shown that in the United States the consumption of milk, cheese, and beef contributed to a daily dietary intake of about 400 mg of branched-chain fatty acids.

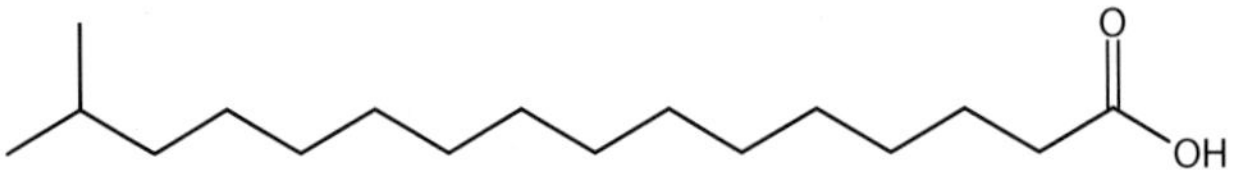

FIGURE 2.2   *iso*-Heptadecanoic acid (15-methyl hexadecanoic acid).

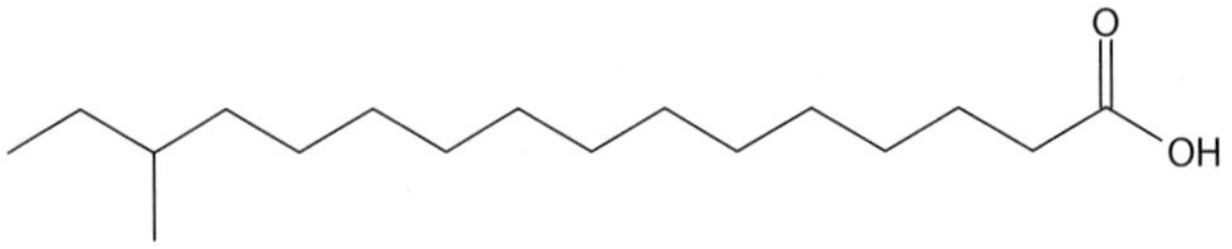

FIGURE 2.3   *anteiso*-Heptadecanoic acid (14-methyl hexadecanoic acid).

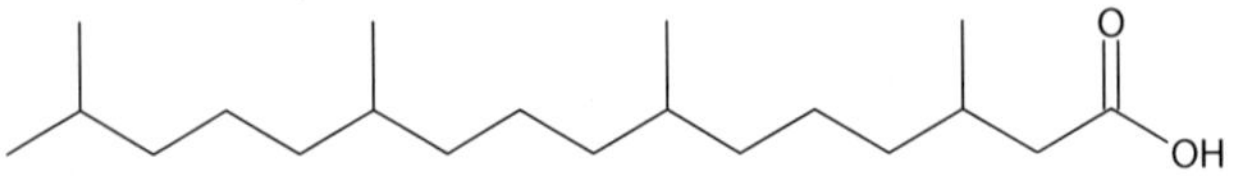

FIGURE 2.4   Phytanic acid.

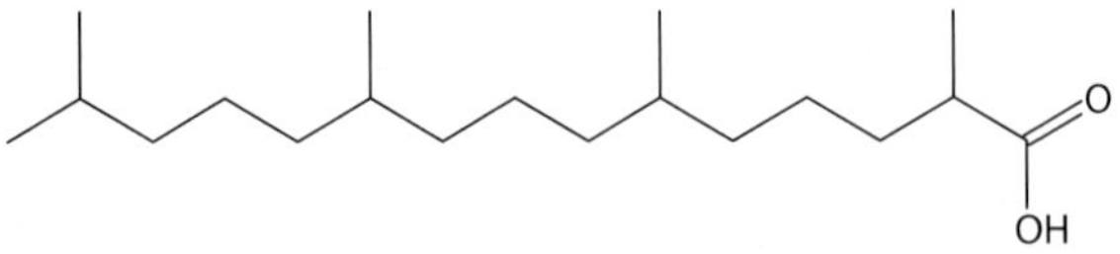

FIGURE 2.5   Pristanic acid.

2.2.2   UNSATURATED FATTY ACIDS

To describe the unsaturated fatty acids, besides the vernacular names, it is essential to indicate the number of carbon atoms of the chain, the number of double bonds, and their position in the chain. In the nutritional field, the nomenclature suggested by R.T. Holman in 1964, currently the most used by physiologists, numbers the double bonds starting from the final methyl group, this order being unmodified by the metabolic lengthening of the chain toward the carboxyl group. This nomenclature also determines the metabolic family, noted by *n-x* (*n* being the number of carbons in the chain, *x* being the position of the first double bond, at the methyl end). The position of the other double bonds is obtained by adding 3 to the preceding one. Three principal series will be thus differentiated, the *n*-9, *n*-6, and *n*-3 fatty acid series, the first double bond being respectively with the ninth, sixth, and third carbon starting at the methyl group. One can also use the expression ω-x (omega-x) instead of *n*-x.

In the chemistry and biochemistry books, the carbon atoms are almost always counted starting from the carboxyl group (C1 carbon). The linoleic acid may thus be named 18:2 *n*-6 or 18:2 ω-6 (or omega 6) by the physiologist or 9,12-octadecadienoic acid by the chemist. In the absence of any indication, the double bonds are considered in the *cis* position (or Z bond), on the reverse, they will have the *trans* position (or E bond).

2.2.2.1   *n*-9 Fatty Acids

The most important fatty acid of this series is the oleic acid (Figure 2.6; 18:1 *n*-9), discovered in 1823 by M.E. Chevreul in animal fats. It is especially abundant in olive oil (50%–80% of the total).

Other fatty acids of this series are present in the diet, but they have no nutritional significance because of their weak concentration in all common foods. The *trans* derivatives of oleic acid will be examined further.

2.2.2.2   *n*-6 Fatty Acids

Linoleic acid (18:2 *n*-6; Figure 2.7), isolated in 1844 from linseed oil, is without doubt the major representative of the *n*-6 fatty acids. That unsaturated fatty acid is mainly provided by foods containing plants (leaves, stems, fruits, and nuts) or their by-products (oils). Linoleic acid is the metabolic precursor of all fatty acids of the *n*-6 series, transformed by successive steps of desaturation and elongation as well in human as in some plants (Section 3.2.1.3.2).

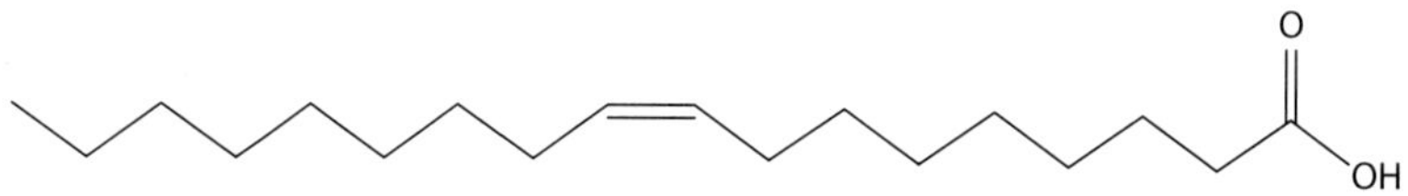

FIGURE 2.6   Oleic acid (9-octadecaenoic acid).

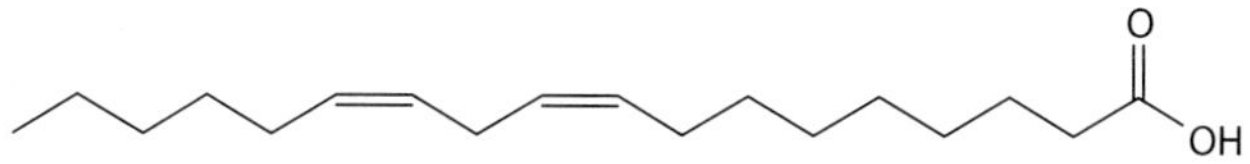

FIGURE 2.7   Linoleic acid (18:2*n*-6) (9,12-octadecadienoic acid).

γ-Linolenic acid (18:3 *n*-6; Figure 2.8) is the first derivative of linoleic acid formed by the action of a desaturase. Discovered in seeds of *Oenothera* (evening primrose), this fatty acid is currently extracted from several plants (seeds of borage and blackcurrant) to produce cosmetics and pharmaceutical compounds.

Dihomo-γ-linolenic acid (20:3 *n*-6; Figure 2.9), a natural derivative of the precedent acid by elongation, is provided in the human consumption only by the breast milk (maximum 0.01 g/100 g). This fatty acid can be metabolized into eicosanoids (series 1) that contribute to the protection of heart and arteries, stimulate immunity, and have anti-inflammatory effects.

Arachidonic acid (20:4 *n*-6) (Figure 2.10), derived from the previous one by desaturation, is an important constituent of membrane phospholipids, giving rise to a bunch of bioactive molecules (eicosanoids such as prostaglandins, isoprostanes, leukotrienes, and hydroxylated derivatives). Poorly represented in plants, except in some microalgae and fungi, it is abundant in animal tissues and without any doubt one of the most physiologically important compounds of the *n*-6 fatty acids. This fatty acid is essential in carnivorous animals such as the felines (cat, lion), which cannot made its biosynthesis from linoleic acid.

2.2.2.3 *n*-3 Fatty Acids

α-Linoleic acid (18:3 *n*-3; Figure 2.11), discovered in 1887 in hemp oil (*Cannabis sativa*), is the precursor of all the representatives of the *n*-3 series produced by successive steps of unsaturation, elongation, and oxidation (Section 3.2.1.3.2). As for linoleic acid, these *n*-3 fatty acids must be found necessarily by all animals in crop productions (leaves, fruits, and vegetal oils). Linolenic acid is a characteristic of linseed oil (*Linum usitatissimum*) (Section 2.3.1.12), amounting up to 50% of the total amount of fatty acids. The common dietary sources for humans are mainly rapeseed oil (Section 2.3.1.3) and very incidentally walnut oil (Section 3.2.1.3.2).

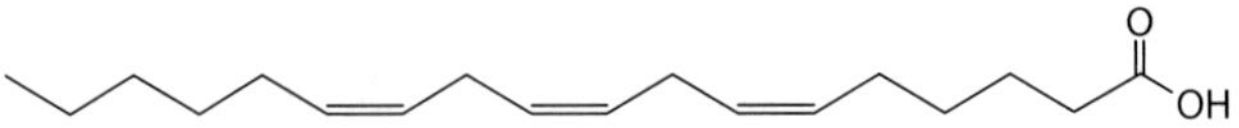

FIGURE 2.8   γ-Linolenic acid (18:3 *n*-6) (6,9,12-octadecatrienoic acid).

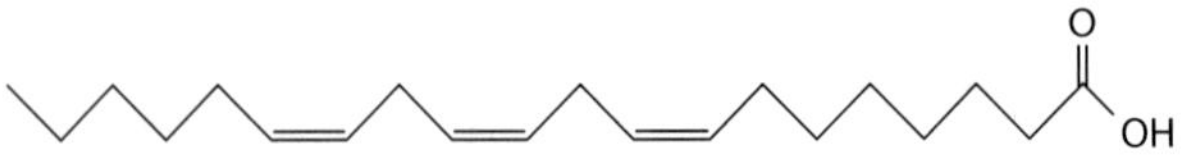

FIGURE 2.9   Dihomo-γ-linolenic acid (20:3 *n*-6) (8,11,14-eicosatrienoic acid).

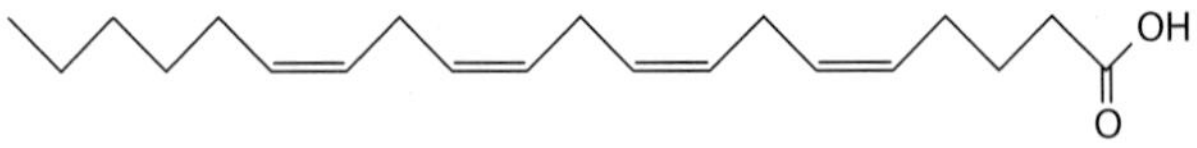

FIGURE 2.10   Arachidonic acid (20:4 *n*-6) (5,8,11,14-eicosatetraenoic acid).

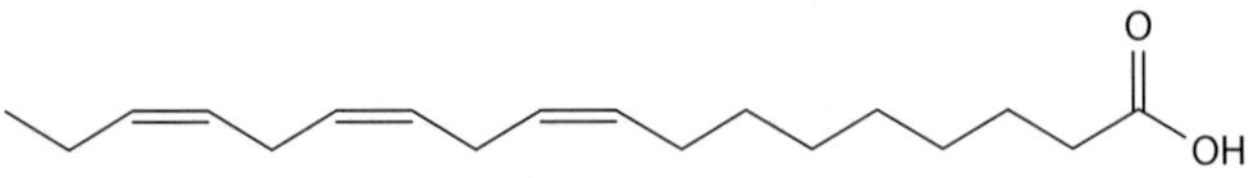

FIGURE 2.11   Linolenic acid (18:3 *n*-3) (9,12,15-octadecatrienoic acid).

The stearidonic acid (18:4 *n*-3; Figure 2.12) is a desaturation product of linolenic acid and is sometimes concentrated in some vegetal (blackcurrant seeds) and animal (fish) oils.

Fish oils are well known for the presence of long-chain polyunsaturated fatty acids in their triacylglycerol molecules, such as eicosapentaenoic acid (EPA) or 20:5 *n*-3 (Figure 2.13) and docosahexaenoic acid (DHA) or 22:6 *n*-3 (Figure 2.14). They are also present in membrane phospholipids of fish tissues. The increasing scarcity of the fishing resources and the contaminations of fish by many toxic substances led to consider a future mass production of these fatty acids starting from transgenic plants, microalgae, and even marine bacteria.

The needs and the food sources of these fatty acids are exposed in Section 3.2.1.3.2 and their incidence in the health problems in Section 4.2.4.

2.2.3 TRANS FATTY ACIDS

Besides unsaturated fatty acids with their double bonds mainly in the *cis* conformation, the natural lipid sources can contain a reduced proportion of fatty acids with one or more double bonds with the *trans* conformation. The difference between these two conformations lies in the distribution of two hydrogen atoms either on the same side of the plan of the ethylene bond (*cis* form) or on both sides (*trans* form) (Figure 2.15).

The *trans* fatty acids of natural origin are mainly monounsaturated (mainly *trans* 11-18:1, vaccenic acid), all are oleic acid isomers. They originate from a bacterial action in the digestive tract of the ruminants and are found in major dietary products such as milk and meat. Even so, some *trans* molecular species are of artificial origin, produced by physico-chemical treatments used for industrial productions. The *trans* configuration has an impact on the physico-chemical and functional properties of the monounsaturated fatty acids, bringing them closer to the corresponding saturated

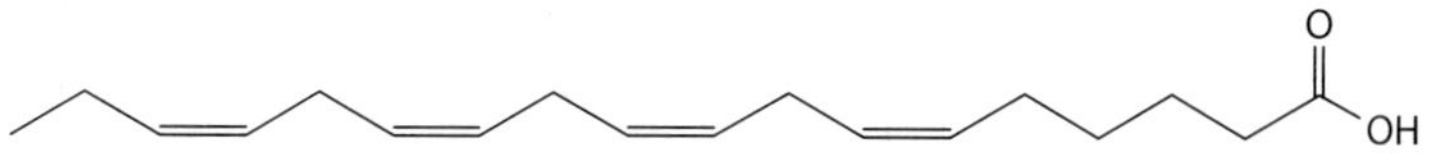

FIGURE 2.12   Stearidonic acid (18:4 *n*-3) (6,9,12,15-octadecatetraenoic acid).

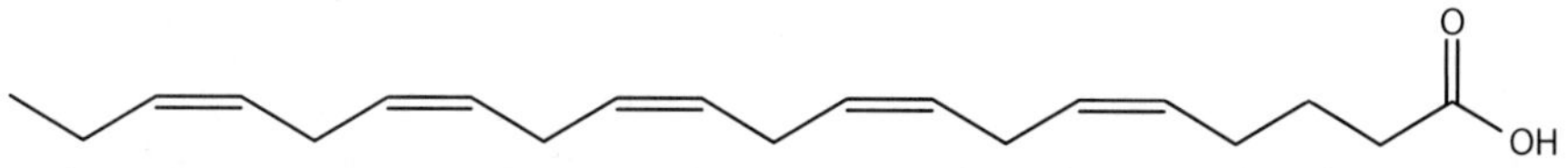

FIGURE 2.13   Eicosapentaenoic acid or 20:5 *n*-3 (5,8,11,14,17-eicosapentaenoic acid).

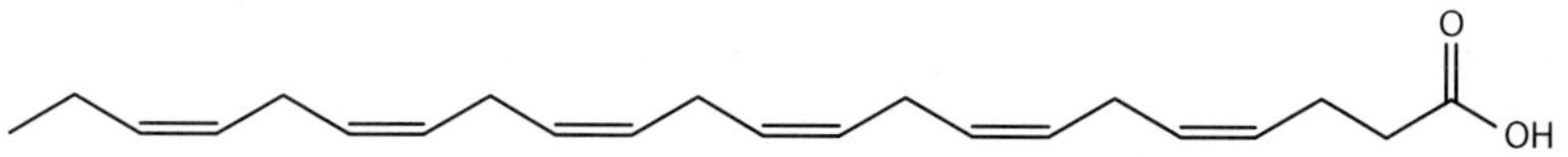

FIGURE 2.14   Docosahexaenoic acid or 22:6 *n*-3 (4,7,10,13,16,19-docosahexaenoic acid).

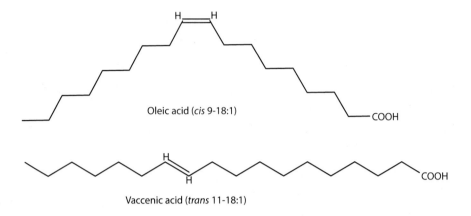

Oleic acid (*cis* 9-18:1)

Vaccenic acid (*trans* 11-18:1)

FIGURE 2.15 Comparative structures of oleic acid (*cis* 9-18:1) and vaccenic acid (*trans* 11-18:1).

fatty acids. Thus, the melting point of oleic acid (18:1) is 13°C, that of its *trans* isomer (*trans* 18:1) is 44°C, and that of stearic acid (18:0) is 70°C.

Besides these monoenes, one can also find very low amounts of *trans* di-unsaturated fatty acids deriving from linoleic acid (18:2 *n*-6) or tri-unsaturated fatty acids deriving from linolenic acid (18:3 *n*-3). The distribution and the food supply of *trans* fatty acids are exposed in Section 3.2.2.1 and their physiological effects in Section 4.2.5.1.

2.2.4 CONJUGATED LINOLEIC ACIDS

Among the *trans* linoleic acid isomers, some have the two conjugated double bonds, one being in a *trans* configuration. All are named with the initials "CLA" for conjugated linoleic acid. All the isomers belonging to this group are characterized by a specific position of the two double bonds on the 18-carbon chain. Indeed, these double bonds are said to be conjugated because they are separated by a simple bond, =C–C=, instead of a methylene bond, =C–C–C=, as for linoleic acid belonging to the fatty acid group known as isolenic. Only the conjugated isomers of linoleic acid are taken into account in that review because of the presence of only trace amounts of conjugated isomers of linolenic acid in natural products. Furthermore, these compounds have at the moment no known nutritional incidence.

The chemical rules show that 56 conjugated isomers are possible for linoleic acid, but only about 20 have been identified in food products. Among these compounds, one finds species having two *cis*, *cis* double bonds; other species with *cis*, *trans* double bonds; and some have *trans*, *trans* double bonds. Rumenic acid (*cis* 9, *trans* 11-18:2) (Figure 2.16) is the most abundant (nearly 90% of the total CLA) in the dairy products and beef.

On the other hand, one isomer of rumenic acid, the *trans* 10, *cis* 12-18:2 (Figure 2.17), is as abundant as the precedent in the vegetal oils treated industrially by alkaline substances and also in the marketed preparations of CLA.

The sources and the food supplies of these fatty acids are exposed in Section 3.2.2.2.

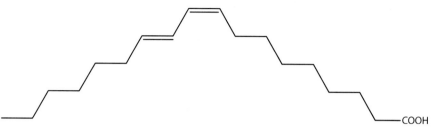

FIGURE 2.16   Rumenic acid (*cis* 9, *trans* 11-18:2).

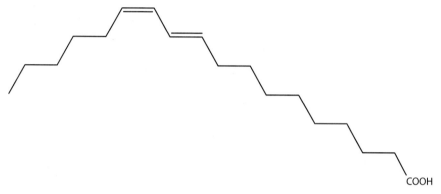

FIGURE 2.17   *trans* 10, *cis* 12-Linoleic acid.

Conjugated fatty acids with three double bonds, such as the α-eleostearic acid (*cis* 9, *trans* 11, *trans* 13-octadecatrienoic acid), a linolenic acid isomer synthesized or of vegetal origin, were studied for their anticancer properties in cultured cells. Similarly, conjugated isomers of EPA, DHA, and arachidonic acid prepared by alkaline isomerization have shown the same properties. To date, none of these products has been the object of clinical applications.

2.3   TRIACYLGLYCEROLS

These major dietary lipids are chemically triesters of glycerol. For a long time they were called triglycerides, but that term has been discarded since 1976 when new rules were decided by the international chemical authorities (International Union of Pure and Applied Chemistry and International Union of Biochemistry [IUPAC-IUB]). Their structure has been established for the first time in 1823 by M.E. Chevreul. This founder of the science of lipids was the first to demonstrate that oils and fats are made of three fatty acids coupled to a glycerol molecule by ester bonds (Figure 2.18). M. Berthelot, 30 years later, confirmed that important discovery by making the synthesis of some triacylglycerol species. His doctorate thesis, defended in 1854 and the first of that kind in the world, was entirely devoted to the structure and synthesis of fats. These discoveries carried out by these two great French chemists were going to become the bases of several sciences, such as analytical chemistry, biochemical synthesis, and lipochemistry. Originally, chemists have considered that these triacylglycerol molecules contained only one type of fatty acid. It was necessary to wait more

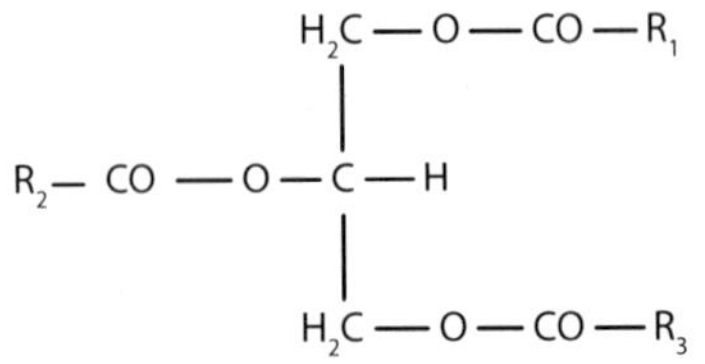

$$H_2C - O - CO - R_1$$
$$R_2 - CO - O - C - H$$
$$H_2C - O - CO - R_3$$

FIGURE 2.18 General structure of a triacylglycerol molecule (R1, R2, and R3 are the fatty acid carbon chains).

than one century, after the development of novel methods, so that the famous German chemist T.P. Hilditch corrected that concept in 1964. Indeed, he showed that natural oils and fats were generally made of mixed triacylglycerol molecules, with two or three types of different fatty acids (Figure 2.18; R1, R2, and R3), saturated or unsaturated. Now, we know that each oil or each fat may contain, thanks to a great number of fatty acid combinations, a great diversity of molecular species. The observed number of these species has been estimated to reach 200 in milk fat. Actually, with a given number of fatty acids, the number of triacylglycerol molecular species observed in natural fats is largely lower than the number of theoretically possible combinations.

The use of the two historical names, oils and fats, has risen from the difference in their fatty acid compositions, oils being more unsaturated than fats and are also more liquid at the same temperature. This difference in physical property (melting point) mainly depends on the unsaturation degree of fatty acids themselves, but also on the nature and the variety of the molecular species of the triacylglycerols present in the mixture. Thus, cocoa butter has a melting point slightly lower than 37°C because of the presence of two major molecular species, POS and SOS, combining two saturated fatty acids, palmitic acid (P) and stearic acid (S) with one monounsaturated fatty acid, oleic acid (O). The value of that melting point takes part in the pleasure of tasting a good chocolate melting slowly in the mouth. The sapidity of lipids is thus demonstrated! The melting point of palm oil is close to that of cocoa butter, but it results from a high proportion of palmitic acid present in two major species, POP and OOP.

Further details will be given on the composition of the lipids of vegetal oils (Section 2.3.1) and animal fats (Section 2.3.2), which are prevailing in the human diet. The physiological importance of the quantity and the quality of these lipids will be exposed in Section 3.2.1 and their impact on our health in Section 4.2.

2.3.1 Crop Productions

Many plants, some of which being cultivated since the antiquity, were first used by humans to match their energy requirements. Some of these plants could have also been incidentally used for one recent time to isolate poorly concentrated lipids such as sterols, carotenoids, and vitamins.

More than 2000 oleaginous plants were indexed on earth, but not more than 13 are the object of large scale productions and international trade, providing more than 90% of the worldwide production of vegetal oils.

The production in 2009/2010 of these most important oils has risen on the whole to approximately 144 million tons and may be classified in three groups, given in

TABLE 2.1

Annual Worldwide Production (2009/2010) of Vegetal Oils (Million Tons) (USDA, Foreign Agriculture Service)

Group 1		Group 2		Group 3	
Palm tree	45.9	Cotton	4.6	Sesame	0.8
Soybean	38.8	Palm kernel	5.5	Linseed	0.7
Rapeseed	22.5	Groundnut	4.6		
Sunflower	11.6	Olive	3		
		Coconut	3.6		
		Corn	2.3		

Table 2.1. The first group is that of the vegetal sources exceeding 10 million tons, the second of those ranging between 2 and 5 million tons, and the last of those with less than 1 million tons but being of some food interest.

It should be noticed that palm oil, with nearly 46 million tons, and soybean oil, with nearly 39 million tons, are the two main crop products, representing together approximately two-thirds of the worldwide oil production. Moreover, it should be noticed that the culture of both plants, palm tree and soya, is in constant increase.

The first group of four major oils (palm, soya, rapeseed, and sunflower) thus represents approximately more than 83% of the total production, the eight other oleaginous sources constituting the remainder (17%), 16% for the second group, and approximately 1% for the third. The study of the distribution of the worldwide productions (in percentage of the total) of the four major oils displays a great geographical diversity, except for palm oil that is collected almost only in Southeast Asia, Indonesia, and Malaysia (Figure 2.19). These four oils are the object of international exports equivalent to approximately 59 billion dollars per year, half being concerned by palm oil.

Vegetal oils are often classified in two groups according to their origin: pulp oils (palm tree and olive tree) and seed oils (all other vegetal sources).

One will find below, by decreasing order of production, a short description of the 12 more important vegetal oils produced in the world.

2.3.1.1 Palm Oil

Origin: Palm oil is extracted, in the form of a red liquid, from the pulp of the fruit of palm tree (*Elaeis guineensis*). This tree, native plant from New Guinea, was introduced in Africa, in Southeast Asia, and in Latin America, in the fifteenth century. The oil extracted from the fruit kernel has a different composition, it is called kernel oil (Section 2.3.1.6). The pulp contains approximately 40% of lipids. The palm tree is the most productive oleaginous source as approximately 6000 L may be collected from 1 ha of plantation, which amounts to 13 times more than for soya and approximately 6 times more than for groundnut, rapeseed, and sunflower. These yields explain the development of the palm tree culture.

Composition: The fatty acid composition of palm oil (Table 2.2) is variable according to the source but is characterized by the prevalence of two fatty acids, oleic and palmitic acids, and of two major triacylglycerol molecular species (OOP:

TABLE 2.2

Fatty Acid Composition of Palm Oil

Fatty Acid	Weight (%)
14:0	0–15
16:0	22–46
16:1 (*n-7*)	0–2.5
18:0	0.5–5
18:1 (*n-9*)	36–68
18:2 (*n-6*)	2–20
18:3 (*n-3*)	<1
20:0	<0.5

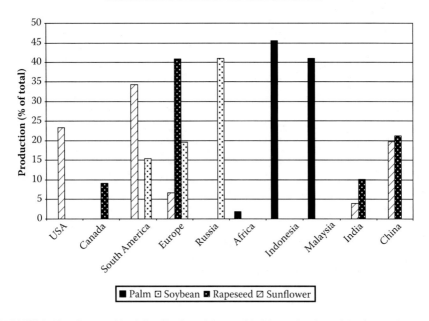

■ Palm □ Soybean ▣ Rapeseed ▨ Sunflower

FIGURE 2.19 Geographical distribution of the worldwide production of the four major vegetal oils (FAOSTAT, 2008).

34%, POP: 12%). The *sn*-2 position is thus occupied at the level of 86% by an unsaturated fatty acid and of 13% by palmitic acid. The red color of crude oil comes from its high concentration in carotene (about 0.8 g/kg) and also in vitamin E (about 0.24 g/kg of α- and γ-tocopherol and 0.56 g/kg of α- and γ-tocotrienol), this composition depending on the refining process. The sterol content may reach 2.5 g/kg of oil, the three quarters being β-sitosterol.

Production: Indonesia and Malaysia provide about 82% of the worldwide production of palm oil (Figure 2.19). The trade of this oil was internationally spread at the beginning of the twentieth century, this increase being linked to reduced production cost and its recent use in the manufacture of a petroleum substitute (biofuel).

Consumption: Palm oil is used up to 80% for human consumption (margarine, edible oil or frying oil, specialized fats, etc.). The domestic consumption in 2013 was 1200 million tons in the United States; in Europe (EU 27), it was about 5700 million tons. French people consume about 4.5 kg of palm oil a year per capita for a consumption of about 36 kg of total lipids. The manufacture of several derivatives with industrial uses (emulsifying agents, soaps, inks, and resins) is carried out with the 19% of the worldwide production, and 1% of the total are transformed into biofuel. In front of the pressure exerted by biodiversity defenders and nutritionists, all opposed to the use of saturated oil in human nutrition, many food industries began to gradually decrease the use of palm oil or are committed themselves to using only sustainable palm oil.

2.3.1.2 Soybean Oil

Origin: Soybean oil is extracted from the seeds of a leguminous plant, soya (*Glycine soya*), native plant in China, which was initially introduced in France in 1740. Soya is regarded as one of the most early cultivated plants (soya has been quoted in China as early as 2800 BC). Soybeans contain between 17% and 20% of oil.

Composition: Soybean oil is very rich in unsaturated fatty acids, particularly in linoleic acid (L, 50%–62%) and linolenic (Ln, 4%–10%) and has low concentration of saturated fatty acids (Table 2.3). The presence of linolenic acid often leads to classify the soya oil in the group of siccative oils. It is characterized by a great diversity of triacylglycerol molecular species, five of them exceeding 7% (LLL: 14%, OLL: 13%, LLP: 13%, LLLn: 8.6%, and OLLn: 7%).

Crude soybean oil contains also phospholipids (lecithin) (up to 22 g/kg), which must be isolated before marketing and can be used as food supplement in pharmaceutical industry (Section 3.2.5), for lubrication, or as emulsifying agent in food or pharmaceutical industries. The tocopherol content is about 1.4 g/kg, 60% being γ-tocopherol, tocotrienols being 100 times less abundant. The sterol content may reach 3 g/kg, half being β-sitosterol.

Production: Latin America (especially Brazil), the United States, and China are the principal soybean oil producers (Figure 2.19). France produces only 2% (~61,000 t in 2008) of the soybean oil of European origin.

TABLE 2.3
Fatty Acid Composition of Soybean Oil

Fatty Acid	Weight (%)
16:0	8–13
18:0	2–5
18:1 (*n*-9)	17–26
18:2 (*n*-6)	50–62
18:3 (*n*-3)	4–10
20:0	<1
20:1	<0.4
22:0	<0.5

Consumption: This oil is mainly used in human consumption (oil and mayonnaise) but also in industry. The enormous increase in its consumption (more than 1000 times) in the United States during the twentieth century was responsible for an increase of 2.6 times of the dietary intake of linoleic acid. The high concentration of linolenic acid in soybean oil leads up to use it in cooking only at room temperature or after moderate heating. In the United States, the domestic consumption is approximately 19 kg of soybean oil per capita per year, in Europe (EU 27) about 3.5 kg per capita per year, and in France about 8 kg per capita per year.

2.3.1.3 Rapeseed Oil

Origin: Rapeseed oil is extracted from seeds of cruciferous (*Brassica napus, Brassica campestris*), native plants from India and was introduced in Europe in 1360. Used until the nineteenth century for lighting and lubrication, it is currently the most important oil for human consumption in the Western countries. In France, the rapeseed oil constitutes 46% of consumed vegetable oils, approximately 1.4 million tons per year (source: CETIOM). Its use, either in the food or in the nonfood industries, has been constantly increasing. Since 1999, it is the first domestic oil consumed in France, and its market share continues to progress.

The seed contains approximately 40% oil, the remainder constituting an oil cake very rich in proteins and largely used in the chicken, pig, and cattle breeding.

Composition: Rapeseed oil is characterized originally by a high percentage of erucic acid (40%–50% of 22:1 *n*-9) and of its metabolic precursor the 20:1 *n*-9 (Table 2.4). Since 1980, the demonstration of a certain toxicity of the erucic acid in the rat led to develop new varieties (Canola) low in fatty acid. This effort survived the demonstration in 2000 of an absence of any toxicity in humans. This example shows that nutritionists should be prudent in generalizing to human the data obtained from experiments carried out in rat or mouse. Studies of other plant varieties tend to reduce the content of saturated fatty acids as well as linolenic acid for the benefit of the amount of lauric (12:0), oleic, and even γ-linolenic (18:3 *n*-6) acid. Among the many molecular species of triacylglycerols, four generally exceed the threshold of 6% (OOO: 23%, OLnO: 11%, OLL: 8%, and OLLn: 6%).

More than half of the α-linolenic acid amount is in the internal position (*sn*-2) in triacylglycerol species, thus resulting in an increase of its bioavailability compared

---

TABLE 2.4
Fatty Acid Composition of Rapeseed Oil (Weight %)

Fatty Acid	Original Oil	Canola
16:0	3–4	3.5
18:0	1–2	1.5
18:1 (*n*-9)	9–16	60
18:2 (*n*-6)	11–16	20
18:3 (*n*-3)	7–12	9.6
20:1 (*n*-9)	7–13	1.4
22:1 (*n*-9)	41–52	0.2

to other sources as soybean oil where only one-third of this fatty acid is in *sn*-2 position (Section 3.2.1.4).

The tocopherol content is about 80 mg/kg, γ-tocopherol being the major species, tocotrienols being absent. The sterol content may reach 8 g/kg, half being β-sitosterol and a quarter being campesterol.

During these last 10 years, the nutritional value of rapeseed oil (standard Canola) was emphasized with reason by nutritionists because of its low content in saturated fatty acids and especially by the low value (close to 2) of its *n*-6 to *n*-3 fatty acid ratio. This ratio makes it possible to balance the essential fatty acid contributions in agreement with the recommendations of the ANSES fixing the limiting value of that ratio at 5 (Section 3.2.1.3.2). Among most common oils containing *n*-3 fatty acids, rapeseed oil presents thus a *n*-6 to *n*-3 fatty acid ratio much nutritionally convenient than that of soybean oil (~7), and that of walnut oil (~4).

Production: The rapeseed oil is produced mainly in northern Europe (more than a third of the worldwide production), before China, India, and Canada (Figure 2.19). France produces approximately the quarter of the rapeseed harvested in Europe, corresponding to approximately 1.6 million tons of oil in 2009.

Consumption: In the United States, the domestic consumption of rapeseed oil in 2013 was about 1.9 million tons and in Europe (EU 27) about 9.2 million tons. In France, people consume on average 13.5 kg of rapeseed oil per capita per year. The rapeseed varieties with low levels of erucic acid (22:1 *n*-9) are used for food industry products (sauces, margarines, and salad oil), the varieties with high levels of erucic acid being used in industry, mainly as lubricant. Moreover, this oil will have likely great developments in the production of biofuel, a renewable substitute for petroleum.

2.3.1.4 Sunflower Oil

Origin: Sunflower oil is extracted from a Compositae (*Helianthus annuus*) native from Central and South America from where it was brought back to Europe during the sixteenth century. The seed contains from 36% to 44% oil and provides about 25% protein used for cattle breeding.

Composition: This oil is characterized by a high unsaturation degree due to a high percentage of linoleic acid (more than 60%), while being deprived of linolenic acid (Table 2.5). The three main triacylglycerol molecular species contain linoleic acid,

TABLE 2.5
Fatty Acid Composition of Sunflower Oil

Fatty Acid	Weight (%)
16:0	5–7
16:1 (*n*-7)	<0.5
18:0	4–6
18:1 (*n*-9)	15–25
18:2 (*n*-6)	62–70
20:0	<1
20:1	<0.5
22:0	<1

oleic acid, and palmitic acid: LLL, 24.5%; OLL, 21.4%; and LLP, 12%. A variety of sunflower rich in oleic acid has been developed in the year 1990. That oil contains up to 90% of oleic acid, 9% of saturated fatty acids, and 13% of linoleic acid. Another not-well-spread variety of sunflower produces an oil containing up to 14% of stearic acid, which is being used for industrial applications.

The tocopherol content (mainly α-tocopherol) is on average of 0.8 g/kg, tocotrienols being poorly represented. The sterol content (mainly β-sitosterol) varies from 2.4 to 5 g/kg.

Production: Russia (including Ukraine) is the largest world producer with more than 30% of the total world production, the remainder being carried out in Latin America and Europe (Figure 2.19). France produces approximately 22% of the sunflower harvested in Europe, corresponding to 568,000 t of oil (2009).

Consumption: In Europe (EU 27), the domestic consumption of sunflower oil in 2013 was about 3.8 million tons, in the Russian federation about 2.2 million tons, and in the United States about 185,000 t. French people consume on average 6.3 kg of sunflower oil per capita per year. This oil is of everyday consumption for salad dressing but is also largely used for the manufacture of margarines (Section 2.3.1.13).

2.3.1.5   Cottonseed Oil

Origin: Cotton oil is extracted from cottonseeds of various Malvaceae (*Gossypium hirsutum* and *Gossypium barbadense*), plants known 5000–7000 years BC in India and in Mexico. At the end of the nineteenth century, oil became a by-product of the production of a vegetal fiber, the most important in the world (~90% of all exploited fibers). The seed contains between 13% and 20% of oil, which contains a toxic polyphenol, the gossypol. That compound must be eliminated from oil before its use as food product.

Composition: Cottonseed oil has a composition close to that of sunflower oil, with a high unsaturation degree caused by 34%–60% of linoleic acid (Table 2.6) and a low content in stearic acid (less than 3%). The four major triacylglycerol species are LLP: 29%, LLL: 17%, OLL: 14%, and PLO: 12%.

Besides gossypol, cottonseed oil contains unwanted cyclopropenic acids (sterculic and malvalic acids) that must be eliminated during the refining processes. The tocopherol content amounts to a maximum of 0.9 g/kg, more than 80% being in

TABLE 2.6
Fatty Acid Composition Cottonseed Oil

Fatty Acid	Weight (%)
14:0	<1
16:0	17–31
16:1	<1
18:0	1–3
18:1 (*n*-9)	13–21
18:2 (*n*-6)	34–60
18:3 (*n*-3)	<1

the form of α-tocopherol. Tocotrienols are practically absent. The content of sterols is on average of 4.6 g/kg, 85% being β-sitosterol.

Production: Cotton is cultivated on five continents, but the most productive oil areas are China (1.5 million tons), India (0.7 million tons), Pakistan (0.5 million tons), and the United States (0.4 million tons).

Consumption: Cottonseed oil is traditionally used in the producing countries for domestic consumption, either directly after refining or after hydrogenation for the manufacture of margarines. In China, the domestic consumption in 2013 was about 1.5 million tons, in India about 1.3 million tons, and in the United States about 220,000 t.

2.3.1.6 Palm Kernel Oil

Origin: Palm kernel is extracted from the almond of the fruit of the palm tree (*Elaeis guineensis*). This almond contains from 3% to 6% of oil on a wet weight basis.

Composition: This oil is, as well as coconut oil, a kind of oil known as lauric oil because it contains 46%–52% of lauric acid (La, 12:0) besides the 15%–17% of myristic acid (M, 14:0) (Table 2.7). The high amount of saturated fatty acids gives it the property to be solid at room temperature (20°C). The four major triacylglycerol species are LaLaLa: 22%, LaLaM: 17%, LaMM: 9%, and LaLaO: 6%.

The contents of tocopherols and tocotrienols of palm kernel oil are extremely low (<30 mg/kg). The sterol content is on average of 1.1 g/kg.

Production: The 5.5 million tons of palm kernel oil produced in the world comes mainly from Malaysia (46%), Indonesia (36%), and Africa (0.5% in Nigeria).

Consumption: Agriculture marketing of palm kernel oil is large because it can replace coconut oil (obtained from copra, Section 2.3.1.9) in many applications, mixed with other oils, used directly as a substitute of the cocoa butter, and manufacture of margarines and soap. This oil is also an important source for the industry of detergents (lauric alcohol synthesis). In Indonesia and Malaysia, the domestic consumption of palm kernel oil in 2013 was about 1.9 million tons, in Europe (EU 27) about 600,000 t, and in the United States about 280,000 t.

TABLE 2.7
Fatty Acid Composition of Palm Kernel Oil

Fatty Acid	Weight (%)
8:0	3–4
10:0	3–7
12:0	46–52
14:0	15–17
16:0	6–9
18:0	1–3
18:1 (*n*-9)	13–19
18:2 (*n*-6)	0.5–2

2.3.1.7 Peanut Oil

Origin: Peanut oil is extracted from seeds of a plant of the Fabaceae family, *Arachis hypogea*. This plant is probably native of Mexico, being cultivated by the Incas more than 5000 years ago. It appeared in Europe only after the discovery of America.

Composition: The seed contains approximately 45% of oil characterized by the dominant presence of two fatty acids, oleic and linoleic acids (Table 2.8). The composition of the triacylglycerol molecular species is very diversified, the four most important species being OLL: 16.2%, OLO: 14.3%, OOO: 11.8%, and OLP: 8.1%.

Peanut oil is moderately unsaturated, therefore it has good heat stability, and hence it is used for cooking even with high temperatures as in frying preparations. The tocopherol content is approximately 0.1 g/kg oil. The groundnut oil contains approximately 1.9 g/kg of sterols.

Production: The 4.6 million tons of peanut oil produced in the world comes from three large producing countries: China (44%), India (24.4%), and Nigeria (7%).

Consumption: In China, the domestic consumption in 2013 was about 2.7 million tons, in India about 1.2 million tons, in the United States about 97,000 t, and in Europe (EU 27) about 91 million tons. French people consume little groundnut oil (0.2 kg per capita per year in 2001), this value dropping since 1969 (10 kg per capita per year) following the development of the local production of sunflower and rapeseed.

2.3.1.8 Olive Oil

Origin: Olive oil is extracted mainly from the pulp of the fruit of an Oleaceae, the olive tree (*Olea europea*). The olive tree, characteristic of the Mediterranean countries, has been known for more than 3000 years in Europe but has owed its expansion during the Roman Empire.

Composition: The olive fruit contains approximately 30% of its weight of an oil characterized by high percentage of oleic acid (from 53% to 80%) but a very low percentage of stearic acid (18:0) (Table 2.9). Three molecular triacylglycerol species dominate are OOO: 39%, OOP: 21%, and OLO: 10%.

The contents of tocopherols and tocotrienols are very low (<0.1 g/kg). The sterol content is also very low (0.1 g/kg on average), β-sitosterol being the major species.

TABLE 2.8
Fatty Acid Composition of Peanut Oil

Fatty Acid	Weight (%)
16:0	8–13
16:1	<0.3
18:0	1–4
18:1 (*n*-9)	35–66
18:2 (*n*-6)	14–41
18:3 (*n*-3)	<0.3
20:0	<2
20:1	<2
22:0	2–5

TABLE 2.9
Fatty Acid Composition of Olive Oil

Fatty Acid	Weight (%)
16:0	8–21
16:1	1–4
18:0	1–6
18:1 (*n*-9)	53–80
18:2 (*n*-6)	2–24
18:3 (*n*-3)	1–2
20:0	<0.5
20:1	<0.5
22:0	<1

Its fatty acid composition and high antioxidant content (tyrosol derivatives) confer a great thermal stability to that oil, allowing it to be used in food even with heating.

Production: The European Community contributes to approximately 80% in the worldwide production of olive oil (3 million tons in 2009), the principal contributors being Spain (32%), Italy (19%), and Greece (14%). France has produced only 7000 t olive oil in 2009.

Consumption: In Europe (EU 27), the domestic consumption in 2013 was about 1.95 million tons and in the United States about 300,000 t. French people consume approximately 1.5 kg of that oil per capita per year. The olive oil is also employed in cosmetic (ointments and soaps) and pharmaceutical industries.

2.3.1.9 Coconut Oil

Origin: Coconut tree (*Cocos nucifera*) is a plant of the Arecaceae family. Coconut oil is extracted from crushed copra, itself obtained by desiccation of the coconut almond. The oil cake is used in the animal feed. The content of lipids of copra is approximately 65%, which corresponds to an oil content of 23% for the totality of the collected fruit.

Composition: Besides the large contents in myristic acid (M, 14:0) and caprylic acid (Cy, 10:0), the high content in lauric acid (La, up to 54%) places that oily product, with palm kernel oil, in the category of lauric oils (Table 2.10). Coconut oil is made up of many molecular species, the four more important being MLaCy: 15%, LaLaCy: 13%, PLaCy: 8%, and LaOCy: 8%.

The content of the coconut oil in tocopherols and tocotrienols is very low (lower than 50 mg/kg). It contains from 0.4 to 1.2 g of sterols per kg of oil, the main ones being campesterol, stigmasterol, and β-sitosterol.

Production: The principal producing countries are Indonesia (30%), Philippines (23%), and India (21%).

Consumption: The high contents of saturated fatty acids and antioxidants in coconut oil explain its traditional use for cooking in the Southeast Asia and the Pacific. It is also used in occidental countries as cooking oils, but is also used after hydrogenation as a basic product for the manufacture of Végétaline, margarines, and many other products created by the food industry. In Philippines, the domestic consumption in 2013 was about 840,000 t, in Europe (EU 27) about 715 million tons, and in the United States about 520,000 t.

2.3.1.10 Corn Oil

Origin: The oil of corn germ (*Zea mays*, Poaceae) is a by-product of the grinding of maize plant seeds, treatment in addition providing a flour rich in starch and proteins. The corn is native from Central America and was brought back in Spain only in 1493 by Christopher Colombus. The seed contains approximately 12% oil, whereas the germ contains from 30% to 40% oil.

Composition: The fatty acid composition of corn oil is characterized by high linoleic acid content (between 55% and 62%), as sunflower oil (Table 2.11).

The tocopherol content is approximately 1.5 g/kg of oil, the majority being made up of γ-tocopherol. This oil contains very few tocotrienols. Corn oil has high phytosterol content (up to 15 g/kg), nearly 77% consisting in β-sitosterol. This high content in antioxidants compensates for its elevated unsaturation and makes it stable to heat, therefore being useful as cooking oil.

Production: The worldwide production is approximately 2.3 million tons; the principal producing countries are the United States (52%), Europe (9.5%), and China (5%).

TABLE 2.10
Fatty Acid Composition of Coconut Oil

Fatty Acid	Weight (%)
6:0	<1
8:0	9–10
10:0	5–10
12:0	40–54
14:0	15–23
16:0	6–11
16:1 (*n*-7)	<2
18:0	1–4
18:1 (*n*-9)	4–11
18:2 (*n*-6)	1–2

TABLE 2.11
Fatty Acid Composition of Corn Oil

Fatty Acid	Weight (%)
16:0	8–13
18:0	1–4
18:1 (*n*-9)	24–32
18:2 (*n*-6)	55–62
18:3 (*n*-3)	<2

Consumption: Besides its use as food (salad sauces and margarines), corn oil is used as a basis for pharmaceutical and cosmetics industries.

2.3.1.11 Sesame Oil

Origin: Sesame oil is extracted from the seeds of *Sesamum indicum* (a plant of the Pedaliaceae family). Sesame oil has been used by the Persian for more than 6000 years as a medicinal and cosmetic compound, from there it penetrated in India, China, and Japan.

Composition: The sesame seed contains approximately 50% of oil rich in linoleic acid (39%–47%), but also in oleic acid (37%–42%) (Table 2.12). Among the many molecular triacylglycerol species, the four more abundant are OLL: 17%, OLO: 15%, LLL: 11%, and OOO: 10%.

Sesame oil is particularly stable during heating despite its strong unsaturation, thanks to the presence of powerful antioxidants such as sesamine and sesamoline, phenolic compounds of the lignan group.

Production: The production of sesame extends from Asia, in India and to the Central America, but the quarter of the worldwide production (800,000 t) comes from China; India producing only 12% of the total.

Consumption: Apart from its use in the diet in the countries of production, the sesame oil is transformed into soaps and various cosmetic products.

2.3.1.12 Linseed Oil

Origin: Linseed oil is extracted from linseed (or flax) (*Linum usitatissimum*), a plant of the Linacae family, used for a very long time as source of fibers for the manufacture of clothes.

Composition: The seeds contain approximately 34% of oil particularly rich in linolenic acid (Ln, 45%–70%) (Table 2.13). The four more important molecular triacylglycerol species are LnOLn: 14.5%, LnLnLn: 14%, LnLLn: 12%, and LnLnP: 9%.

Production: The worldwide production of linseed oil in 2007 was 700,000 t, 21.5% being harvested in China, 21.4% in the United States, and 18.8% in Belgium.

Consumption: As linseed oil is highly unsaturated and has siccative properties when reacting with oxygen, it was poorly used in human consumption, except as food supplement to raise the intake of *n*-3 fatty acids (authorization given in 2006 by the AFSSA). By a decree of July 12, 2010, the marketing in France of the virgin

---

TABLE 2.12

Fatty Acid Composition of Sesame Oil

Fatty Acid	Weight (%)
16:0	8–11
18:0	4–6
18:1 (*n*-9)	37–42
18:2 (*n*-6)	39–47
18:3 (*n*-3)	<0.5
20:0	<1

TABLE 2.13
Fatty Acid Composition of Linseed Oil

Fatty Acid	Weight (%)
16:0	4–10
18:0	2–8
18:1 (n-9)	10–20
18:2 (n-6)	12–24
18:3 (n-3)	45–70

linseed oil is now authorized as edible oil. That oil is largely used for industrial applications (manufacture of paints, varnish, inks, and even plastics).

2.3.1.13   Processed Products: Margarines

Origin: Margarine is currently manufactured starting from vegetable oils, but at the origin the process started from an emulsion of milk and beef fat. The manufacture process of that kind of butter substitute was patented in 1872 by Hippolyte Mège-Mouries, following a contest launched by Napoleon III for the discovery of a fat similar to butter. In 1875, vegetal lipids (especially from coconut) were included in the preparation, but later they were replaced by hydrogenated whale oil around 1911, following the discovery of the hydrogenation reaction of fatty acids by W. Normann in 1901. This process primarily consists of reacting oils with hydrogen in the presence of a catalyst (nickel and palladium) under strictly controlled physical conditions. From the reaction, it results a transformation of the fatty acids by the removal of double bonds, thus producing saturated fatty acids. The hydrogenation process contributes to raise the melting point of the product and thus to harden it to a previously intended plasticity. Around 1950, whale oil was replaced by fish oils and later by vegetal oils that became more available and cheaper.

Currently, margarine is an emulsion of fats (80% minimum) and an aqueous phase often containing milk (20%). The type of fat and oil entering the composition of a margarine is very variable and the nutritional characteristics of the finished product depend on it. The majority of margarine brands currently on the market are made from vegetal oils (mainly from palm tree, rapeseed, sunflower, soybean, and groundnut) more or less hydrogenated. Vegetal oils are thus transformed into saturated fats, but, unfortunately, they may also contain *trans* fatty acids (Section 3.2.2.1). However, the manufacturing processes have been gradually improved to ensure that the consumers will ingest a minimum amount of these *trans* fatty acids, compounds suspected to cause several pathologies (Section 4.2.5.1). These *trans* fatty acids are generally at a concentration lower than 2 g for 100 g of finished product. The margarine made by food industries has a *trans* 18:1 content frequently higher than that found in domestic margarines. The most recent manufacturing processes used fats obtained by interesterification (Section 3.2.1.4) and not by hydrogenation, thus allowing the marketing of margarines practically without *trans* fatty acids. After mixing with skimmed milk, the whole is crystallized by cooling, and then mixed to obtain the desired plasticity. The labeling of the marketed products makes it possible to know if hydrogenation is part of their manufacturing process. Unfortunately, so far, the law does not make obligatory the indication of the *trans* fatty acid content.

Composition: The fatty acid composition of margarine depends closely on the vegetal oils employed for their manufacture. The average composition of a margarine manufactured from sunflower and rapeseed oils is given in Table 2.14. It should be noted that the concentration of the polyunsaturated fatty acids of the margarine (14.3% of linoleic and linolenic acid) is quite lower than that of original oils, the difference being the result of the hydrogenation process.

In recent decades, margarine or table spreads have gone through many developments to improve their healthfulness. Thus, formulas with specific fatty acid compositions have been firstly put on the market, either enriched in *n*-3 fatty acids from linseed oil or in oleic acid from olive oil. These specifications were intended to prevent or to fight against cardiovascular disease (Section 4.2.4.1). This tendency to a *n*-3 fatty acid supplementation with fats from vegetal sources is, in France, a consequence of the AFSSA recommendation in June 2003 to tend progressively toward an *n*-6 to *n*-3 fatty acid ratio lower or equal to 5. Other margarine formulas enriched in phytosterols (Section 3.2.3.2) are present on the market; they are intended to fight the intestinal absorption of cholesterol (Section 4.3.2). These products were authorized by the European Commission in 2000. Margarines with a fat content lower than 62% (low-calorie margarines) may be also included in the group of "healthy margarines"; they are naturally proposed within the scope of fighting obesity.

Production: Several countries have legislated on manufacturing and marketing of margarine as well as spread.

In Canada, margarine is an emulsion of fat with water or vegetal oil, but without milk, and contains not less than 80% fat. A supplementation with vitamins A and D is accepted. Low-calorie margarines must contain not less than 40% fat. In the United Kingdom, the use of any partially hydrogenated oils in the margarine manufacture is forbidden.

In the European Union, a directive of 1994 standardized names for the marketing of spread and margarine replacing butter. Thus the name "margarine" was reserved for products containing at least 80% of lipids. Spread that contains 60%–62% of fat may be called "three-quarter-fat margarine" or "low-fat margarine." Spread that contains 39%–41% of fat may be called "half-fat margarine," "low-fat margarine,"

TABLE 2.14

Fatty Acid Composition of One Margarine

Fatty Acid	Weight (%)
12:0	11.7
14:0	29.5
16:0	11.7
18:0	7.6
18:1 (*n*-9)	66.4
18:2 (*n*-6)	12.2
18:3 (*n*-3)	2.1

or "light margarine." A spread with any other percentage of fat is called "fat spread" or "light spread." For all these products, the final content of milk lipids should not exceed 3%. No legislation was promulgated by the European authorities about the addition of vitamins. The examination of the labels on commercial products attests that the addition of vitamins A and E is often carried out, similarly for carotene.

In 2012, the world production of margarine amounted to about 9.4 million tons including the share of Europe (EU 27) representing about 2.44 million tons. The most important European producing countries are Poland with 404,000 t, Germany with 383,000 t, the United Kingdom with 321,000 t, the Netherlands with 247,000 t, and Belgium (+Luxembourg) with 243,000 t. France produced only 78,000 t but 123,000 t are imported. From 2000 to 2008, the production of margarine dropped by 13%, but the offered products evolved from the "traditional" margarine to the "healthy margarine." The importance of this last category increased from 8% to 63% of the market.

Consumption: In the United States, the domestic consumption of margarine in 2010 was estimated to 1.6 kg per capita and exhibited a persistent downward trend since the early 1990s, but remaining unchanged during the last 10 years. In 2010, the consumption was only 31.6% of the amount recorded in 1980 on a per capita basis. In Canada, the consumption remained at the level of about 3.7 kg per capita since about 7 years. In Europe (EU 27), the margarine consumption was slowly decreasing and is actually about 4.2 kg per capita per year.

The domestic margarine consumption is estimated in France at 3.2 kg per capita per year. The French study INCA2 (2006–2007) has indicated an average consumption of 8.1 g/day, for example, 2.9 kg by year; this figure is about 4.3 kg in Europe.

2.3.2 Livestock Productions

Animal fats of commercial importance are by-products of the production of meat (carcasses of cattle) or fisheries (fish oil). In the terrestrial animals, these fats are present mainly in specific depots (fat cells) localized in subcutaneous or peritoneal tissues. To these sources, it is necessary to add fats derived from dairy products (cream and butter). The total amount of animal fats produced in the world in 2007 was approximately 24.4 million tons, equivalent to 16% of the total production of lipids.

Besides the fats marketed as such and coming from fat tissues, the nutritionist must take into account the fats included in meats (also in fat cells) (Section 2.3.2.5), the latter being important foodstuffs bringing also most of our dietary proteins.

2.3.2.1 Tallow

Origin: Tallow is recovered during the preparation of the cattle meat (8%–25% of the raw meat), but the distinction between its ovine and bovine origin is not made at the production level. It is not a fat of common use, but it is incorporated as an ingredient in many products resulting from food industry.

Composition: Tallow is a solid fat at room temperature; its fatty acid composition depends much on the anatomical localization and of the animal species. In general, tallow is characterized by a high level of saturated fatty acids (palmitic and stearic acids) (40%) while containing moderate amounts of oleic acid (35%–50%) and linoleic acid (1%–5%). The ruminant tallow contains also on average 5% (3.4%–6.2%)

trans fatty acids derived from oleic acid, the vaccenic acid (*trans* 11-18:1) being the most abundant (about 62% of the total *trans* fatty acids) (Wolff et al 1998). These *trans* fatty acids are naturally originating in the rumen from the biohydrogenation of fatty acids produced during the digestion of dietary plants (Section 3.2.2.1). The *trans* fatty acid intake via tallow is relatively reduced because of the low human consumption of that product. The *sn*-1 position in triacylglycerol species is mainly occupied by palmitic acid, whereas the *sn*-2 position is by oleic acid. The *sn*-3 position is occupied by palmitic or stearic acid, as in the POP and POS species that are the most abundant. Tallow contains from 80 to 140 mg cholesterol for 100 g fat.

Production: The worldwide production of tallow in 2007/2008 has been approximately 8.7 million tons.

Consumption: Used in human food but mainly for animal feeding, tallow is used predominantly for the manufacture of soaps, detergents, and candles. The fatty acids of tallow are also used for the manufacture of many chemical derivatives such as pesticides, dispersion agents, plastics, paints, and varnishes.

2.3.2.2 Lard

Origin: Lard originates mainly from the subcutaneous fat of pig. Lard is obtained after melting the various sources of adipose tissue (skin, visceral fat around kidney) by heating. Lard may be rendered by different processes to separate pure triacylglycerols from other cellular fractions. In some countries, lard may be hydrogenated to improve stability at room temperature. It is also often treated with bleaching and deodorizing agents, with emulsifiers and antioxidants.

Composition: Its fatty acid composition, especially in linoleic acid, varies much according to the animal diet, but is generally close to that of tallow. In contrast, the *sn*-2 position is mainly occupied by palmitic acid (~58%), whereas the *sn*-3 position is occupied by unsaturated fatty acids. The *sn*-1 position is diversely occupied as in SPO, OPL, and OPO species that are the most abundant. Lard contains from 300 to 400 mg cholesterol/100 g.

Production: The worldwide production of lard in 2007/2008 has been approximately 8.3 million tons. In France, the production of lard varies from 20,000 to 50,000 t a year.

Consumption: Like tallow, lard is used for the manufacture of human and animal foods, but also as lipid bases for oleochemistry (mainly biofuel and soaps). Unfortunately, one practically does not have reliable information on the uses of pig fats, generally, directly used by butchers and manufacturers according to their needs, only the remainder (lard) is quantified statistically by the official organizations.

2.3.2.3 Butter

Origin: It has been proved that preparations similar to butter were known in the Stone Age as early than 6500 BC. Sumerian tablets discovered in ancient Mesopotamia dating from 2500 BC illustrated butter making. Butter is a fatty product obtained primarily from cow's milk by mechanical treatments, 25 L of milk providing approximately 1 kg of butter. It is made by churning fresh or fermented cream or milk. This operation gathers the fat drops in suspension and thus separates butter from the remaining watery cream, called buttermilk. Cow's milk contains on average 40 g of lipids per liter in the form of triacylglycerol droplets surrounded by a complex membrane made of proteins and phospholipids. Butter can also be manufactured from the milk of other mammals, including sheep, goat, buffalo, and yak.

Composition: Butter is the most complex natural fat because more than 400 fatty acids could be characterized, more than 60% of them being saturated. Its composition is specific and variable but with a high percentage of short- and medium-chain fatty acids (from 4 to 12 carbons), including capric acid (C, 10:0) and a typical component, butyric acid (B, 4:0) (Table 2.15). The structure of the molecular species is very diversified, with a preferential localization of palmitic acid in the *sn*-1 and especially *sn*-2 positions (~32%). The molecular species PPB, PPC, and POP are the most abundant.

According to regulations, the term *butter* is reserved to products with a milk fat content of not less than 82% but less than 90% and with a maximum of 16% water in the form of small drops. These regulations are similar in the United States and in Europe.

Besides the *cis* fatty acids, butter is well known to contain a large variety of *trans* fatty acids (Section 3.2.2.1). The latter have 14–20 carbons with a *trans* double bond in various positions on the carbon chain. Butter contains on average 3.7 g/100 g of *trans* fatty acids derived from oleic acid (18:1 *n*-9). The most abundant isomer is the *trans* 11-18:1 (vaccenic acid) (~50% of the total *trans* fatty acid content). Qualitative and quantitative variations in the composition of these *trans* fatty acids are observed according to the type of animal diet (ensilage or pasture).

Butter also contains isomers of linoleic acid (*cis* 9, *cis* 12-18:2), the CLA, fatty acids with two double bonds in a conjugated situation (Section 2.2.4). These double bonds are sometimes both in position *cis*, *cis*, but only one or both may be in the *trans* position, moreover they can have various positions on the carbon chain. Butter contains on average 0.45–0.8 g/100 g fat of CLA, 90% of them being represented by rumenic acid (*cis* 9, *trans* 11-18:2; Figure 2.16).

Butter naturally contains cholesterol (to a maximum of 350 mg/100 g) and high amounts of vitamins A (~0.7 mg/100 g) and E (~3 mg/100 g), but little vitamin D (~1 mg/100 g). β-Carotene is also found and some manufacturers add that natural pigment to ensure a more attracting yellow color.

TABLE 2.15
Average Fatty Acid Composition of Butter

Fatty Acid	Weight (%)
4:0	2.9
8:0	0.6
10:0	2.1
12:0	2.6
14:0	9.4
16:0	24.1
16:1	2.6
18:0	10.9
18:1 (*n*-9)	28.2
18:2 (*n*-6)	2.1

Clarified butter is a kind of butter without its water, leaving almost pure butterfat. This product is made by heating butter to its melting point and then allowing it to cool, the fatty layer being separated by density. It thus represents the milk fat in the purest state. This product offers economic benefit as for its use, its transport, and its storage. This practice also makes it possible to tolerate, for a short time, strong temperatures during cooking, up to 180°C, without blackening.

Production: The worldwide production of butter is approximately 7 million tons (2007/2008). At first rank is India that contributes more than one-fifth to this production; second in production is the United States (522,000 t). The European production of butter was estimated at 2.1 million tons in 2008. France produces 406,000 t of butter, the second rank in Europe (EC 27) after Germany (445,000 t).

Low-fat butters are present recently on the market. With a lower fat content and a similar taste, they are useful especially as they are easier to spread, but they support cooking very badly. These butters are manufactured from low fat and pasteurized cream. Some additives such as starch are sometimes added. Low-fat butter must have a fat content equal or lower than 65%, whereas in France, butter known as "light butter" contains a maximum of 40% fat.

Another kind of easily spreadable butter has also been put on the market. To obtain it, it was necessary to melt butter, to cool it slowly, and to collect the soft fraction when maintained at the refrigerator temperature. This fraction is finally mixed with normal butter to give it physical softness. Despite this property, this is a traditional butter with a fat content reaching 82%. The consultation of the labels on the packaging is thus essential to appreciate the potential fat intakes.

In some areas, salt is, for a long time, added to butter, without exceeding 3%, to allow a longer conservation.

Consumption: Butter is consumed preferably uncooked because heating beyond 130°C degrades it by giving rise to toxic compounds. The domestic butter consumption in 2012 in the United States was about 2.5 kg per capita per year, in Canada about 2.8 kg, and in Europe (EU 27) about 4 kg.

Butter constitutes the largest contribution of lipids in the French diet, that is to say 12.1% of the total lipids consumed by an adult man (Razanamahefa et al. 2005). Similarly, butter is the first food supply of saturated fatty acids, for example, 17.3% of the total ingested saturated fatty acids (INCA Study 1999). INCA2 investigation (2006–2007) has indicated that the average consumption of butter was 11.6 g/day, equivalent to 4.2 kg per capita per year. The INRA organization determined that butter consumption by French people decreased more than half between 1969 (8.3 kg per capita per year) and 2001 (3 kg per capita per year); this fall being partially compensated by an increase in the consumption of the vegetal margarines (Section 2.3.1.13). The Interprofessional National Center of the Dairy Economy (CNIEL) and the services of the Ministry for Agriculture considered that the French consumption of butter is 7.6 kg per capita and per year. These differences in estimates originate in the various sources of consumption data, the different types of consumer surveys, and in the production and import statistics. Moreover, the share of the quantities of the "hidden" butter in the commercial meals remains very badly known.

The butter oil is largely employed in food industry for the manufacture of chocolate, pastry, confectionery, and ice creams, foodstuffs seldom perceived as containing butter.

2.3.2.4 Fish Oils

Origin: Fish oils are generally the by-products of the preparation of fish meals starting from species not directly consumed by humans. The composition of these oils is
thus very dependent on the species used, the period, and their capture area. The fatty
fish such as herring (*Clupea harengus*), anchovy (*Engraulis* spp., *Anchoa* spp.), and
menhaden (*Brevoortia patronus*) are the species generally used to prepare these oils.

Composition: Fish oils contain more than 90% triacylglycerols, the unsaponifi-
able fraction (from 1% to 3%) being primarily made up of cholesterol, hydrocarbons
(squalene), and fatty alcohols (with 14–22 carbons). Globally, they have a fatty acid
composition largely more varied than that of the majority of lipid sources. Thus,
more than 50 fatty acid species have been described in the oils of seawater fish,
8 fatty acid species representing more than 80% of the total. The fatty acid composition of the two more important sources of fish oil is given in Table 2.16.

Fish oil composition is variable but characterized by a high content in polyunsaturated fatty acids of the *n*-3 series: 20:5 *n*-3 (EPA) and 22:6 *n*-3 (DHA). The concentrations of 22:1 *n*-11 (cetoleic acid, an isomer of erucic acid) and 20:1 *n*-9 (gadoleic
acid) are dependent on the food ingested by fish. The complexity of the composition
in polyunsaturated fatty acids is, even still today, at the origin of our ignorance of
the triacylglycerol molecular species present in these oils. On the other hand, the
study of the average stereospecific distribution of the fatty acids in the triacylglycerol
species has shown that the *n*-3 fatty acids are mainly acylated at the *sn*-2 position
of glycerol. The saturated and monounsaturated fatty acids (except for the 22:1) are
preferentially acylated at the *sn*-1 position.

An example of the fatty acid distribution in herring oil is given in Table 2.17.

TABLE 2.16
**Fatty Acid Composition (Weight %) of Oils Prepared
from Two Fish Species, Menhaden (*Brevoortia
tyrannus*) and Herring (*Clupea harengus*)**

Fatty Acid	Menhaden	Herring
14:0	7–12	5–8
16:0	15–26	10–19
16:1 (*n*-7)	9–16	6–12
18:0	2–4	1–2
18:1 (*n*-9)	8–14	9–25
20:1 (*n*-9)	—	7–20
20:5 (*n*-3)	11–16	4–15
22:1 (*n*-11)	<1	7–30
22:6 (*n*-3)	5–14	2–8

TABLE 2.17

Distribution in Weight (%) of the Main Fatty Acids of Herring Oil on Each Carbon Position

Position	14:0	16:0	16:1	18:0	18:1	18:2	20:1	22:1	20:5	22:5	22:6
sn-1	6	12	13	1	16	3	25	14	3	1	1
sn-2	10	17	10	1	10	3	6	5	18	3	13
sn-3	4	7	5	1	8	1	20	48	4	1	1

Source: From Brockerhoff, H. et al., Lipids, 3, 24–9, 1968.

The oil extracted from menhaden (*Brevoortia tyrannus*), a fish abundantly caught in North Atlantic, has a similar structure. These results are important in human nutrition if one considers the preferential bioavailability of unsaturated fatty acids in the central position (*sn*-2) of glycerol carbon chain (Section 3.2.1.4).

Production: The worldwide production of fish oil has decreased during the last decade and is currently estimated at 1 million tons a year, this value corresponding to a daily ration of no more than 500 mg oil per capita. About 10 countries contribute to fish oil production. The first one is Peru (31%), and the second being Chile (15%).

Except the increasing supply difficulties, one of the main challenges that industry will have to overcome is the elimination of the numerous pollutants accumulated in the lipid reserves of fish (Tocher 2009). Treatments by distillation of oils have been exploited to eliminate the very small quantities of mercury, arsenic, dioxane, polychlorinated biphenyls, and other organic pollutants, but with a considerable incidence on the selling price. The development of a controlled aquaculture for fish or algae as sources of *n*-3 fatty acid-rich oils and of new genetically modified crops will be probably the solutions to the problem of an increasing environmental pollution (Nichols 2010).

Consumption: Approximately one-tenth of the worldwide fish oil production is used as food supplement for humans, the remainder being used in animal feeding (aquaculture), and by chemical and pharmaceutical industries.

2.3.2.5 Meat-Included Fats

The amount of lipids included (or "hidden") in muscle tissues of the livestock (mammals, birds) depends on the species, of the analyzed muscles and also largely of the breeding conditions (food composition). Despite that variability, the average values for the different anatomical localizations corresponding to the most usually consumed pieces are given in Tables 2.18 (beef and calf), 2.19 (lamb and pork), and 2.20 (chicken, duck, and rabbit).

It is clearly seen that, in the same animal, various anatomical localizations lead to various nutritional contributions in lipid amount but also in fatty acid intake. Beef and lamb provide the highest quantities of saturated fatty acids but have a fatty acid profile with an *n*-6 to *n*-3 fatty acid ratio between 2 and 4. Calf meat has lower lipid content than the previous ones. It has also less saturated fatty acids, but the *n*-6 to *n*-3 fatty acid ratio is less satisfactory. Pork, rabbit, and poultry supply less lipids than the other animals, chicken and rabbit being the richest in *n*-3 fatty acids.

TABLE 2.18
Total Lipid Amounts (g/100 g) and Fatty Acid Composition of Some Pieces of Beef and Calf (mg/100 g)

	Beef				Calf	
	Steak	Filet	Entrecote	Ground beef	Cushion	Shank
Lipid content	3.4	6.7	17	5	2.6	2.6
Saturated	800	2741	8061	1643	799	1289
16:0	474	1613	4293	933	442	724
18:0	226	761	2676	501	232	347
n-9	785	2588	6014	1766	887	1762
n-6	116	141	242	178	312	346
n-3	62	77	93	79	23	24

Source: Data from "Valeurs nutritionnelles des viands—CIV—INRA, 2006–2009."

TABLE 2.19
Total Lipid Amounts (g/100 g) and Fatty Acid Composition of Some Pieces of Lamb and Pork (mg/100 g)

	Lamb				Pork	
	Leg	Neck	Saddle	Chop	Loin	Ham
Lipid content	5.1	13.7	18.4	18	3	2.9
Saturated	1967	5794	7942	6361	1194	850
16:0	986	2799	3842	2983	777	292
18:0	548	1674	2203	1771	360	558
n-9	1685	4491	5876	4716	1521	1100
n-6	318	568	795	646	270	320
n-3	80	128	172	136	9	20

Sources: Data for lamb (Valeurs nutritionnelles des viandes—CIV—INRA, 2006–2009) and for pork (Gandemer, G. et al., *Journées Rech. Porcine Fr.*, 22, 101–10, 1990.)

TABLE 2.20
Total Lipid Amounts (g/100 g) and Fatty Acid Composition of Some Pieces of Chicken, Duck, and Rabbit (mg/100 g)

	Chicken (Thigh)	Duck (Magret)	Rabbit (Back)
Lipid content	2.8	2.1	2
Saturated	792	681	780
16:0	557	491	558
18:0	235	157	150
n-9	776	1034	559
n-6	661	371	559
n-3	123	13	115

Sources: Data for chicken (Lessire, M., *Inra Prod. Anim,* 14, 365–70, 2001.), duck and rabbit (Chartrin, P. et al., *Anim. Res.*, 55, 231–44, 2006.)

2.4 PHOSPHOLIPIDS

Phospholipids are present in food in the form of glycerophospholipids or sphingo-sylphospholipids. They are integral part of meat and dairy products consumed by humans. Although constitutive of the cellular membranes, they are also present in the form of reserves in milk and egg yolk. Plants also contain phospholipids but their low amount in food (maximum 0.2%) reduced enormously their nutritional interest, except in the form of supplementation products prepared from some seeds. Their interest is based as well from the point of view of their fatty acid content as from their nonlipidic component (polar head). These phospholipids will be regarded here as only able to sup-ply choline to the organism. Choline is a water-soluble quaternary amine essential for the building of all cell membranes and especially for the performance of the nervous system (Section 3.2.5.1). Choline is present in the polar head of a glycerophospholipid, the phosphatidylcholine and of a sphingosylphospholipid, the sphingomyelin.

The phosphatidylcholine (Figure 2.20), as all the glycerophospholipids, contains a common core, the diacylglycerophosphate (or phosphatidic acid), to which choline is linked. Palmitic acid (16:0) is generally acylated at the *sn*-1 position (R1), whereas oleic acid (18:1 *n*-9) or linoleic acid (18:2 *n*-6) is acylated at the *sn*-2 position (R2).

One still employs the historical term *lecithin* (of the Greek lecithos, egg yolk) given in 1845 by the French chemist, N.T. Gobley, to name the phosphorylated lipid that he had just discovered in egg yolk. Phosphatidylcholine is the most abundant phospholipid in animal and vegetable tissues (~50% of the total phospholipids), con-taining approximately 14% of choline. By extension, the term *lecithin* is now used in food industry (*Codex Alimentarius*) to name a mixture rich in phospholipids (at least 60%) composed mainly of phosphatidylcholine and with a content of neutral (or simple) lipids lower than 40%. Lecithin is extracted from various animal (egg yolk) or vegetal (soybean or others) sources. The raw soybean oil has the highest content of lecithin (from 1% to 3%) among all vegetal oils.

Phosphatidylcholine contained in our food (Section 3.2.5.1) is digested like the other glycerophospholipids in the intestine by giving different metabolites such as fatty acids, a glycerophosphate molecule, and choline. The latter is the original element of choline-containing phospholipids and it must be supplied to a suitable amount to the organism. A nutritional supplementation in phosphatidylcholine is recommended in various path-ological situations, but without differentiating the role of choline from that of the other parts of the molecule (Section 4.5.1). It has been shown that lysophosphatidylcholine,

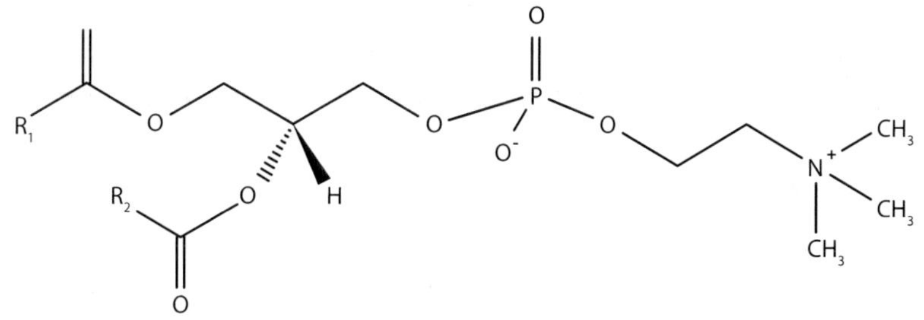

FIGURE 2.20   Phosphatidylcholine (R1 and R2 are the carbon chains of the two fatty acids).

resulting from the hydrolysis of phosphatidylcholine after removing one fatty acid, may be directly absorbed by the intestinal mucosa before being transformed back into the initial phospholipid and if necessary being transported in the blood.

Sphingomyelin (Figure 2.21) is the most abundant sphingolipid present in our food, this family being formed by lipids containing a long-chain amino alcohol, such as sphingosine. It is a very heterogeneous lipid family because it brings together sphingomyelin and many glycolipids. All these lipids have in common a ceramide core formed by a fatty acid linked to the –NH$_2$ group of sphingosine. If one adds one phosphocholine group to the primary –OH group of the sphingosine, it results in a molecule of sphingomyelin rather close structurally to phosphatidylcholine. The fatty acid linked to the amino alcohol is generally saturated and have from 18 to 24 carbon atoms.

Sphingomyelin is present in all the animal cellular membranes, including the external wall of the lipid globules of milk. It constitutes approximately 10% of the nervous tissue lipids and 3%–6% in other tissues, whereas practically absent in plants. Dietary sphingomyelin, with its content in choline (14%) contributes, like phosphatidylcholine, to the regular supply of that quaternary ammonium salt considered as an essential nutrient and usually grouped within the B-complex vitamins. Like other phospholipids, sphingomyelin supplied by our food (Section 3.2.5.2) is hydrolyzed in the intestine being mainly transformed into a ceramide, which even can be hydrolyzed into sphingosine and fatty acid. These products may have a direct action on the intestinal mucosa (Section 4.6.2) or may be recycled by enterocytes to form other sphingolipids.

Sphingomyelin is no longer regarded as an inert membrane component, but as a source of bioactive lipids (ceramide, ceramide phosphate, and sphingosine) known to be involved in cellular signaling.

Phosphatidylserine (Figure 2.22) is constituted by a phosphatidic acid linked to an amino acid, the serine. It is a minor constituent of cellular membranes, but the

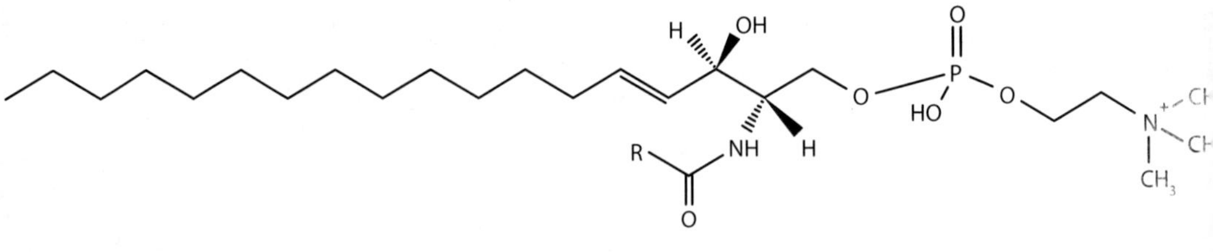

FIGURE 2.21 Sphingomyelin (R is the fatty acid carbon chain with 20–24 carbon atoms).

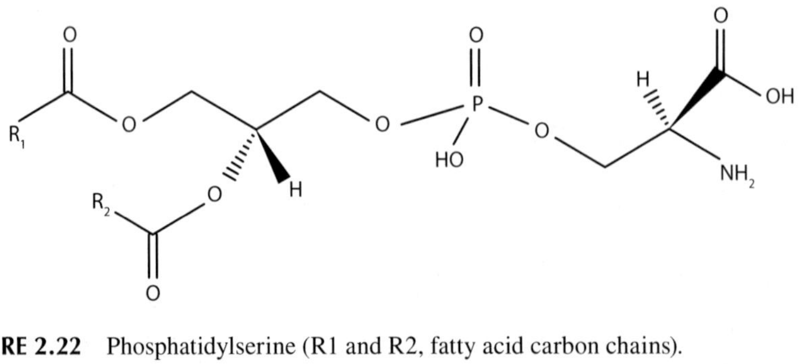

FIGURE 2.22 Phosphatidylserine (R1 and R2, fatty acid carbon chains).

nervous system, especially the white matter, contains great quantities of this phospholipid (up to 18%). One of its characteristics is to have a highly unsaturated fatty acid (often DHA) at the *sn*-2 position (R2). Its exact structure has been elucidated only in 1948 by J. Folch.

Phosphatidylserine can be regarded as a bioactive lipid because it has been shown to be involved in several physiological mechanisms, such as the activation of protein kinase C and the initiation of blood coagulation. Some nutritional studies suggest that this phospholipid could have an interest in the improvement of brain function (Section 4.5.2.1) or physical performances (Section 4.5.2.2).

2.5   GLYCOLIPIDS

Although several types of glycolipids are present in animal and vegetal cells, only the glucosylceramide (Figure 2.23), classified in the group of glycosphingolipids and among the most represented, has been the target of nutritional research. This glycolipid, discovered in plants (wheat flour) in 1954, can be considered as the simplest one in plants as well as in animals. The compound has a hydrophobic part represented by a ceramide, as in the sphingomyelin molecule but linked to a glucose moiety. This glycolipid is often classified in the lipid group named cerebrosides.

Besides this glycolipid, food of animal origin contains a huge number of glycosphingolipids having a more or less diversified polar head (globosides, sulphated, or phosphorylated glycolipids), the most complex being the gangliosides (Section 4.6.1). In plants, the variety is not so wide and represented by mono- and oligohexosylceramides containing one or more carbohydrate moieties (glucose, galactose, mannose, and inositol), some of them being phosphorylated. Thus, a phosphoceramide bound to a tetrasaccharide (sometimes called phytoglycolipid) constitutes from 5% to 12% of the soya lecithin.

Glycolipids are, like phospholipids, digested in the intestine giving firstly a ceramide and then a fatty acid and an amino alcohol. All these metabolites will be absorbed by the enterocytes to be again transformed into more or less complex sphingolipids (Vesper et al. 1999). Like sphingomyelin, these lipids are regarded on the nutritional level mainly as ceramide donors (Section 3.2.6) being able to act directly on the intestine (Section 4.6.2) or to be integrated in more complex lipids in all the human body.

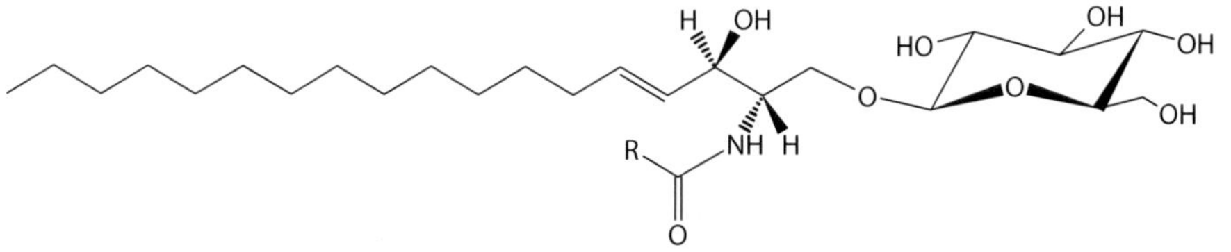

FIGURE 2.23   Glucosylceramide (R, fatty acid carbon chain).

2.6   CHOLESTEROL AND PHYTOSTEROLS

Cholesterol (Figure 2.24) and phytosterols (vegetal sterols) belong to the vast group of the steroids. They derive metabolically from squalene, a common terpene in plant and animal kingdoms and all have a four-cycle skeleton, the sterane nucleus (cyclopentanoperhydrophenanthrene). The steroid group includes many other parent compounds, such as hormonal steroids, biliary acids, and vitamin D.

Cholesterol is the major sterol in animals where it takes part in the construction of the cellular membranes, ensuring them certain rigidity. It is more particularly abundant in the suprarenal glands, the nervous system, the liver, and the gallstones. It is in these crystalline concretions formed within the gallbladder that it was isolated for the first time in 1770 by F. Poulletier de la Salle. Later, in 1815, it was recognized in animal fats by M.E. Chevreul who gave it the name *cholesterine* (from the Greek kholé = bile and stereos = solid). M. Berthelot, taking into account the alcohol function, modified in 1859 the name into cholesterol. This lipid is also found esterified by a fatty acid, the cholesterol ester, the preferred form to be transported in plasma within lipoproteins. Except for some exceptions, cholesterol is absent in plants.

The implications of cholesterol in the human diet are exposed in Section 3.2.3.1 and its importance for health is exposed in Section 4.3.1.

Phytosterols are the plant sterols, equivalent to cholesterol in animals. They can be divided into two groups: the true phytosterols or "Δ5-sterols" with a double bond in position 5 on the sterol nucleus, and the stanols or "5α-sterols" with an entirely saturated sterol nucleus.

Among more than about 40 different forms of phytosterols, the most important is the β-sitosterol (Figure 2.25). It is present in the membranes of the vegetal cells but

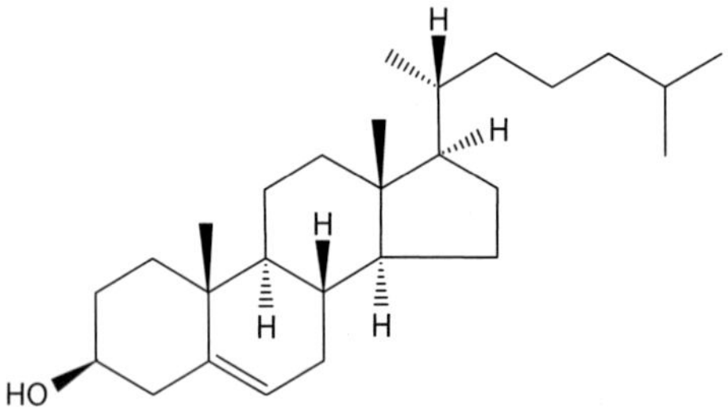

FIGURE 2.24   Cholesterol.

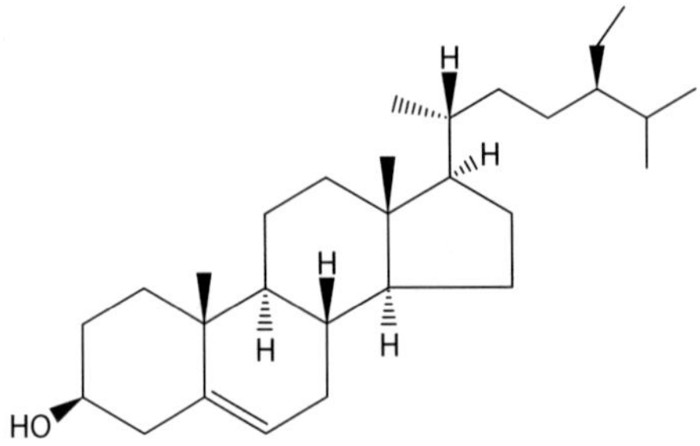

FIGURE 2.25   β-Sitosterol.

also in many oils (corn and wheat germ). One also finds other types of phytosterols such as campesterol in rapeseed oil and ergosterol in yeasts and fungi.

Concurrently to these phytosterols, plants also contain their saturated derivatives, the stanols. The most abundant is sitostanol, the saturated derivative of β-sitosterol, and campestanol. Like all phytosterols, they are often combined with one fatty acid to give stanol esters. Cholestanol has been synthesized for nutritional experiments.

The involvement of phytosterols and phytostanols in human nutrition is exposed in Section 3.2.3.2, and its importance for health is exposed in Section 4.3.2.

2.7 LIPOSOLUBLE VITAMINS

The liposoluble vitamins (vitamins A, D, E, and K) form a heterogeneous group of compounds that may be considered as true lipids. They were all discovered during the twentieth century as nutritional factors at the origin of deficiency diseases. Some of them must be imperatively supplied by the diet; others may be biosynthesized by humans after ingestion of precursors or by cutaneous photoreaction. Their functions are very diverse and their supplementation proves gradually useful for important therapeutic actions, some of them still remaining to be confirmed.

2.7.1 VITAMIN A AND CAROTENOIDS

2.7.1.1 Vitamin A

Vitamin A is a generic term for several substances chemically related to retinol (Figure 2.26) and with similar biological activities, such as retinol esters, retinal, and retinoic acid. Retinol may be transformed into retinal after substitution of the alcohol function by an aldehyde function or into retinoic acid after substitution by a carboxyl functional group. They all belong to the group of the retinoids. They may be considered also as apocarotenoids because they are formed from a carotenoid molecule by the loss of a fragment at one end of the carbon chain. They have a double vitamin activity in animals: photoreception (retinol and retinal) and regulation of cell multiplication (retinoic acid). The physiological and nutritional properties of vitamin A are exposed in Section 3.2.4.1, and its importance for health is exposed in Section 4.4.1.

Paul Karrer (1889–1971) discovered that carotenoids, and especially β-carotene (Figure 2.27), were the metabolic precursors of retinoids (provitamin A). This famous Swiss chemist was awarded the Nobel Prize in 1937. Retinol is present in plasma and tissues of animals (especially in the liver) in an esterified form after combining with oleic acid or palmitic acid. These esterified forms must be hydrolyzed

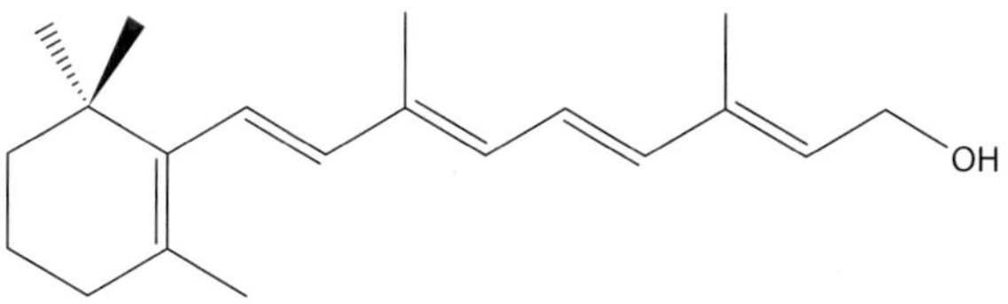

FIGURE 2.26 All-*trans*-Retinol.

to recover a physiological activity. The natural sources of vitamin A for humans are the retinoids contained in the ingested animal products (meats and dairy products) and some carotenoids (provitamin A) of vegetal origin. Among nearly 600 known carotenoids, less than 10% can be sources of vitamin A (Bendich et al. 1989).

2.7.1.2 β-Carotene

β-Carotene (Figure 2.27) is the most known carotenoid, since its isolation from carrot and the separation of its two other isomers (α and γ-carotene) in 1931. Its structure was elucidated by P. Karrer in 1930. This compound gives an orange-red color to many vegetables and fruits but is masked by chlorophyll in all the green plants. As for all carotenoids, its principal function is to protect vegetal cells against the attacks exerted by the oxygenated reactive species formed by the primary action of light energy. One of the oxidation metabolites of β-carotene, the β-cyclocitral, is a messenger involved in the mechanisms of defense to oxidative stress (Ramel et al. 2012). Carotene may be obtained by chemical synthesis but is currently and progressively prepared from raw palm oil and alfalfa, an important forage crop, *Medicago sativa* (Fabaceae). Fungi and marine microalgae are also important sources of carotene in some countries.

Carotene is industrially used as food coloring (E160a) in various edible products (butter, margarine, sweet pastries, drinks, and confectioneries) and animal foods. It is also used as provitamin A as a vitamin supplement. The physiological properties and the nutritional importance of carotenoids are discussed in Section 3.2.4.1, and its importance for health is discussed in Section 4.4.1.

2.7.1.3 Lutein and Zeaxanthin

Lutein (Figure 2.28) is a carotenoid of the xanthophyll group, oxygenated analogues of β-carotene. Lutein is present in some food of vegetal origin in free (95%) or esterified form. Zeaxanthin is a *cis* isomer of lutein, both hydroxyl groups being in the same spatial plan. Several other stereoisomers are present in the human serum and retina. Lutein (E161b) and zeaxanthin (E161h) are authorized as food additives.

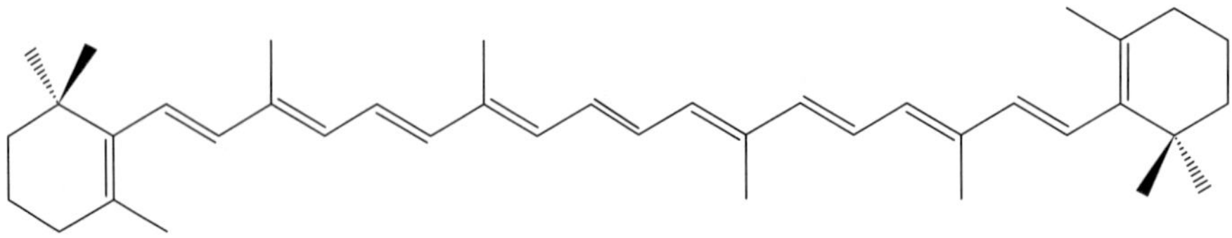

FIGURE 2.27 β-Carotene.

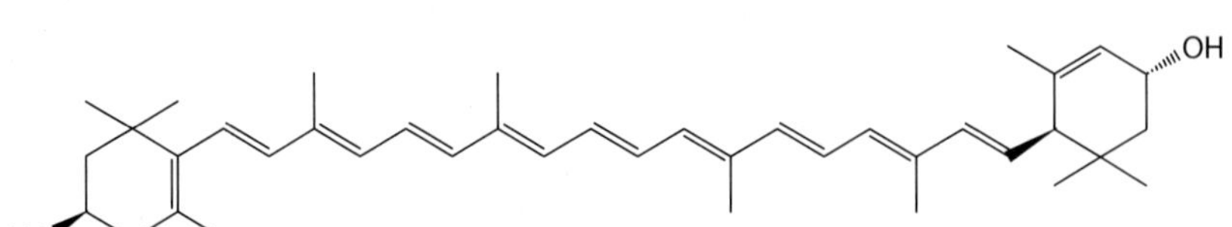

FIGURE 2.28 Lutein.

The food sources and the possible dietary requirements of lutein and zeaxanthin are exposed in Section 3.2.4.1, and its importance for health being exposed in Section 4.4.1.

2.7.1.4 Astaxanthin

That carotenoid (Figure 2.29) pertains by its chemical structure to the group of xanthophylls and is present in many marine animals consumed by humans: shellfish, mollusk, and fish (Maoka 2011). This compound is the origin of the blue color of lobster, because of its combination with a specific protein. Astaxanthin is responsible for the orange-red color of salmon muscles, shrimps, and other shellfish present in fish shops. In humans, its dietary intake is also done by the consumption of many livestock (poultries and fish), themselves fed food enriched with carotenoids.

Astaxanthin has unquestionable powerful antioxidant properties, which would exceed even those of β-carotene and vitamin E.

The food sources of astaxanthine are discussed in Section 3.2.4.1, and its importance for health is discussed in Section 4.4.1.

2.7.1.5 Lycopene

Lycopene (Figure 2.30) is the first metabolite deriving directly from phytoene by desaturation, this last compound being the precursor of all carotenoids. Its red color rises from the presence of 11 conjugated double bonds. It is authorized as food additive (E160d).

Lycopene is the carotenoid having the most powerful antioxidant action enabling the elimination of free radicals. The food sources of lycopene are exposed in Section 3.2.4.1, and its importance for health is exposed in Section 4.4.1.

2.7.2 VITAMIN D

Vitamin D is a group of several molecules deriving from sterols by opening of one cycle (secosteroids). The most abundant form of vitamin D in the human body is cholecalciferol (vitamin D_3; Figure 2.31) that may be compared to a provitamin D.

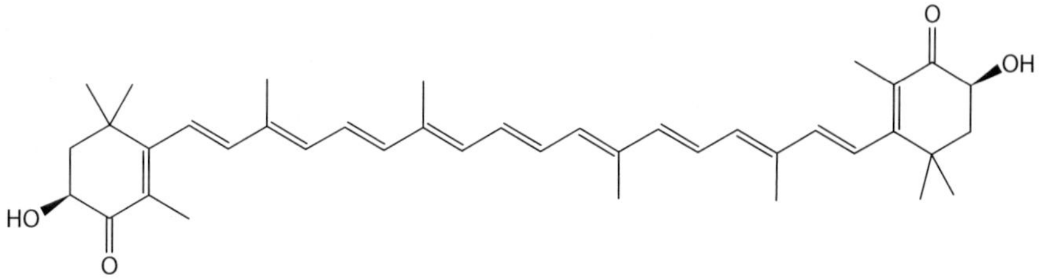

FIGURE 2.29 Astaxanthin.

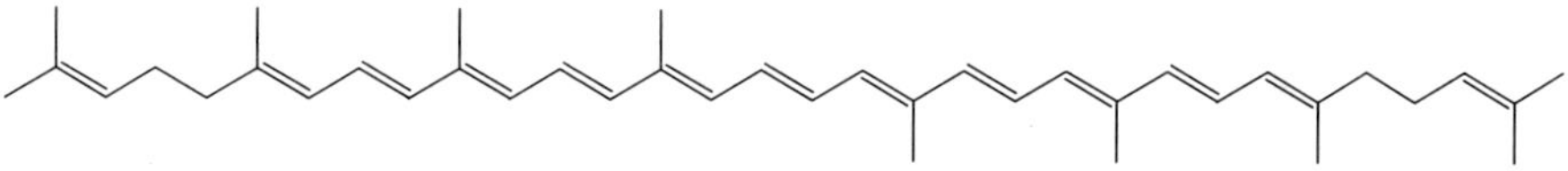

FIGURE 2.30 Lycopene.

This compound is formed by ultra violet (UV) irradiation (290–320 nm, UV-B) of a hydroxylated derivative of cholesterol localized in the skin. The vitamin D_3, from cutaneous or dietary origin, is metabolized and stored in the liver in the form of 25-hydroxycholecalciferol (or calcidiol, Figure 2.32). This last compound is then transported in plasma toward the kidney where it is transformed into an active form, calcitriol or 1,25-dihydroxycholecalciferol (Figure 2.33). This last metabolic step is controlled by the parathyroid hormone, secreted in response to hypocalcaemia. To a lesser extent, skin is known to carry out all the previous reactions. A natural calcitriol metabolite, the 1,25-dihydroxy-3-epi-cholecalciferol, has also a vitamin D activity, but its exact role remains to be defined (Molnar 2011). The complex cellular metabolism of vitamin D leads to consider this bioactive molecule as more of a hormone than as a vitamin

In parallel, UV irradiation of ergosterol contained in mushrooms and some vegetal oils is able to produce ergocalciferol (vitamin D_2). This derivative with a biological activity lower than that of vitamin D_3 is poorly present in the diet but is sometimes used in some countries in pharmaceutical preparations for vitamin D supplementation.

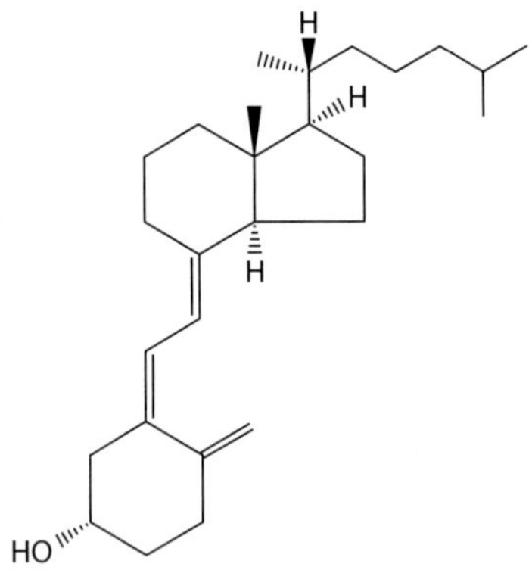

FIGURE 2.31   Vitamin D_3: Cholecalciferol.

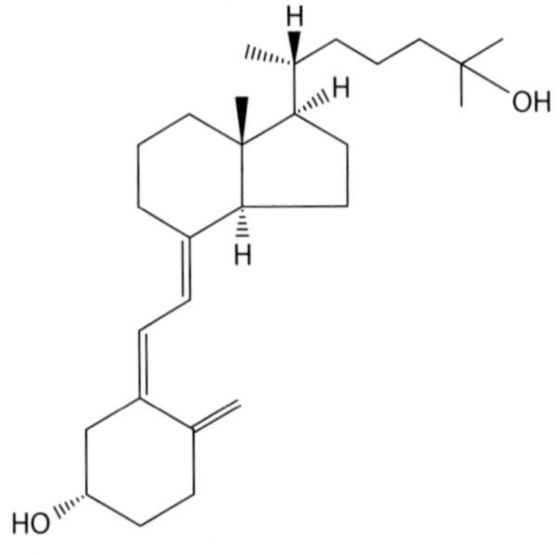

FIGURE 2.32   25-Hydroxy-vitamin D_3: Calcidiol.

The food sources of vitamin D, its synthesis during sun exposure, and the dietary reference intake in humans are exposed in Section 3.2.4.2. Its importance for health is exposed in Section 4.4.2.

2.7.3 VITAMIN E

Vitamin E is a generic term, which refers to a group of eight different molecules but with close structures (tocochromanols). All these compounds may be distributed into two groups: four tocopherols, vitamin E with a saturated side chain (Figure 2.34), and four tocotrienols, vitamin E with an unsaturated side chain (Figure 2.35). All these compounds are specific of the vegetal kingdom and are essential for humans (Section 3.2.4.3).

Tocopherols (Figure 2.34) have a phytyl chain (made of three isoprenoid units) entirely saturated. They are represented by four forms (vitameres) differentiated by the number and the position of the methyl groups on the phenolic nucleus: α-, β-, γ-, and δ-tocopherol.

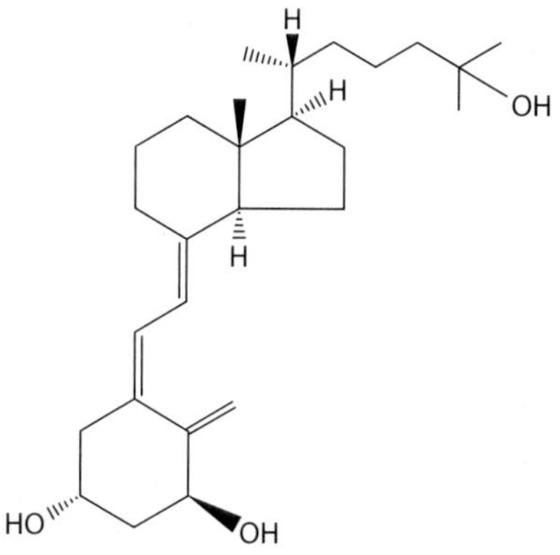

FIGURE 2.33 1,25-Dihydroxy-vitamin D_3: Calcitriol.

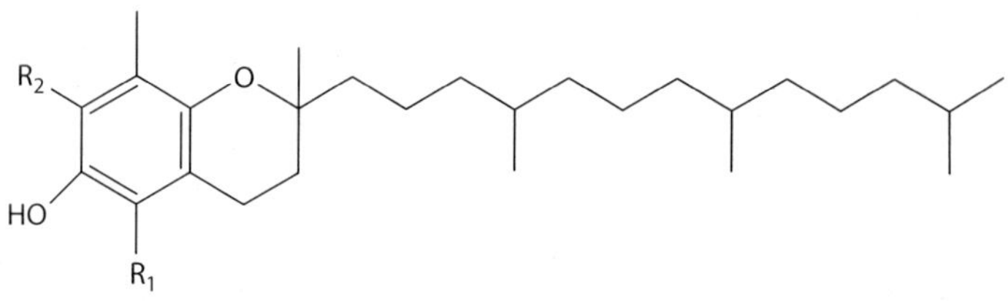

FIGURE 2.34 Tocopherols

R_1	R_2	
CH_3	CH_3	α-Tocopherol
CH_3	H	β-Tocopherol
H	CH_3	γ-Tocopherol
H	H	δ-Tocopherol

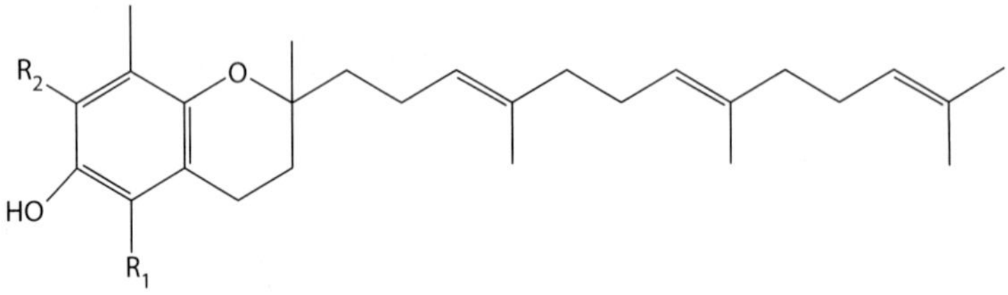

FIGURE 2.35   Tocotrienols

R$_1$	R$_2$	
CH$_3$	CH$_3$	α-Tocotrienol
CH$_3$	H	β-Tocotrienol
H	CH$_3$	γ-Tocotrienol
H	H	δ-Tocotrienol

α-Tocopherol is the most known form, often used to define chemically and physiologically the "vitamin E" complex.

Tocotrienols (Figure 2.35) have a phytyl chain with three double bonds. As for tocopherols, they are represented by four forms differentiated by the number and the position of the methyl groups on the phenolic nucleus: α-, β-, γ-, and δ-tocotrienol.

Besides their role in animal reproduction that was the source of their discovery, all the compounds forming the complex of vitamin E protect lipids and prevent to various degree the oxidation of polyunsaturated fatty acids present in cell membranes and circulating lipids.

The food sources of vitamin E and the dietary reference intake in humans are exposed in Section 3.2.4.3, and its importance for health is exposed in Section 4.4.3.

2.7.4   VITAMIN K

As for vitamin E, vitamin K refers to a group of molecules belonging to the group of prenylated quinones, all having an antihemorrhagic activity.

The most important form of the vitamin K complex is phylloquinone or vitamin K$_1$ (Figure 2.36), which is provided by plants. This molecule is a methylated naphthoquinone with a monounsaturated side chain of four isoprenoid units.

The other forms are the menaquinones or vitamin K$_2$ (Figure 2.37), which are different by the length and the unsaturation degree of the side chain. These molecules are also methylated naphthoquinones but with a polyunsaturated side chain with 4–13 isoprenoid units. They are named menaquinone-n (or MK-n), n being the number of isoprenoid units. In humans, these menaquinones are synthesized mainly by the gram-positive bacteria living in the posterior intestine.

The food sources and the dietary reference intake of vitamin K in humans are exposed in Section 3.2.4.4, whereas its importance for health is exposed in Section 4.4.4.

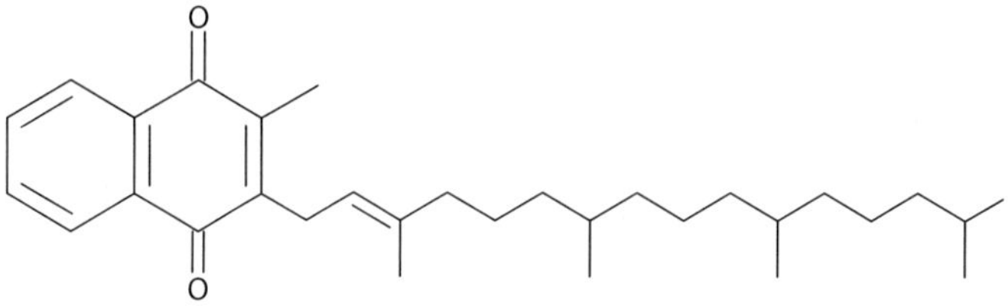

FIGURE 2.36 Vitamin K1 (Phylloquinone).

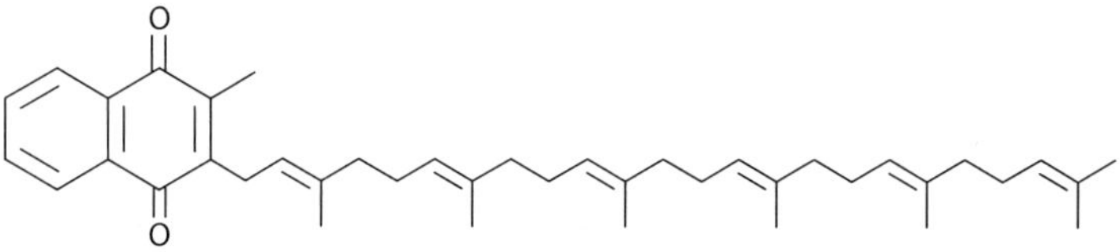

FIGURE 2.37 Vitamin K2 (Menaquinone-6).

2.8 LIPID SUBSTITUTES

To fight against obesity, the manufacturers tried, with clinicians, to launch artificial fats with low energy contents. Several lipid substitutes have been synthesized but little are on the market because of a lack of toxicological studies in humans.

2.8.1 LOW-ENERGY LIPIDS

Among the lipids used in low-calorie diets, there are various triacylglycerols known as "structured triglycerides" obtained by interesterification of different acylglycerols with synthetic glycolipids (acylated sugars).

2.8.1.1 Structured Triglycerides

Salatrim has been synthesized by using oils rich in short-chain fatty acids (2, 3, or 4 carbons) and hydrogenated vegetal oils, thus containing mainly stearic acid. Salatrim is the acronym corresponding to "short- and long-chain acyl triglyceride molecules." Typically, that product contains from 30% to 67% of short-chain fatty acids in the *sn*-1 and *sn*-2 positions of glycerol and from 33% to 70% of long-chain fatty acids in the *sn*-3 position. It should not contain more than 0.5% free fatty acids and not more than 2% monoacylglycerols. Salatrim has an energy value of 21 kJ/g (5 kcal/g). It should not be used for frying, but it can replace cocoa butter. It is marketed under the brand Benefat® by Pfizer and Nabisco.

Figure 2.38 is an example of one molecular species abundantly present in this product.

The manufacturer has launched about 10 Salatrim mixtures differing by the nature of the short chains and the vegetal oil used for their synthesis. The used numbering consists in indicating with figures the number of carbons in the short chains and with letters the nature of the oil providing the long chains. Thus, Salatrim 43SO is made by interesterification of tributyrin (four carbons), tripropionine (three carbons), and hydrogenated soybean oil. Moreover, the manufacturer indicates in the specification sheet the molar ratio of the used oils; here the mixture is 11 parts of tributyrin for 1 part of tripropionine and 1 part of soybean oil.

Caprenine is constituted by the fatty acids 8:0 at the *sn*-1 position, 10:0 at the *sn*-2, and 22:0 at the *sn*-3 position (Figure 2.39), and it has an energy value of 16.7 kJ/g (4 kcal/g), therefore the half energy than natural triacylglycerols. This molecule is marketed by Procter & Gamble, Cincinnati, Ohio. It may be used in the manufacture of chocolate.

Neobee® (more exactly Neobee® MCTs), produced by Stepan Lipid Nutrition, Northfield, Illinois, is a triacylglycerol with only medium-chain fatty acids (with 8 and 10 carbons) being provided by coconut oil and palm tree oil. That structured triglyceride is used in human nutrition, mainly in children and sportsmen. It is also used in the manufacture of some food preparations (confectionery, dried fruits, and drinks) (Figure 2.40).

The physiological effects of these structured triacylglycerols are exposed in Section 4.7.1.

2.8.1.2 Nondigestible Glycolipids

The only product still marketed under the name of Olean® by Procter & Gamble is Olestra, a mixture of saccharose molecules bound to six to eight fatty acid chains of vegetal origin (soya and cotton) (Figure 2.41). Molecules containing only one to three fatty acids are used as food additive (sucroester, E473). The chemical structure of Olestra makes it insensitive to intestinal enzymes. As it contains fatty acids, Olestra is able to dissolve the liposoluble vitamins and carotenoids present in food, but these compounds are excreted with the undigested molecules of Olestra.

The physiological effects of this lipid substitute are exposed in Section 4.7.1.

$$H_2C-O-CO-CH_3$$
$$H_3C-(CH_2)_{16}CO-O-C-H$$
$$H_2C-O-CO-(CH_2)_2-CH_3$$

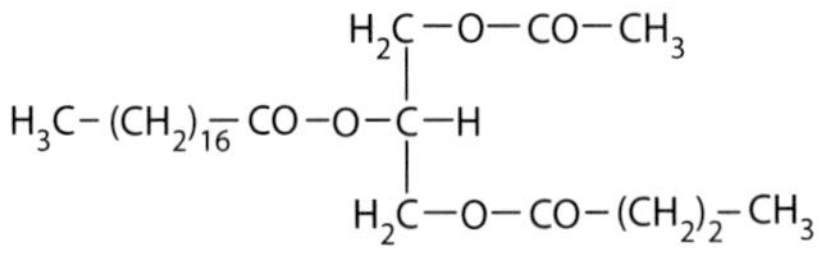

FIGURE 2.38   One of the molecular species present in Salatrim.

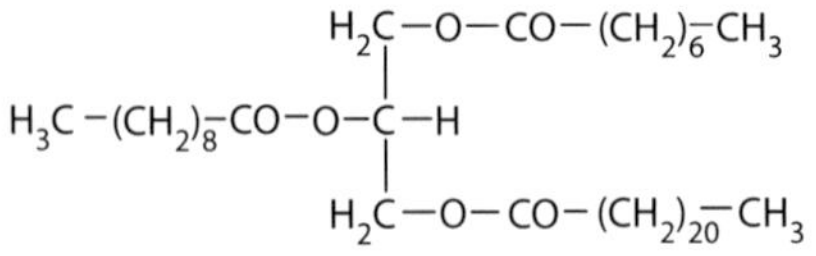

$$H_2C-O-CO-(CH_2)_6-CH_3$$
$$H_3C-(CH_2)_8-CO-O-C-H$$
$$H_2C-O-CO-(CH_2)_{20}-CH_3$$

FIGURE 2.39   Caprenine.

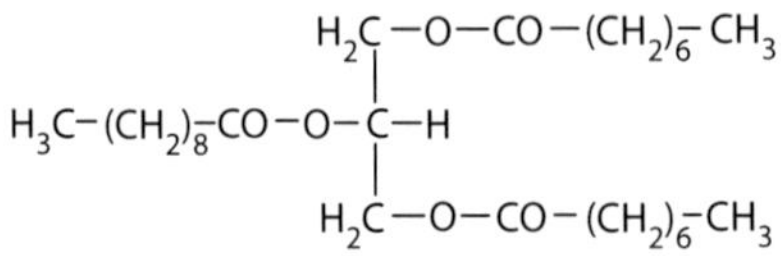

$$H_2C-O-CO-(CH_2)_6-CH_3$$
$$H_3C-(CH_2)_8-CO-O-C-H$$
$$H_2C-O-CO-(CH_2)_6-CH_3$$

FIGURE 2.40   Triacylglycerol species present in Neobee.

2.8.2 DIACYLGLYCEROLS

The diacylglycerols are natural components of marketed vegetal oils, but their content is maintained to relatively low levels. Their presence in natural oils is mainly due to hydrolysis of triacylglycerol molecules, a high content being a sign of poor quality. They are in general the object neither of investigations nor directive for standards of composition. According to the literature, the contents are from about 1% for rapeseed and soybean oils to 5% for palm and olive oils.

Diacylglycerols are present under the structure of two isomers, 1,3-diacyl-sn-glycerol (1,3-DAG, Figure 2.42) and the 1,2-diacyl-sn-glycerol (1,2-DAG, Figure 2.43). In commercial refined oils, the 1,3-DAG form is prevalent (~70%), as a consequence of a natural migration of fatty acids until that equilibrium resulting from high-temperature treatments.

The oils marketed under the name of "diacylglycerol oils" contain approximately 80% of diacylglycerols (mixture of 1,3- and 1,2-diacylglycerols with a ratio of 7:3), 20% or less of triacylglycerols, 5% or less of monoacylglycerols, emulsifying agents, and antioxidants. The main fatty acids comprising these oils are the oleic, linoleic, and linolenic acids. The vegetal diacylglycerol oil may be used in a ratio of 1:1 (weight/weight) to replace liquid vegetable oils in all their uses.

These oils are manufactured by esterification of fatty acids, derived from rapeseed and soybean oils of food quality, with either monoacylglycerol or glycerol. The esterification, catalyzed by a purified mould lipase, provides a mixture that will be

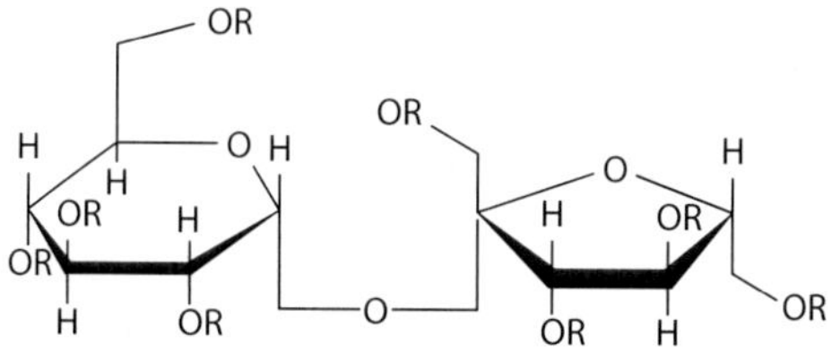

FIGURE 2.41 One molecular species contained in Olestra (R, fatty acid).

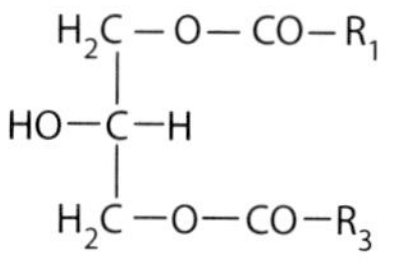

FIGURE 2.42 1,3-Diacyl-sn-glycerol (R1 and R3, fatty acid carbon chain).

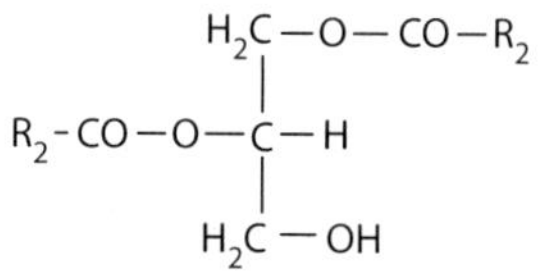

FIGURE 2.43 1,2-Diacyl-sn-glycerol (R1 and R2, fatty acid carbon chain).

further refined, washed, and deodorized. Some allowed additives (antioxidant and emulsifying agents) will be added before marketing.

The nutritional effects of diacylglycerols are examined in Section 4.7.2.

REFERENCES

Bendich, A., Olson, J.A. et al., 1989. Biological actions of carotenoids. *FASEB J.* 3:1927–32.

Brockerhoff, H., Hoyle, R.J. et al., 1968. Positional distribution of fatty acids in depot triglycerides of aquatic animals. *Lipids* 3:24–9.

Chartrin, P., Bernadet, M.D. et al., 2006. Effect of genotype and overfeeding on fat level and composition of adipose and muscle tissue in ducks. *Anim. Res.* 55:231–44.

Gandemer, G., Pichou, D. et al., 1990. Effects of rearing system and genotype on the chemical composition and sensory quality of the longissimus dorsi muscle of pigs. *Journées Rech. Porcine Fr.* 22:101–10.

Lessire, M., 2001. Matières grasses alimentaires et composition lipidique des volailles. *Inra Prod. Anim.* 14:365–70.

Maoka, T., 2011. Carotenoids in marine animals. *Mar. Drugs* 9:278–93.

Metzger, J.O., 2009. Fats and oils as renewable feedstock for chemistry. *Eur. J. Lipid Sci. Technol.* 111:865–76.

Molnar, F., Sigüeiro, R. et al., 2011. 1a,25(OH)2-3-epi-vitamin D3, a natural physiological metabolite of vitamin D3: Its synthesis, biological activity and crystal structure with its receptor. *PloS ONE* 6:e18124.

Nichols, P.D., Riep, B. et al., 2010. Long-chain omega-3 oils—An update on sustainable sources. *Nutrients* 2:572–85.

Ramel, F., Birtic, S. et al., 2012. Carotenoid oxidation products are stress signals that mediate gene responses to singlet oxygen in plants. *Proc. Natl. Acad. Sci. USA* 109:5535–40.

Ran-Ressler, R.R., Devapatla, S. et al., 2008. Branched chain fatty acids are constituents of the normal healthy newborn gastrointestinal tract. *Pediatr. Res.* 64:605–9.

Razanamahefa, L., Lafay, L. et al., 2005. Consommation lipidique de la population française et qualité des données de composition des principaux groupes d'aliments vecteurs. *Bull. Cancer* 92:647–57.

Tocher, D.R., 2009. Issues surrounding fish as a source of ω3 long-chain polyunsaturated fatty acids. *Lipid Technol.* 21:13–6.

Vesper, H., Schmelz, E.M. et al., 1999. Sphingolipids in food and the emerging importance of sphingolipids to nutrition. *J. Nutr.* 129:1239–50.

Wolff, R.L., Precht, D. et al., 1998. Occurrence and distribution profiles of trans-18:1 acids in edible fats of natural origin. In *"Trans fatty acids in human nutrition,"* pp. 1–33. (Edited by J.L. Sebedio and W.W. Christie, Oily Press, Dundee, Scotland, United Kingdom.)

3 Lipids and Human Nutrition

3.1 INTRODUCTION

3.1.1 History

To better understand the state of our current nutritional requirements, it is important to consider the historical evolution of the human diet since the appearance of the genus *Homo* during the Paleolithic Age (approximately 2.5 million years ago). A glance at the past allows to clarify the present situation and foresee the future. It is now well acquired that the nutritional characteristics of the modern man result from a long evolutionary process during which the environmental pressure has modeled slowly its genetic inheritance. It should not be forgotten that we are preceded by only 2 or 3 generations that knew a sophisticated diet, increasingly influenced by food engineering; 10 generations during the industrial era; approximately 500 generations that lived almost exclusively from agriculture; and more than 100,000 generations that only knew gathering plants, fishing, hunting, and scavenging wild animals (hunter-gatherers). Given the slowness of the evolutionary processes, natural selection could have had only very few repercussions since the appearance of agriculture, about 10,000 years ago, and still much less since the Industrial Revolution, not more than one century ago. Thus, except for total energy requirements, which depend closely on body weight and physical exercises, everything indicates that the modern man remains very close to his ancestors of the Paleolithic Age in terms of his genetic equipment and thus his metabolic and nutritional capacities.

PALEOLITHIC DIET

This diet is mainly based on the elimination of two among the four great groups of food, for example, dairy products and cereals (pasta, bread, and rice). Globally, it is a low-carbohydrate diet (22%–40%), with 19%–35% proteins and 28%–47% lipids. Green vegetables and leguminous plants are allowed but cooked by steam vapor. The massive consumption of crude vegetables and fruits is encouraged. Proteins must be ingested preferably in the course of only one meal.

Even if the composition of this diet was debatable, Prof. J. Seignalet could get undeniable beneficial results on the health of patients suffering from various chronic and inflammatory diseases. Clinical investigations with large cohorts are desirable before concluding an unquestionable interest for definite patients.

This concept, now well accepted, was analyzed and developed in 1985 by the uncontested specialist of Paleolithic nutrition Prof. S.B. Eaton in the United States (Eaton et al. 1997). He defined that the meat-based diet of our ancestors *Homo sapiens* of the Paleolithic Age was primarily based on small and medium-sized mammals, especially birds and fish, until approximately 20,000 years ago, the whole diet containing no more than 20%–25% of fats (on an energy basis). Cereals were consumed only later, around 15,000 years ago, and dairy products, except breast milk, appeared only with the Neolithic period, around 10,000 years ago. During the last century, there occurred in the Occident a large food diversification and especially a trend toward the continuous increase in fat content of the common diet. In 1998, Prof. Eaton specified the most probable fatty acid composition of the average Paleolithic diet (Section 3.2.1.3.2). These Paleo-anthropology data have been adapted to advise patients eager to lose weight. The name "Paleolithic diet" used by many nutritionists had some success as more than 15,000 hits may be obtained with an Internet search engine using that expression (146,000 using "stone old diet"). The most known promoters of this diet are Prof. J. Seignalet of the Montpellier University, France, and Prof. L. Cordain, the United States. According to these nutritionists, the virtues of these "ancestral diets," bringing mainly carbohydrates and proteins, but eliminating dairy products and cereals, could even be extended to the prevention of autoimmune diseases (multiple sclerosis and arthritis). Of course, no serious study on a great scale came to support these assertions and, moreover, the effectiveness of these diets is far from winning unanimous support, among nutritionists as well as users.

Research on the stature, physical activity, and most probable diet of Paleolithic men led to an estimate of their energy intake to about 3,000 kcal (12,555 kJ) per day, with 20%–25% of that value coming from lipids. Among almost all people, the late introduction of mechanization has reduced the energy requirements of humans to 40%–50% of the basic value that had been necessary 20,000 years ago. This ancestral energy intake would correspond to an intake not exceeding 1500 kcal (6,277 kJ)/day today, not only in townsmen but also in the rural ones. This value is largely lower than that estimated for present-time populations, but it is possible to think that it corresponds to a healthy status for a sedentary individual.

3.1.2 Total Energy Requirements

Before describing the amounts of lipids essential for a healthy performance of the body, it is necessary to define the total energy that is required to satisfy the needs in humans.

The World Health Organization (WHO) defined in 1996 the energy requirements of an individual as "the quantity of energy necessary to compensate for the expenditure and to ensure a size and a body composition compatible with the long-term maintenance of a good health and a physical-activity adapted to the economic and social context." This means that the balance between energy expenditure and energy intake allows the maintenance of a stable body weight, thus contributing to the conservation of a healthy status. When the intake is lower than the expenditure, there is weight loss (reduction of fat depots and muscular proteins); conversely, when the intake is higher than the expenditure there is weight increase, the surplus being stocked in the form of fats, primarily in adipose tissue.

The energy expenditure of an individual may be split into four sectors: (1) basal metabolism (60%–75% of the total), (2) physical activity (15%–30% of the total), (3) diet-induced thermogenesis induced by food transformation (7%–13% of the total), and (4) possibly growth (2%–7%). The basal metabolism is represented by the minimal energy expenditure compatible with life, but with the body in a total rest. Although specific to each individual, this value is appreciated by taking into account an individual's sex and body weight. The most recent equations making it possible to consider basal metabolism were proposed in 2011 by the Scientific Advisory Committee on Nutrition in the United Kingdom (Henry 2012). For individuals from 30 to 60 years, the formulas are as follows:

Man: BM (kcal/day) = 14.2 × weight (kg) + 593 or 1,658 kcal/day for a 75-kg man
Woman: BM (kcal/day) = 9.7 × weight (kg) + 694 or 1,421 kcal/day for a 75-kg woman

where BM refers to basal metabolism.

The daily average value of the total energy requirements has been estimated to be 2,000 kcal (8,370 kJ) for women and 2,500 kcal (10,462 kJ) for men. These basic values obviously vary according to sex, age, weight, physiological state, and level of physical activity, as shown in Table 3.1. The values given in the table have been calculated by taking into account the basal metabolism, the level of physical activity, and, if needed, the growth.

The daily energy requirements may thus fluctuate between 1360 and 2900 kcal for children, 2100 and 3000 kcal for men, and 1800 and 2150 kcal for women. One may find on the Internet automatic calculators, allowing one to appreciate these values starting from personal data (http://www.logidiet.com/fderad/DEJ.htm). A complete report on the energy requirements of the human population can be consulted on the Food and Agriculture Organization (FAO) site (http://www.fao.org/docrep/007/y5686e/y5686e01.htm#TopOfPage).

To more precisely appreciate the energy requirements of adult men and women, it is possible to use formulas taking into account the major variation factors: sex, age, weight (in kilograms), size (in meters), and physical activity. These formulas make it possible to specify the needs for a determined subject to adapt the total energy requirement to specific physiological situations if necessary.

TABLE 3.1
Average Daily Energy Requirements in Men and Women According to Age and Physical Activity

Age and Activity	Daily Energy Requirements in Men	Daily Energy Requirements in Women
Children from 1 to 3 years old	1,364 kcal (5,708 kJ)	1,364 kcal (5,708 kJ)
Children from 13 to 15 years old	2,900 kcal (12,136 kJ)	2,490 kcal (10,420 kJ)
Adults		
Low activity	2,100 kcal (8,788 kJ)	1,800 kcal (7,533 kJ)
Average activity	2,700 kcal (11,300 kJ)	2,000 kcal (8,370 kJ)
Intense activity	3,000 kcal (12,555 kJ)	2,200 kcal (9,207 kJ)
Pregnancy	—	1,990–2,150 kcal (8,328–8,998 kJ)

TABLE 3.2
Values of Physical AC

	Sedentary	Low Activity	Average Activity	High Activity
Man	1	1.11	1.25	1.48
Woman	1	1.12	1.27	1.45

Estimated energy requirements in a man (MER) in kilocalories per day:

$$MER = 662 - (9.53 \times age) + AC \times [(15.91 \times weight) + (539.6 \times height)]$$

Estimated energy requirements in a woman (WER) in kilocalories per day:

$$WER = 354 - (6.91 \times age) + AC \times [(9.36 \times weight) + (726 \times height)]$$

where weight is in kilograms and height is in meters.

AC is a physical activity coefficient that may be evaluated for four activity levels (Table 3.2).

As an example, taking into account the average values of weight and size of the French population between 18 and 65 years (INSEE 2009 survey), 77 kg and 1.75 m for men and 63 kg and 1.63 m for women, the values of energy requirements amounted to 2,545 kcal (10,651 kJ) and 1,920 kcal (8,035 kJ) for men and women, respectively.

To simplify the text, the value of 2,500 kcal (10,462 kJ) will be adopted as the single reference in that work. The last survey of the French population (INCA2 2006–2007) has appreciated the average daily energy requirement of 2,500 kcal in adult men and 1,855 kcal (7,763 kJ) in women.

3.1.3 Nature of Dietary Lipid Intakes

The dietary lipids (oils or fats) provided by plants are mainly constituted of triacylglycerols (Section 2.3.1), those provided by animals consisted of triacylglycerols (Section 2.3.2) but also of phospholipids (Section 2.4), the latter reaching about 15% of the total lipid content in beef fillet (Table 3.3). Among food, egg is an exception as it contains the highest phospholipid amount (up to nearly 30% of the total lipids). Other lipids are ingested with triacylglycerols and phospholipids but in very small amounts, various simple lipids such as sterols (Section 2.6), carotenoids and liposoluble vitamins (Section 2.7), and low amounts of complex lipids such as glycolipids (Section 2.5). It should be stressed that cholesterol is absent in crop products and remains a characteristic of animal products (meat, eggs, and dairy products) (Table 3.3). Its concentration does not exceed 70 mg by 100 g of meat, but hen egg represents a considerable cholesterol supply (about 200 mg in one egg) (Section 2.6).

Thus, it should be stressed that among whole lipids, fatty acids share an important part of the dietary energy supply (95%–98% of the total lipids), whether they come from triacylglycerols or phospholipids. The fatty acids constitute on average 96% of the triacylglycerol weight and 71% of the phospholipid weight and are therefore in the first rank in dietary lipids. The quantitative aspect is not of course the only criterion to be considered, as some fatty acids play a significant role in many cellular mechanisms (Section 4.2).

The ingested triacylglycerols come from the meat (muscles) and the fat depots (adipocytes) in animals, as specific reserves in oleaginous plants. In the latter, oil may be localized in seeds (soybean, groundnut, rapeseed, sunflower, walnut, etc.) or in the pulp of a fruit (olive, palm tree, coconut, etc.). The phospholipids come from the cellular membranes in vegetables and animals and also from lipid reserves, such as egg yolk.

Vegetable oils and animal fats (lard, fish oil, butter, and creams) are practically only made up of lipids, but some animal products (meats, eggs, and dairy products) contain only from 4% to 25% lipids. On the other hand, plants usually found in human diet contain no more than 1% lipids. Obviously, fruits and seeds may contain high amounts of lipids (from 10% to 70%). The average lipid contents of the principal unprocessed foods consumed by humans in occidental countries are given for plants (Table 3.4) and animals (Table 3.5). It is obvious that these values cannot be regarded as being fixed and definitive, because they depend on the culture and breeding conditions of different places.

In food prepared by industry as well as in processed food derivatives prepared at delicatessen, the lipid contents are much more various and in general much higher than in the original products. Thus, lipids added during manufacture ("hidden lipids") constitute 1%–17% for preparations containing meats, 1%–13% for fish, 0%–17% for pasta, 2%–56% for charcuterie, 9%–34% for cheeses, 1%–9% for yoghurts, and 6%–35% for pastry. The high lipid contents in many commercial culinary preparations are undoubtedly at the root of excessive fat consumption that leads to overweight and obesity in people with a positive energy balance (intake higher than expenditure). This too-elevated lipid consumption in Western countries rises partly from the significant role lipids play in food palatability. In animals, it has been shown that fats promote the search for energy-rich foods, behavior that can be considered to be favorable to species' survival. Similar investigations cannot be undertaken in humans, but everybody may perceive at a precise time a preference for dishes containing sauces made of butter, milk cream, or goose fat! This well-known preference for fatty food, in animals as well as in humans, may be related to the specific perception of fats at the level of taste buds. Knowledge

TABLE 3.3
Amount of Lipids in Some Foods of Animal Origin (Grams per 100 g Fresh Weight)

Foodstuff	Total Lipids	Phospholipids	Cholesterol
Chicken (fillet)	9.7	0.4	0.065
Beef (fillet)	6.4	1.0	0.070
Salmon (fillet)	14.0	0.6	0.035
Hen's egg (whole)	9.9	2.4	0.380
Cow's milk (whole)	4	0.04	0.012

Note: Data are from the CIQUAL 2008 tables (ANSES). The total lipids are made up mainly of triacylglycerols, phospholipids, and cholesterol.

TABLE 3.4
Lipid Content of Unprocessed Crop Products (Grams per 100 g Fresh Weight)

Cereals	%	Vegetables	%	Seeds	%	Fruits	%
Rice	0.8	Cauliflower	0	Green pea	0.2	Pineapple	0.12
Wheat	3.5	Potato	0.1	Kidney bean	0.3	Pear	0.12
Quinoa	5	Cucumber	0.1	Lentil	0.4	Grape	0.16
		Tomato	0.1	Corn	1.3	Watermelon	0.2
		Asparagus	0.1	Chickpea	2	Peach	0.25
		Mushroom	0.2	Hazelnut	36	Orange	0.2
		Green beans	0.2	Walnut	52	Apple	0.2
		Lettuce	0.3	Almond	54	Banana	0.3
		Red pepper	0.3			Strawberry	0.6
		Carrot	0.3			Kiwi	0.6
						Raspberry	0.65
						Chestnut	5
						Avocado	14
						Coconut	35

TABLE 3.5
Lipid Content of Unprocessed Animal Products (Grams per 100 g Fresh Weight)

Meat	%	Fish	%	Milk	%
Chicken fillet	1	Scallop	0.2	Cow's milk	3–4
Turkey	2.4	Crab	0.5	Breast milk	4
Veal cutlet	2.5–5	Blue whiting	0.6		
Rump steak	3.5	Cod	0.6		
Veal liver	5	Oyster	1.2		
Rabbit	5.6	Shrimp	2		
Beef fillet	6.5	Trout	3-6		
Lamb leg	9	Sardine	4.5		
Entrecôte	12	Tuna	6-16		
Veal chop	15	Salmon	14		
Roasted pork	15				
Lamb rib	17				
Pork belly	27				

Note: Data are from the CIQUAL 2008 tables (ANSES).

in humans is still fragmentary and is a matter of behavior more than physiology, and investigations in laboratory rodents open up new horizons. In 1997, T.A. Gilbertson in the United States reported that long-chain fatty acids (≥14 carbons), released by triacylglycerols under the action of lingual lipase, were detected by the tongue. This property is likely based on the inhibition of potassium channels on taste bud cells. The importance of this effect still remains to be determined, but very recent studies in the rat have shown that lipids would be an element of taste perception, as salty, sour, bitter, sweet, and umami. This perception would lie in the existence of chimioreceptive proteins at the level of cellular membranes, such as CD36 protein (Gaillard et al. 2008).

Triacylglycerols are concentrated reserves of energy (on average 9 kcal/g or 38 kJ/g), twice as much as for proteins and carbohydrates. In an organism, they are mainly localized in fat cells that are found among other cells in several tissues (muscle and conjunctive tissue) or in adipocytes clusters (adipose tissue) in various body places (under the skin, around muscle and internal organs, and breast). The accumulation of this adipose tissue is rather visceral in man and subcutaneous in woman. These fat depots may easily reach 20% and 25% of the body weight in man and woman, respectively. These amounts represent a considerable energy reserve (nearly 30 kg of lipids for an individual of 120 kg).

Moreover, triacylglycerols may be converted into cholesterol, phospholipids, and other kinds of lipids, if necessary. They are the major source of unsaturated fatty acids that almost all animals cannot synthesize when ingesting other nutrients (essential fatty acids). The phospholipids participate, with cholesterol, in the constitution of cellular membranes. The polyunsaturated fatty acids present in phospholipids play a role in membrane structure and also in the formation of many bioactive lipids (prostaglandins, leukotrienes, etc.) known to be essential for cellular signaling. The cholesterol originating from animal sources is ingested at a rate of approximately 350 mg/day, and then it may be integrated directly into cellular membranes or metabolized into bile acids, essential to lipid digestion, as well as steroid hormones and vitamin D.

In short, more than 95% of the energy provided by lipids comes from the fatty acid moieties of triacylglycerols (oils and fat) contained in food. These fatty acids are also stored in adipose tissue to be used as reserves in the event of insufficient food intake or energy supplies or intense demand. The majority of the vegetables we eat contain less than 5% of lipids, whereas the meat and fish may contain up to 27% of lipids. The modern diet places at our disposal culinary preparations containing up to 56% of lipids; these lipid-rich foods being likely selected by an unconscious consumer demand.

3.2 METABOLISM AND DIETARY REQUIREMENTS

Until the discovery in 1849 by Claude Bernard (Figure 3.1) of the digestion of fats by pancreatic juice, scientists thought that fats contained in food did not undergo any transformation in the body. This important discovery has established that the lipid digestion in the intestine is followed by the absorption of free fatty acids through the intestinal mucosa. The fate of these fatty acids and the biosynthesis of body lipids (lipogenesis) caused bitter academic battles starting from 1840 between the French school of the famous chemist Jean-Baptiste Dumas (Figure 3.2) and the German school managed by the famous Baron Justus von Liebig (Figure 3.3). As Jean-Baptiste Boussingault (Figure 3.4) wrote in 1845, "The question of the production of fats during the nutrition process raised a biting controversy, it was much discussed and very little experimented."

FIGURE 3.1   Claude Bernard (1813–1878).

FIGURE 3.2   Jean-Baptiste Dumas (1800–1884).

FIGURE 3.3   Justus von Liebig (1803–1873).

Whereas Dumas supported that animals got the lipids only by eating plants and could carry out only limited oxidations, Liebig supported that body fats could be formed starting from carbohydrates by a simple chemical reduction. Finally, contrary to Dumas, Boussingault carried out experimental investigations in pigs and geese and proved brilliantly and definitively that animals are able to synthesize fats starting from carbohydrates, without eliminating a direct tissue incorporation of dietary fatty acids. The later stage, which was also the most important, was the

FIGURE 3.4 Jean-Baptiste Boussingault (1802–1887).

investigations done in 1904 by F. Knoop in dogs, studying the metabolism of fatty acids carrying a phenyl group. This artifice enabled him to determine that the degradation of fatty acids was done by successive removals of fragments containing two carbon atoms (β-oxidation). These fragments could be used later on for other biosyntheses or as energy source. The advent of the isotopes of hydrogen, carbon, and phosphorus initiated a large wave of rational exploration of lipid metabolism. Only outstanding research in this field was retained here. Shortly after the discovery in 1932 of deuterium, a natural isotope of hydrogen, by the American chemist H.C. Urey (he got the Nobel Prize in 1934), the German biochemist R. Schoenheimer, emigrant in 1933 in the United States, carried out in 1936 a fundamental experiment for understanding the metabolism of fatty acids. By using the heavy isotope of hydrogen (D or 2H) in the form of heavy water, this scientist demonstrated that fat depots were used not only during periods of food shortage, as it was accepted at the time. He brought irrefutable evidence that fatty acids are in constant renewal. The concept of "turnover" had just been born. He also discovered that palmitic acid (16:0) could be transformed into stearic acid (18:0), the latter being able to be desaturated into oleic acid (18:1). As labeled linoleic acid was never formed in Schoenheimer's experiments, he confirmed the essential character of this fatty acid, as it had been shown before by the American G.O. Burr during nutritional experiments. The in vivo oxidation of lauric acid (12:0) was confirmed in 1948 by the American R.P. Geyer by using trilaurine labeled with ^{14}C and introduced into blood circulation. The metabolic steps of the synthesis of fatty acids were specified in the United States in 1945 by D. Rittenberg, starting from acetic acid labeled with ^{13}C, and in 1950 by R.O. Brady, starting from short-chain fatty acids labeled with ^{14}C. In consequence of the complexity of the mechanisms concerning the catabolism of fatty acids, 50 years went by before the German biochemist F. Lynen (1911–1979) (Figure 3.5) summarized all the steps and all the involved enzymes. He was awarded the Nobel Prize in 1964 for the whole of his fundamental work on the biochemistry of fatty acids.

Concurrently to recognizing their contribution to the construction of membranes and their direct catabolism, fatty acids were recently recognized to be important actors in communication mechanisms and cellular regulations. These mechanisms involve the fatty acids themselves, which are able to modify the expression of several genes and their oxygenated derivatives, and the prostanoids, whose relationship with fatty acids was established by the Swedish S.K. Bergström in 1962. In 1977, the American P.K. Flick discovered that the fatty acid composition of food modulated the activity of hepatic lipogenesis. Since then, it became increasingly obvious that fatty acids control the expression of not only many genes involved in the metabolism of lipids, but also of carbohydrates (Pegorier et al. 2004).

3.2.1 TRIACYLGLYCEROLS AND NUTRITION

3.2.1.1 Absorption: Digestion

After ingestion, food lipids are hydrolyzed mainly in the duodenum by pancreatic lipase. The American biochemist F.H. Mattson established in 1964 that this enzyme, highly specific of the fatty acids esterified in the external positions of glycerol (*sn*-1 and *sn*-3), catalyzes the formation of *sn*-2 monoacylglycerols and free fatty acids. The hydrolysis speed varies according to the chain length of the fatty acids, the short chains being more quickly hydrolyzed than the longer chains. The whole released fatty acids will be absorbed by the intestinal epithelium, which will form again by sequential ester-ifications of triacylglycerols, phospholipids, and cholesterol esters. These products will be transported in the form of complex structures (chylomicrons) (Section 3.2.1.2) in the lymph and then, by the thoracic channel, in the general circulation toward the liver (for about 30%), adipose tissue (for about 30%), and various organs (for about 40%). In con-trast, the fatty acids with a chain not exceeding 10 carbons, ingested mainly with dairy products, are quickly absorbed and directly transported bound to albumin by the portal vein to the liver. Besides the short time taken for them to be absorbed by the intestine, they have a lower energy value than the longer fatty acids. The shortest fatty acids (up to six carbons) are able to control the absorption of water and sodium by the intestinal epithelium. They behave rather like simple carbohydrates. The fatty acids with a 6- to 10-carbon chain are used mainly as immediate energy source without storage in adi-pose tissue, the property being at the origin of their clinical interest for weight control (St-Onge and Jones 2002). They have also been used in patients unable to absorb longer fatty acids and in athletes who have fast and important energy requirements.

The triacylglycerols synthesized in the liver from other sources (mainly carbo-hydrates) will be exported to the peripheral tissues in the form of very-low-density lipoproteins (VLDLs). The distribution and metabolism of fatty acids with 12 and

FIGURE 3.5   Feodor Lynen (1911–1979).

more carbons vary according to their structure. Saturated and monounsaturated fatty acids are used especially for storage, the former also being able to be unsaturated at the liver level. The n-3 polyunsaturated fatty acids, such as eicosapentaenoic acid (EPA) and docosahexaenoic acid (DHA), will be either oxidized (as for the linolenic acid) or incorporated into phospholipids of cellular membranes. The precursors of the n-6 fatty acids (linoleic acid) will be elongated and desaturated to give mainly arachidonic acid, a membrane component and also an important source of oxygenated mediators (prostaglandins, thromboxanes, and leukotrienes).

Far from the meals (a kind of fasting time), the fatty acids coming from adipose tissues will be delivered in free form or complexed with albumin to the peripheral tissues where they will be oxidized or reconverted if necessary into complex lipids (triacylglycerols and phospholipids).

3.2.1.2 Blood Transport

The presence of a milky lymph after a fatty meal was noted shortly after the discovery of intestinal lymph vessels (or chyliferous vessels) in 1622 by the Italian surgeon G. Aselli and the description of the anatomy of the lymphatic system in 1651 by J. Pecquet, doctor of King Louis XIV of France. Large fatty droplets were recognized in these vessels and named chylomicrons in 1924 by the American S.H. Gage. In blood, the first description of protein-bound lipids has been done in 1901 by J. Nerking in Germany. All later works in this field arose from the fundamental discovery in 1929 of a serum lipoprotein complex by the French doctor and biochemist M. Macheboeuf. This complex contained 59% proteins and 41% lipids, including 23% phospholipids and 18% cholesterol; it corresponds to what is currently called high-density lipoprotein (HDL). Thereafter, improvement of techniques, mainly ultracentrifugation, allowed the description of five principal lipoprotein types playing a role in the transport of lipids in human plasma.

From a structural point of view, lipoproteins are globular particles of high molecular weight with a membrane made of a phospholipid and cholesterol monolayer and a core containing nonpolar lipids (triacylglycerols and cholesterol esters) and apoproteins. The latter are used for the recognition of lipoproteins by specific receptors and enzymes; furthermore, they determine the function and metabolic fate of the lipoprotein particle.

The most important lipoproteins are the following:

- Chylomicrons are synthesized by the intestine. They have a variable diameter from 800 to 5000 Å and a density of 0.93 and are composed of about 86% triacylglycerols, 3% cholesterol esters, 2% cholesterol, 7% phospholipids, and 2% proteins.
- VLDLs are synthesized and secreted by the liver. They have a diameter varying from 300 to 700 Å and a density from 0.95 to 1.010, and they are composed of about 55% triacylglycerols, 12% cholesterol esters, 7% cholesterol, 18% phospholipids, and 8% proteins.
- Intermediate-density lipoproteins (IDLs) originate from the VLDLs. Their size and density are intermediate between those of VLDL and low-density lipoprotein (LDL), for example, from 270 to 300 Å and from 1.008 to 1.019,

respectively. They contain approximately 23% triacylglycerols, 29% cholesterol esters, 9% cholesterol, 19% phospholipids, and 19% proteins.

- LDLs, discovered in 1950 by J.L. Oncley originate from the IDL. Their size is approximately 220 to 272 Å and their density varies between 1.019 and 1.060. They contain approximately 6% triacylglycerols, 42% cholesterol esters, 8% cholesterol, 22% phospholipids, and 22% proteins.
- The HDLs are secreted by the liver and the intestine and are derived from triacylglycerol-rich lipoproteins (chylomicrons and VLDLs). Several forms have been described according to their increasing density. Phospholipids are their main lipidic components. The heaviest have a size from 70 to 90 Å and a density from 1.125 to 1.210 and contain about 3% triacylglycerols, 13% cholesterol esters, 4% cholesterol, 25% phospholipids, and 55% proteins.

The discovery in 1974 by the Americans M. Brown and J. Goldstein (Nobel Prize for 1985) of LDL receptors has renewed all former knowledge on the mechanisms of blood lipid transport.

The metabolism of the lipoproteins and their roles in the transport of lipids (fatty acids and cholesterol) are complex and based on the function of several receptors and enzymes. To simplify, this metabolism may be divided into three parts: the exogenous pathway (starting from the intestine toward other tissues), endogenous pathway (from the liver to other tissues), and reverse cholesterol transport (from the tissues to the liver) (Figure 3.6).

3.2.1.2.1 Exogenous Pathway

Its function is to bring dietary lipids from the intestine to peripheral tissues for energy production, storage, or biosynthesis of new molecules. The dietary lipids are first of all hydrolyzed mainly in the small intestine, absorbed by the intestinal epithelial cells that reconstitute them into chylomicrons, secondarily secreted in the lymph and exported into blood circulation. The free fatty acids formed by the hydrolysis of triacylglycerols at the level of vessels, muscles, and adipose tissue will be later stored or used directly for energy production. Because only the triacylglycerols are hydrolyzed, the chylomicron remnants will be enriched in cholesterol esters. These remnants will be collected by the liver, via receptors shared with LDL.

3.2.1.2.2 Endogenous Pathway

Few hours after a meal, the fatty acid requirements of peripheral tissues are satisfied by VLDLs transporting the lipids synthesized by liver. These VLDLs will be hydrolyzed in the capillaries, and the released fatty acids will be collected by tissues and used as an energy source. The triacylglycerols remaining in the VLDL residues, IDLs, will also be hydrolyzed, thus leading to LDLs strongly enriched in cholesterol esters and cholesterol ("bad cholesterol"). At the same time, many exchanges and transformations of lipids have been described between various lipoprotein classes (exchanges of cholesterol esters with triacylglycerols and interesterification of cholesterol between HDLs, chylomicrons, or IDLs). The LDL will slowly disappear from circulation at the level of the liver, after recognition by specific receptors. Their average half-life is normally about 3 days. One of the main functions of HDL is to transport cholesterol ("good cholesterol") from peripheral tissues to the liver (reverse transport of cholesterol). Indeed,

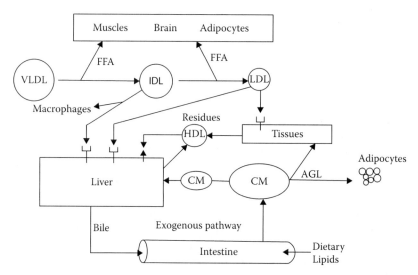

FIGURE 3.6 Summary of the transport pathways of various lipoproteins. FFA, free fatty acids; CM, chylomicrons.

HDL uptakes cholesterol and cholesterol esters from other lipoproteins, especially from liver, steroidogenic tissues, kidneys, intestines, and macrophages by receptor-mediated endocytosis. Free cholesterol will be esterified, thus leading to exchanges with triacylglycerols contained in other HDLs and also to transfers toward LDLs and VLDLs, lipoproteins that will be brought back to the liver. In the liver, cholesterol will be transformed into bile acids excreted in the bile fluid. In steroidogenic tissues cholesterol will be transformed into steroid hormones, and in the skin it will be the precursor of vitamin D. The rise of cholesterol level in the liver inhibits its endogenous synthesis.

3.2.1.3 Dietary Intakes of Lipids and Fatty Acids

Until the Second World War, no country has made recommendations for the quantity or quality of the lipids absorbed by consumers, to prevent any pathology. It can be easily understood that at this time the health priority was not the recommendation of low-calorie diets to afflicted populations. In the 10 years following the war, contrary to what happened in Europe, U.S. society knew an economic and commercial expansion accompanied by an increased consumption of foods varied and rich in energy. This period of food abundance has continued to this day, as it has been estimated that the average energy intake of Americans evolved from 1900 kcal/day in 1950 to 2660 kcal in 2008, for example, an increase in 760 kcal in 58 years (Grotto and Zied 2010).

The concomitant increase in cardiovascular diseases was quickly connected to these nutritional changes by the famous American physiologist Ancel Keys whose memory will remain linked to the Mediterranean diet and the K-ration of troops during the Second World War. Officially, it was in 1961 that the Central Committee of the American Heart Association launched its first recommendations for a reduction of lipid intake (from 40%–45% to 25%–35% of the total energy intake) and the proportion of saturated fats.

From 1968 to 1977, 18 medical and scientific organizations (North America, Europe, Australia, and New Zealand) proposed a framework based on clinical investigations defining the limits of lipid consumption. Several committees advised a reduction of not only the total energy intake, but also the amount of saturated fats and cholesterol. All agreed to recommend a lipid ration lower than 35% of the total energy intake (the role of fats and edible oils in human nutrition, FAO/OMS Report, 1977). In the United States, the work of the senatorial committee on human nutrition and needs (McGovern Committee) led for the first time, between 1968 and 1977, to the definition of precise nutritional recommendations concerning the amount and quality of the lipids consumed by Americans (U.S. Senate Select Committee on Nutrition and Human Needs 1977). Thus, the senatorial council recommended a lipid contribution not exceeding 30% of total energy, composed of three equivalent parts (each one 10% of total energy), saturated, monounsaturated, and polyunsaturated fatty acids, the cholesterol contribution not exceeding 300 mg/day.

Later research and many epidemiological investigations in several countries made it possible to specify various recommendations in terms of quantitative and qualitative lipid intake, and consequently of fatty acids, for human consumption. The diffusion of these recommendations to health professionals and the public ought to prevent various long-term pathologies such as obesity, cardiovascular diseases, and some cancers.

Several countries have established standards about essential nutrients. These standard values may vary from one country to another according to various factors like climate and dietary and cultural habits. Thus, a few years ago Canada and the United States brought their dietary rules in line, rules established by the experts of the National Academy of Sciences and intended for the general public and health professionals. These standards are gathered under the name of dietary reference intake (DRI) (Food and Nutrition Board, Institute of Medicine [IOM], National Academy of Sciences). The DRI is a system of nutritional recommendations constituting four types of reference values, each one having its own definition and use.

The current DRI is thus composed of the following:

- Estimated average requirement (EAR), which is expected to satisfy the needs of 50% of the population in one definite age group.
- Recommended dietary allowance (RDA), which is defined as the daily dietary intake of a nutrient considered sufficient to meet the requirements of 97.5% of healthy subjects (calculated in each life stage and sex group). This value is about 20% higher than EAR.
- Adequate intake (AI), used where no RDA has been established.
- Tolerable upper intake level. This value is defined to prevent excessive intake of nutrients (like vitamins or sterols) that may be harmful in large amounts.

The member countries of the European Community reexamined in 2009 the standards followed by each country and finally succeeded in creating the recommended daily intakes named AJR for *Apports Journaliers Recommandés* (the European Food Safety Authority [EFSA] in 2010).

The French agency of food health and security, French Agency for Food, Environmental and Occupational Health & Safety (ANSES), defined its own standards, which are translated into French population reference intakes (ANCs for *Apports Nutritionnels Conseillés*) (Martin 2001). The definitions of these standards are not different from those of the U.S. DRIs. One important point is that the ANC may vary according to countries and are periodically updated, taking into account new research results. Thus, in December 2010 the ANC for vitamin D was increased in the United States by three times for definite group of subjects. Presently, the French ANCs have in general lower values than their American counterparts (DRIs).

How was the ANC defined by the ANSES? The ANC retained for each type of nutrient and for a well-defined group of people (e.g., sex and age) corresponds to intake levels that make it possible to cover the needs of the individuals of that group. Therefore, by definition any individual whose intake of a nutrient is higher than the ANC covers his needs. On the other hand, if the contribution is lower than the ANC there is a risk of deficiency. The ANC is determined in studying statistically a group of representing individuals, but the average value is increased by 30% (two standard deviations of the data distribution). This statistical safety margin takes into account the individual variability and enables to meet the needs for most of the population (97.5% of the whole). Thus, the ANC is generally selected on the basis of 130% of the average need of a definite nutrient. It should not be forgotten that ANCs are intended to meet the needs practically for the whole population and do not correspond to an individual standard.

The ANC must be distinguished from the AJR, which corresponds to the average needs for the population, values that are used mainly for the labeling of food products and easier to reach than the ANC. These standards have also been used in the United States as DRI. They are established as single values for each nutrient and do not take into account the differences related to age, weight, and gender. They are harmonized at the European level and have regulation values. The AJR value is based on the needs of a healthy adult of median age. It is obvious that an individual whose intakes are equivalent to the AJR has little risk to be deficient.

3.2.1.3.1 Whole Lipid Intakes

What is the evolution of the amount of dietary lipids since the prehistoric times? If one evokes the average food diet of our ancestors, known as Paleolithic diet (Section 3.1.1), specialists admit that it consisted of 65% of food of animal origin and 35% of food of vegetable origin. The most recent analysis of various foods of this diet, products of hunting and gathering at that time, led S.B. Eaton to evaluate the energy distribution of the basic macronutrients into 37% of proteins, 41% of carbohydrates, and 22% of lipids in 1997.

It should be stressed that in this lipid fraction 41% came from animal fats and were especially from game meat, these foodstuffs containing less lipids (approximately 4% in weight) than the meat marketed nowadays. Among this lipid fraction, the saturated fatty acids constituted of only approximately 6% of the total energy intake, whereas the unsaturated to saturated fatty acid ratio being close to 1.4.

What is the situation today concerning the importance of lipids in our food intake? In 2003, a report of FAO experts (WHO technical report series 916) determined that the inhabitants of 49 out of the 173 studied countries exceeded the threshold fixed at 30% of fats in their daily energy intake, all these countries being in the Occident and in Oceania. At that time, France occupied the first place with an intake of 3597 kcal/day (860 kJ/day), including 41.9% (167 g/day) of lipids. A summary of this study is illustrated in Figure 3.7.

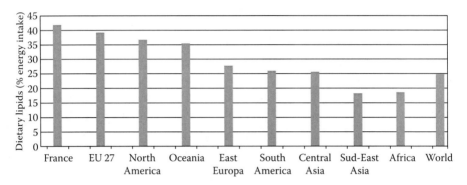

FIGURE 3.7   Importance of lipids in the energy intake in various countries (data for 2000).

Another review published in 2009 by L. Elmadfa, Austria, concerning the diets in 28 countries distributed on all continents showed that the contribution of lipids to the total energy intake is extremely variable not only across the countries (from 11% to 51%) but also within each country. The most important variations were observed in Africa (13%–51%); but they remained significant in America (26%–37%), Asia (11%–36%), and even Europe (28.5%–46%). France had a rank similar to eight other countries (Alaska, Taiwan, New Zealand, Austria, Denmark, Germany, Hungary, and Spain) characterized by a lipid intake ranging from 35% to 40%. These variations may lead only to regional recommendations to modify the importance of lipids in dietary intakes.

In France, one may consider that until the end of the eighteenth century (during the 1780s) the population was in a situation of subsistence economy. The diet could vary, but a balance was established around a daily energy intake of approximately 1750 kcal (420 kJ). The second period, which lasted for about a century until the 1880s, was characterized by innovations in the techniques of agricultural production, but the distribution of energy sources varied little despite an increase in food consumption (Figure 3.8). It has been estimated that in a century the average energy intakes increased from a threshold of survival of 1750 kcal/day (420 kJ/day) to 3000 kcal/day (718 kJ/day). The third phase, known as the phase of nutritional transition, was characterized by a fall in the carbohydrate fraction compensated by an increase in that of the lipids. This phase also lasted 100 years (from 1880 to 1980). One may establish a parallel between that phase of the dietary evolution and the development of the mechanization allowed by engines utilizing steam, electricity, and oil. That phase of nutritional transition was prolonged by a phase of stability (1980–2000) where the total energy intake did not increase anymore and where the proportion of each great nutrient group remained stable. During the next 7 years, the FAO studies and the food consumption surveys carried out by the INSEE and SECODIP showed that the proportion of calories supplied by lipids decreased slightly (39% instead of 42%) without any important change in the proportion of carbohydrates (44%–45%) but with a modest increase in the protein part (15%–17%).

The detailed study by INRA and INSERM on the food and nutrient consumption in France from 1969 to 2001 has specified that the contribution of lipids to the energy intake increased by 3.4% between 1969 and 1980, returning in 2001 to the initial 1969 amount (39.3%).

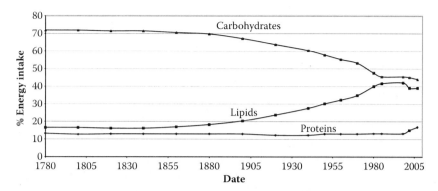

FIGURE 3.8 Evolution of the dietary composition of the French people since 1780. (Data from the Food and Agriculture Organization, INSEE, and SECODIP.)

The average lipid intakes in the current French population were appreciated (INCA 1999 study on nearly 3000 people) at approximately 100 g/day for men (or 36% of an average energy intake of 2500 kcal/day) and at 80 g/day for women (or 37.5% of an average energy intake of 1940 kcal/day). More specifically, 11.5% of the men and 12.5% of the women had total lipid intakes lower than 30% of the energy intake, whereas 33.9% of the men and 32.9% of the women had total lipid intakes equal to or higher than 40% of the energy intake (National Study of Nutrition and Health 2006, http://www.invs.sante.fr/publications/2007/nutrition_enns/). The children (3–17 years old) had an average total lipid intake estimated to 36.2% of the energy intake, the values being similar in boys and girls. In accordance with the objective of the French National Plan for Nutrition and Health (PNNS 2006–2010), 41.6% of boys and 39.4% of girls had total lipid intakes less than 35% of the energy intake. These proportions did not vary with age either for boys or for girls (http://www.ladocumentationfrancaise.fr/rapports-publics/074000748/index.shtml).

Many other consumption studies have been performed in various French areas in the year 1990. The four more important studies (INCA2, SUVIMAX, E3N, and MEDHEA) have confirmed that lipids constituted from 37% to 40% of the total energy intake, including 37%–41% of saturated fatty acids (AFSSA 2003 Report). The INCA 1998–1999 national survey, like the others, does not allow the differentiation of lipid intakes from animals and plants. These studies have taken into account only the data of vegetal oil consumption, the green vegetables being neglected because of their very low lipid content. Their contribution would be between 5.4% and 9.3% of the total lipid ration of the French population.

The national study INCA2 focusing on the dietary habits of nearly 4000 individuals from 2005 to 2007 established the quantity and nature of lipid intake of animal origin. In the adult, this contribution was on average 56 g/day, equivalent to 63% of the total lipid intake. Whatever the age, the contributions are higher in men, 65.0 g/day against 48.6 g/day among women. Whatever the sex and the age, butter constituted the nutrient that brought the main part of the animal lipids, for example, 28% of the ration in men and 32% in women. Cheeses brought 16% of the lipid ration in men and 19% in women, milk bringing 11% of the animal lipids in children but only half of that value in adults. Overall, for the contribution of animal lipids, milk and its derived products amounted to 60% in men and 66% in women. Meats (mainly beef, pork, and poultry) contributed to 30% of the intake in boys and 27% in girls. Eggs contributed on average only from 4% to 5% of the animal lipids, and sea products (fish, shells, and shellfish) contributed 1%–3%.

With respect to the evolution of lipid consumption with age, it appeared that the consumption increased clearly in absolute value in men to reach its maximum between 18 and 30 years old, decreasing slightly to be stabilized thereafter. Then the lipid consumption fell down but only in subjects more than 65 years old. The evolution is the same in girls and in boys; but the consumption stabilizes with adolescence in girls, decreasing slightly in the adult age. From 14 years old, the lipid intakes of girls were definitely lower than those for young boys (INSERM 2000 report, "*For a Nutritional Policy of Public Health in France*," http://www.inserm.fr/thematiques /sante-publique/rapports-publies). The great NutriNet-Health study, which was initiated in 2008 and follows about 260,000 subjects, will soon bring more precise figures on the lipid consumption of the French people.

Despite several surveys and investigations, the estimate of lipid consumption remains very approximate and will be clearly more and more because of the commercial success of the prepared dishes. Indeed, the obscure part of this consumption is what nutritionists call "hidden or invisible fats." They are in opposition to "visible fats," and therefore measurable, which are clearly separated from animal tissues (subcutaneous fat) or milk (butter) and some vegetal parts like seed oils. These hidden fats are mixed with animal meat (mammals, birds, and fish) or vegetal fractions (cereals and leguminous plants). Moreover, the proportion of this hidden component is variable according to the quality and source of the food or to the manufacturing process of industrial dishes. When it exists, the labeling is for the consumer his only source of information to appreciate its lipid content and its contribution to the energy intake.

If one compares the estimated values of average lipid intake in the French population (40%) with the current official recommendations (30%), it is obvious that the French ingest a 25% excess of lipids. This excessive lipid intake is likely related to the well-known tendency of increasing the frequency of overweight and obesity in adults as in children even in the Mediterranean regions, which were until now little concerned with these modifications. Overweight and obesity are defined for adults from 17 to 70 years old by calculating the individual value of body mass index (BMI = weight in kilogram/size in square meter). If the BMI is in the range of 25–30 there is an overweight state, beyond a value of 30 there is an obesity state, and the "ideal" weight is reached when BMI lies between 18 and 25.

The most recent results for France (investigation Obépi-Roche 2012) have indicated that 15% of the adults are obese (or almost 6.8 million people) and 32% are overweight (or approximately 14.4 million people). Furthermore, this phenomenon is expanding as the number of obese has increased by 70% and the average weight increased by 3.1 kg since 1997. In spite of a very recent slowdown, less than one person out of two has a normal BMI.

This evolution may be hardly related to the consumption of lipids as the latter was maintained constant during the same period. Although these body weight values are the lowest within the European countries, they must be connected with the position of the French people in the leading group of lipid consumers. This French lipid consumption rate (40%) is equivalent to that measured in Denmark (37%) or in the United Kingdom (38%) and Germany (40%), but it is more important in Belgium (42%) and weaker in Italy (33%), Finland (34%), Sweden (36%), Ireland (32%), and Portugal (30%). From this global view, France cannot be regarded as a Mediterranean country; perhaps it would be necessary to call this an aspect of the "French paradox"! The question is still in full debate because since 1997 (Hu et al. 1997) most studies agreed to denounce any association between lipid consumption and overweight or obesity. As it was reported in 2006 by R.L. Prentice, United States, it is the same conclusion for the risks of cancer and especially of cardiovascular disease (Section 4.1). More precise studies will be necessary to explain the mechanisms that lead to an increase in the fatty mass with a lipid-rich diet (Pereira-Lancha et al. 2010). It should not be forgotten that body lipid reserves are always much higher than those of carbohydrates and proteins and that contrary to the latter the ingestion of new lipids does not induce automatically their tissue oxidation. Thus, a sudden need for 200 kcal does not induce any uptake from the lipid reserves, whereas an intake of 200 kcal of lipids is practically entirely stored in adipose tissue.

In spite of the poor knowledge of the physiological laws controlling the energy reserves of the human body, specialists around the world generally admit that lipid-rich diets (> 40% of the energy intake) are to be avoided, but there is still debate according to the countries to choose between a threshold lower than 20% and an intermediate value (approximately 30%). It is certain that each individual must appreciate this proportion according to his physical expense but without neglecting the nature of the consumed lipids.

Regarding good food practices, it is currently admitted that for an average energy intake of 2500 kcal/day in a 75-kg adult man the maximum contribution of lipids should be close to 30% of the total energy intake, or about 83 g/day of lipids (scientific report of the ANSES, *EFSA J.* 2010, http://www.efsa.europa.eu/en/efsajournal/pub/1507.htm). This higher limit is unfortunately observed only for 12% of the French population, varying from 15% up to 45% in the diet of the world population.

Because the higher limit of 35% is generally adopted (especially in relation to the existence of moderate physical work), an effort of restriction and modification of dietary intake is advised for most subjects of the Western countries, some frequently exceeding that higher limit. A restriction of the lipid ration would ensure without any doubt a reduction in the frequency of associated pathologies, such as obesity, cardiovascular disease, and early mortality. The recommendation by certain nutritionists of a lipid ration limited to 20% of the total energy intake must be considered with prudence because it may be detrimental to health. Indeed, the compensation of that lipid reduction by a rise in the proportion of carbohydrates is physiologically not advisable and, moreover, this limitation may involve deficiencies of essential fatty acids and liposoluble vitamins, especially in children, pregnant women, and elderly people. It is thus significant that a large study, done in 2006 by B. Howard in the United States on approximately 49,000 women during 8 years, could not show that the replacement of a portion of dietary fats by fruits and vegetables had any effect on the incidence of cardiovascular diseases.

For preterm babies, lipid requirements were established in 2007 by an expert committee from the European Society of Pediatric Gastroenterology, Hepatology and Nutrition (Agostini et al. 2010). This committee recommended in the majority of cases a daily intake of lipids from 4.8 to 6.6 g/kg of weight, this value corresponding to about 40%–55% of the total energy intake.

On request from the European Commission, the EFSA has given in 2013 its scientific opinion on the nutrient requirements and dietary intakes of infants and young children (*EFSA J.* 2013, http://www.efsa.europa.eu/en/efsajournal/pub/3408.htm). For infants aged 0 to less than 6 months, intakes of total fat considered adequate are 21–34 g/day in boys (19–32 in girls) for energy requirements from 360 to 580 kcal/day. The dietary fat corresponds to an average of 52% of the total energy intake. For infants from 6 to less than 12 months, the recommended fat intake must be between 30% and 40% of the total energy intake. There is evidence that a fat intake as low as 25% can be adequate. In infants from 12 to less than 36 months, the recommended fat intake is the same as the previous one.

Summary: in France, the ANSES revised in September 2010 its preceding opinion by recommending that the lipid part must reach a maximum of 35%–40%, but the ANSES specifies that it must be in the absence of excessive energy intake. Overall, this range is respected by most French consumers, but it should be stressed that nearly 34% of the adult men exceed it.

The National Plan Nutrition-Health (PNNS) 2001–2005 defined nine priority nutritional objectives in terms of public health. It is recommended to reduce the average contribution of the total lipid intake to less than 35% of the energy intake. In parallel, a reduction of a quarter of the consumption of saturated fatty acids to a level less than 35% of the total lipid intake is recommended (French delegated minister of health: http://www.sante.gouv.fr/IMG/pdf/1n1.pdf).

3.2.1.3.2   Specific Fatty Acid Intakes

The determination of specific fatty acid intakes is primarily based on the distinction between the essential fatty acids whose synthesis cannot be carried out by the human body and the nonessential fatty acids whose synthesis can be done by the body. Essential fatty acids were discovered by G.O. Burr in 1929 in the United States. This was made possible thanks to the knowledge acquired beforehand concerning the liposoluble vitamins and to the improvement of fatty acid analyses. At that time, the concept of essential fatty acids rose from their effects on body growth and the prevention of dermatitis. In the beginning, the essential fatty acids were compared to vitamin F. Although linoleic acid (18:2 *n*-6) was quickly recognized as an essential fatty acid, the essentiality of linolenic acid (18:3 *n*-3) like that of the other *n*-3 fatty acids was largely recognized only later. Thus, it was only in 1982 that R.T. Holman of the United States described the first case of deficiency in linolenic acid in humans. It is now recognized that *n*-3 fatty acids are quite essential for mammals, but the gravity of the deficiency disorders depends on the ingested quantity of linoleic acid. As underlined by the Canadian S.C. Cunnane (2003), the needs for one fatty acid species can be given only in the presence of the other species. Thus, an optimal *n*-6 to *n*-3 fatty acid ratio equal to 4 was determined in the rat by S. Yehuda in Israel (1993). This physiological characteristic will induce several investigations in human subjects (Section 4.2.3.3).

More recently, the concept of essential fatty acids evolved starting from a definition including at the beginning only linoleic acid to a definition extended to several polyunsaturated fatty acids from the *n*-6 and *n*-3 series (Cunnane 2003). If the precursors, linoleic and linolenic acids, are most concerned, their derivatives (arachidonic acid, EPA, and DHA) are also essential as sources of many bioactive lipids (eicosanoids and docosanoids). All polyunsaturated fatty acids could thus be regarded as essential. It should be noted that many other polyunsaturated fatty acids do not have a clear function at the present time and that those with long chains (20 and 22 carbons) may be essential only at certain periods of development (pregnancy and childhood). To appreciate the measurement of the real nutritional contributions of a food intake including dairy products (Table 3.6), meats (Table 3.7), fish and mollusks (Table 3.8), vegetal oils or fats (Table 3.9), and vegetables (Table 3.10), it is necessary to consult detailed tables of food composition. Since 2012, the new downloadable version of the CIQUAL table of food composition is available (www.anses .fr/TableCIQUAL/ index.htm). Among many information, one can find there the fatty acid composition of almost 1400 foodstuffs with a possibility to search per foodstuff, per nutrient, or per lipid compound (fatty acid and vitamin). Moreover, this table gives the possibility of seeking any food "richest in …" or "the least rich in … ."

3.2.1.3.3 Specific Intakes of Nonessential Fatty Acids

3.2.1.3.3.1 Saturated Fatty Acids   From a purely physiological point of view, the presence of saturated fatty acids in the diet is not necessary. Humans as the other animals are able to make their synthesis from acetate units originating from carbohydrates, especially in case of lipid deficiency. In the presence of high amounts of dietary lipids, the *de novo* synthesis of these compounds is stopped.

Interest was focused on the need for limiting the proportion of dietary saturated fatty acids in the 1950s, physiologists showing that the concentration of serum lipids (cholesterol, phospholipids, and triacylglycerols), in humans as in animals, was directly connected to the saturation degree of the ingested lipids. These demonstrations resulted in a major part of the concomitant explosion of the analytical methods for lipids. Thus, L.W. Kinsell, in the United States, showed in 1953 that patients ingesting vegetal oil (very unsaturated), even in great amounts, had weak cholesterolemia. Conversely, this author showed in 1955 that very saturated lipids from coconut caused a rise in serum cholesterol. In 1957, the works of the famous American

---

TABLE 3.6
Energy Content and Lipid and Fatty Acid Composition of Dairy Products (Kilocalories and Grams per 100 g Fresh Weight)

Dairy	Energy (kcal)	Total Lipids	Saturated	Monounsaturated	Polyunsaturated
Butter	748	82.6	57	21.7	3.1
Camembert (45%)	277	20.7	13.8	5.3	0.5
Milk cream	206	20	12.6	5.7	0.5
Emmental	383	29.6	18.7	7.9	1.2
Roquefort	366	32	21	6.1	1
Cow's milk	70	3.7	2.8	0.8	0.1
Breast milk	69	3.8	1.9	1.4	0.5

TABLE 3.7
Energy Content and Lipid and Fatty Acid Composition of Meats (Kilocalories and Grams per 100 g Fresh Weight)

Meat	Energy (kcal)	Total Lipids	Saturated	Monounsaturated	Polyunsaturated
Lamb leg	226	14	6.4	5.7	0.8
Beefsteak	148	4	1.7	1.9	0.2
Turkey breast	144	2.3	0.5	0.5	0.4
Rabbit	167	9.2	3.4	2.6	2.1
Hen's egg	142	8.5	2.7	4	1.5
Roasted pork	123	4.3	1.6	2	0.3
Chicken fillet	124	3.9	1	1.3	0.8
Veal scallop	151	3	1	1.1	0.4

TABLE 3.8
Energy Content and Lipid and Fatty Acid Composition of Fish and Mollusks (Kilocalories and Grams per 100 g Fresh Weight)

Fish/Mollusk	Energy (kcal)	Total Lipids	Saturated	Monounsaturated	Polyunsaturated
Cod	82	0.8	0.2	0.1	0.3
Blue whiting	92	0.9	0.1	0.3	0.2
Sardine	201	11.2	3	3.6	3.4
Salmon	201	11.8	1.8	4.7	3.3
Trout	119	4.3	0.9	1.4	1.5
Farmed bass	104	4.1	1.14	1.20	1.70
Coalfish	102	0.9	0.15	0.12	0.31
Sole	70	0.4	0.13	0.10	0.16
Mackerel	250	14.2	3.22	4.03	3.57
Red gurnard	100	9	2.72	3.54	2.50
Farmed daurade	76	4.8	1.17	1.46	1.74
Oyster	78	1.5	0.34	0.34	0.52
Octopus	85.5	0.9	0.26	0.17	0.43

Note: Data are from the CIQUAL tables (ANSES: http://www.anses.fr/TableCIQUAL/), F. Médale (2009), and http://www.nutraqua.com/.

Meats are grilled, sautéed or roasted; fishes are steamed; and mollusks are uncooked.

physiologist E.H. Ahrens (1915–2000), founder of the *Journal of Lipid Research*, brilliantly confirmed the preceding results on the close relationships between the dietary unsaturated to saturated fatty acid ratio and the plasma lipids, especially cholesterol. Finally, the epidemiological study, known as the Seven Countries Study, undertaken over 25 years with more than 12,000 people, has shown that the mortality rate of whatever origin was closely related to the amount of dietary saturated lipids (between 4% and 23% of the total energy intake).

TABLE 3.9
Energy Content and Lipid and Fatty Acid Composition of Common Vegetal Oils and Animal Fats (Kilocalories and Grams per 100 g)

Vegetal	Energy (kcal)	Total lipids	Saturated	Monounsaturated	Polyunsaturated
Groundnut	900	100	19.8	45.2	30.1
Rapeseed	900	100	7.6	58.9	29.7
Walnut	900	100	9.3	17	69
Olive	900	100	15.1	77.2	6.9
Sunflower	900	100	11.5	20	64.4
Goose fat	900	100	27.3	57.1	11
Margarine (35%)	330	35	8	9	17.5

Note: Data are from the CIQUAL tables.

TABLE 3.10
Energy Content and Lipid and Fatty Acid Composition of Common Vegetables (Kilocalories and Grams per 100 g)

Vegetables	Energy (kcal)	Total Lipids	Saturated	Monounsaturated	Polyunsaturated
Eggplant	19	0.20	0.030	0.010	0.110
Carrot	26	0.20	0.038	0.004	0.117
Cauliflower	22	0.28	0.042	0.007	0.138
Kale	17	0.90	0.107	0.019	0.485
Spinach	16	0.30	0.030	0.010	0.162
White bean	102	0.32	0.090	0.030	0.150
Green bean	33	0.24	0.071	0.007	0.115
Lentil	107	0.50	0.070	0.080	0.230
Green pea	81	0.48	0.099	0.037	0.297
Potato	70	0.11	0.025	0.002	0.055
Tomato	17	0.21	0.037	0.025	0.100
Apricot	43	0.13	0.010	0.064	0.029
Pineapple	55	0.15	0.016	0.026	0.078
Avocado	221	23.5	1.89	15.48	1.82
Banana	88	0.18	0.057	0.019	0.060
Kiwi	50	0.61	0.030	0.060	0.330
Peach	41	0.11	0.010	0.034	0.044
Apple	54	0.58	0.214	0.023	0.245
Grape	67	0.28	0.066	0.014	0.143

Data are from Souci and others (2000) and the CIQUAL tables.

Recommended intakes of saturated fatty acids: although there remain many controversies concerning the maximum level of dietary saturated fatty acids, it is generally accepted that it should not exceed in the adult 10% of the total energy intake (either a DRI for the United States or an AJR in Europe of 28 g/day for a total intake of 2500 kcal/day). On the basis of observational studies, some nutritionists even recommended to lower this threshold to 8% (22 g/day). The 28-g/day threshold thus corresponds to approximately 40% of the total lipid ration, the remainder being formed of unsaturated fatty acids. It is necessary to note that these recommendations are not the conclusions of precise physiological studies but arose primarily from several observations describing an increase in the cholesterolemia connected to an increase in the amount of dietary saturated fatty acids in humans as in animals.

In the United States, the dietetic advice given in 2005 by the U.S. Department of Agriculture (USDA) indicates that the maximum intake of saturated fats should not exceed 10% of the total energy intake or an equivalent of 28 g of saturated fats for a total intake of 2,500 kcal (http://www.health.gov/dietaryguidelines/dga2005 /document/html/chapter6.htm). This intake should even be lowered to 7% of the total energy intake in subjects having elevated LDL-cholesterol levels.

Several studies have shown a great dispersion of the data making unconvincing conclusions on the possible negative effects of the use of a diet rich in saturated lipids. Moreover, it appears less and less harmless to replace dietary saturated fatty acids by carbohydrates that proved to be able to induce the biosynthesis of other saturated fatty acids (Hudgins 2000). Much remains to be done in this field, especially as long as the biological activities of these fatty acids will remain poorly known (Section 4.2.1). It should be noted that in spite of the colossal research efforts in this field the first mechanism linking saturated fatty acids and cholesterol metabolism was revealed only in 2005 by J. Flax (the United States).

At the present time, although it is premature to fix precise DRI recommendations, it is advisable to maintain a saturated fatty acid intake lower than 12% of the total energy intake. This limit is far from being universal but, undoubtedly, cannot be applied either to newborn babies or to young children, who require much higher levels than adults to ensure proper growth. The high content of saturated fatty acids in the human breast milk (approximately 50% of total fatty acids) shows well that the natural requirements should not be reduced to only one value for every age and for all populations. In addition, the consumption of great amounts of dairy products does not appear related to the incidence of cardiovascular diseases, according to a large Australian study of 1529 adults followed for 14 years (Bonthuis et al. 2010). There even seems to be a slight benefit for people consuming around 340 g of dairy products a day.

Saturated fatty acid intakes: the review in 2009 by L. Elmadfa of the diets in 28 countries distributed on all continents has shown that the contribution of saturated fatty acids to total energy intake varied from 3% to 5% in China, by 10% in Canada, by 11% in the United States, and up to 25% in Nigeria.

The USDA has reported in 2005 that Americans consumed approximately 25.5 g of saturated lipids daily, more of the quarter being supplied by milk and its derived products.

In the European countries, this contribution varied from 8.9% in Portugal to 16.5% in Belgium, France being at the level of 15.6%. Efforts thus remain to be made in the last two countries, as in most countries in northern Europe, to bring back these rates under the threshold of 12%. A large survey realized in Europe in 1999 (the TRANSFAIR study) has made it possible to note a slight increase of saturated lipids in our diet since then. Only the populations of Italy, Portugal, Spain, and Greece now respect the recommended limit of 12% of total energy intake. It is remarkable that in the European populations dairy products contribute 30%–57% (average: 41%) of the total saturated fatty acid intake.

The INCA 1999 survey published by the ANSES noted that the French population still consumed too much saturated fatty acids. According to this report, it appeared that the proportion of saturated fats was identical for both sexes. Saturated fatty acids accounted for 16% of the energy intakes, a proportion close to that noted in Germany (17%). This elevated saturated fatty acid consumption in France is related to a rather high meat and dairy products (butter and cheese) consumption. The investigation thus confirmed that the elevated consumption of saturated lipids without association with an increase in the incidence of cardiovascular pathologies is a characteristic of the French paradox.

The INRA/INSERM study on food and nutrient consumption in France has shown that the contribution of saturated fatty acids to the energy intake increased by 1.4% between 1969 and 1977 and then decreased to 15.5% until 2001. The report emphasized that the saturated fatty acid overconsumption of almost 6% corresponded to a total lipid overconsumption.

The French Nutrition-Health study in 2006 showed that the proportion of saturated fatty acids in total lipids was on average estimated to be 37.8%, higher in men (38.3%) compared to women (37.3%). In addition, only 28.7% of men and 34.4% of women had a saturated fatty acid consumption that was in conformity with the objective recommended by the Nutrition Health National Plan (PNNS) (< 35% saturated fatty acids in total lipids). In children (3–17 years), saturated fatty acids accounted for on average 38.9% of the total lipid intake, identically in boys and girls. In its last report on September 2010, the ANSES recognized that saturated fatty acids are consumed in excess by the French population because they account for on average 16% of the energy intake, although the advised nutritional intake should be lower than 12%. These intakes are mainly made up of lauric, myristic, and palmitic acids, fatty acids that are now regarded as atherogenic when ingested in excess. On the other hand, other saturated fatty acids with shorter or longer chains may have beneficial effects on health.

It appeared increasingly obvious to physiologists that saturated fatty acids cannot be regarded as a homogeneous group because they differ by their nutritional and cellular functions (Legrand and Rioux 2010). Thus, short-chain fatty acids, mainly butyric acid (4:0), and those with a medium chain (6:0 to 12:0) have a metabolism different from that of long-chain fatty acids (14:0 and more) from their absorption to their catabolism (Section 4.2.1.2). The sum of short- and medium-chain fatty acids constitutes approximately 6% of the total fatty acids in breast milk, the single lipid source in the infant, and approximately 13% in cow's milk. In the infant, it may thus be estimated that the absorption of saturated fatty acids (from 4:0 to 12:0) is 1 g/day, including approximately 0.12 g/day of butyric acid. An adult who consumes 1 L of cow's milk is absorbing 13 g/day of saturated fatty acids (4:0 to 12:0), including 1.4 g of butyric acid. It should be specified that the contribution of short-chain fatty acids remains difficult to appreciate because an important proportion results from the bacterial fermentation of plants in the digestive tract. Apart from these estimates, few investigations were devoted to the absorption of these fatty acids by consumers in various countries. In the United States, it has been calculated that men and women consumed 2.6 and 2.1 g/day, respectively, in 2007–2008 (http://www.ars.usda.gov/main /site_main.htm?modecode=12-35-50-00). These short- and medium-chain fatty acids accounted for 8.3% and 9.5% of the total consumed saturated fatty acids in men and women, respectively. The American surveys tended to show that the majority of these compounds are mainly from animal source (58% of the total) in the form of dairy products. In consequence of the specificities of dietary habits, it is for the moment difficult to extrapolate these results to other countries.

Among long-chain fatty acids, some, such as palmitic acid (16:0), are known to deteriorate the cellular metabolism in various tissues. They are able to inhibit cellular glucose uptake, stimulate the inflammatory reactions, and contribute to the rise in cholesterolemia, thus supporting the development of atherosclerosis. Myristic acid (14:0) would occupy a particular place among such acids as it could exert a hypocholesterolemic role even at normal concentrations in the diet (Section 4.2.1). On the other hand, this fatty acid may be atherogenic when its contribution is higher than 4% of the total energy intake. Apart from these harmful effects, biochemists know very well that myristic acid takes part in the mechanism of myristoylation of proteins, an important process in the binding of active proteins on specific membrane targets. It should be specified that the impact of a possible dietary deficiency in myristoic acid is not yet defined.

From the point of view of a physiologist and also a dietician, it is necessary to define in the long-chain fatty acid group the subgroup of fatty acids from 12 to 16 carbons (lauric, myristic, and palmitic acids), those being atherogenic if absorbed in excess. The French ANSES authority had actually established for that subgroup an RDA in the United States or an ANC in Europe of 8% of the total energy intake, for example, 22 g/day for an adult man of 75 kg spending 2500 kcal/day. The second group is that of stearic acid (18:0), which, contrary to other saturated acids, is neither hypercholesterolemic nor atherogenic. This fatty acid was even proposed as a substitute for *trans* fatty acids of industrial origin used to raise the melting point of vegetal oils (Valenzuela et al. 2011). This physiological neutrality could rise from the calcium salt formation in the intestinal lumen that would reduce the intestinal absorption of fatty acids by eliminating them in feces. This mechanism, described in infants but not in adults, was clearly highlighted in animals in 1958 by K.K. Carroll in Canada. This lower absorption of stearic acid has also a consequence on the energy value of the lipids containing this fatty acid. Thus, in 1998 G.S. Ranhotra, the United States, estimated in the rat that tristearine had a food energy of only 3 kcal/g instead of the theoretical 9 kcal/g. A better knowledge of these phenomena should be a priority for nutritionists.

The preferential bioconversion of stearic acid into oleic acid has also been suggested. For these reasons, saturated fats are interchangeable neither in the human diet nor in nutritional experiments in animals. Thus, if one considers the composition of the saturated fatty acid pool, beef fat with 50% of palmitic acid (16:0) and 38% of stearic acid (18:0) cannot be compared except for their energy value with palm oil containing 88% of 16:0 and 9% of 18:0. These two sources have similar contents in saturated fatty acids (about 50% of the total fatty acid pool). Besides these characteristics, it must be emphasized that the differences in structure of the ingested triacylglycerol molecules are also able to influence the fate of the fatty acids (Section 3.2.1.4). This great diversity of composition could possibly induce sometimes opposite effects and thus different recommendations troubling the image of the relationships between lipids and health in the consumers. New research and precise information will be the only means to ensure a reasoned use of the many lipid sources at our disposal.

The European Parliament recognized that "the general public was interested in the correlation between diet and health and in the choice of a suitable diet corresponding to the needs for each one." Also, the parliament adopted on July 6, 2011 a legislative resolution intended to inform the consumers by imposing the addition in the obligatory nutritional declaration printed on each food package containing fats not only the amount of fats and its energy value but also the proportion of saturated fats.

To encourage a reduction in the consumption of saturated fats, several countries have established taxation of food containing these fats. Hungary initiated the movement in July 2011 by deciding on a tax of €0.37 on certain foods with high fat contents. Denmark followed, deciding in October 2011 on a tax of about €2 per kilogram of saturated fats on all products containing more than 2.3% fatty acids (fat tax), but it abolished this tax in November 2012. Among other countries, France could adopt this kind of taxation. An econometric study even showed that a tax policy on saturated fats could be consistent with an incentive of the population to modify the quality of its diet in connection with a beneficial medical effect (Tiffin and Arnoult 2011).

3.2.1.3.3.2 n-9 Fatty Acids   Foodstuffs contain variable amounts of *n*-9 fatty acids, represented mainly by oleic acid (18:1 *n*-9). Its concentration varies between 0.1% and 22% in dairy products and meat products and between 17% and 77% in oils and fats used for cooking (Tables 3.6 through 3.10).

Few studies have been devoted to the physiological effects of monounsaturated fatty acids, but some experiments have suggested that their consumption could be associated with an improvement of the blood markers related to cardiovascular disease but without significant evidence on the incidence of these pathologies.

In the world population oleic acid represents 3.5%–22% of the total energy intake, whereas in Europe this proportion varies according to countries, between 11% (Denmark) and 22% (Greece) (Simopoulos 2010). In France, this value is established at approximately 14%.

To inform consumers, a European Council directive emitted in 1990, related to the nutritional labeling of foodstuffs (Directive 90/496/CEE), advised to include the amounts of monounsaturated fatty acids.

Considering the concentration ranges recommended by nutritionists, the ANSES organization has defined for oleic acid an ANC (or RDA) value between 15% and 20% of the daily energy intake (e.g., from 42 to 55 g/day). The lower limit of this intake must be defined taking into account the risk associated with the replacement of oleic acid by atherogenic saturated fatty acids. The upper limit was suggested by the epidemiological and clinical data concerning cardiovascular disease risk factors (Section 4.2.2).

3.2.1.3.4   Specific Intake of Essential Fatty Acids

3.2.1.3.4.1   n-6 Fatty Acids   Although the concept of essential fatty acids applied to *n*-6 fatty acids was discovered in 1929 when studying the growth and prevention of dermatitis in rats, it was necessary to still wait nearly 20 years to extend that concept to humans. This discovery was made in 1947 by A.E. Hansen in the United States, when examining children suffering from an eczema that was cured by the ingestion of *n*-6 fatty acids of animal origin. A more precise nutritional study, extended to more than 400 children, made it possible in 1963 for the same author to appreciate for the first time the requirements of essential *n*-6 fatty acids in these children at approximately 1% of the total energy intake.

All essential fatty acids with the *n*-6 structure are biosynthesized by a series of desaturation (by specific desaturases) and elongation (elongases) from the precursor linoleic acid (Section 2.2.2.2) (Figure 3.9). Almost all mammals must find these fatty acids in plants or animals included in their diet. These transformations take place in the cytosol and the mitochondria.

Linoleic acid (18:2 *n*-6) is first desaturated into γ-linolenic acid (18:3 *n*-6) and then transformed into dihomo-γ-linolenic acid (20:3 *n*-6) (Section 2.2.2.2). The latter is not very abundant in animal and human tissues, but it may be metabolized by cyclooxygenases mainly into prostaglandins and thromboxanes of the first series. These prostanoids have anti-inflammatory properties, contrary to those derived from arachidonic acid (20:4 *n*-6). This fatty acid, represented well in animal tissues, is the most important of the *n*-6 fatty acids for the constitution of membranes and the cellular signaling operated by its derivatives: prostanoids, leukotrienes, and hydroxylated acids. Arachidonic acid may be further transformed into other polyunsaturated fatty acids with 22 carbons present in the membranes of some cell types.

It is remarkable that the values of ANC (or RDA) recommended for linoleic acid regularly evolved upward since its essential character was specified. It is now well known that the values initially suggested in the adult were overestimated because the first epidemiological studies did not take into account the *n*-3 fatty acid intake. A more precise sight of this problem has been obtained when it was shown in the animal that the addition of linolenic acid (18:3 *n*-3) to the diet reduced the specific needs in linoleic acid.

On the other side, it appears increasingly fundamental to limit the imbalance between the two groups of essential fatty acids, the *n*-6 and the *n*-3 series, by reducing the contribution of linoleic acid especially when the intake of linolenic acid is low, as frequently observed. This imbalance, with a too high amount of *n*-6 fatty acids, may only accentuate the incidence of pathologies affecting the brain and the cardiovascular system. To prevent these disorders, a ratio of linoleic acid to linolenic acid equal to or lower than 5:1 is generally highly recommended (Section 4.2.3.3) and a value of 10:1 or higher is considered as nutritionally inappropriate.

Physiologically, the adipose tissue will store the 18:2 *n*-6 in the form of triacylglycerols and the other tissues will use it as phospholipids to build up their cellular membranes, mainly in the form of highly unsaturated derivatives (e.g., arachidonic acid).

$$18:2\ n\text{-}6 \xrightarrow{\Delta 6} 18:3\ n\text{-}6 \xrightarrow{E} 20:3\ n\text{-}6 \xrightarrow{\Delta 5} 20:4\ n\text{-}6$$

FIGURE 3.9 Metabolism of *n*-6 fatty acids. The Δ sign indicates a desaturation step, the figure indicating the place of the new double bond; the carbon numbering starts from the carboxyl group. The symbol E indicates an elongation by the addition of two carbons.

This compound may also be the source of eicosanoids that then play a significant role in the normal performance of the nervous, cardiovascular, and immunizing systems. On the other hand, it may also generate allergic and inflammatory reactions. When consumed in excess, the n-6 fatty acids may suppress the beneficial effects of the n-3 fatty acids, sources of specific derivatives. The dietary n-6 fatty acids could thus take part in cardiovascular disorders, and also in inflammatory diseases (asthma or arthritis). This last hypothesis, supported by some physiologists, is far from being verified because no serious research could highlight an arachidonic acid accumulation in blood in the case of linoleic acid excess. On the contrary, experiments in humans have revealed a reduction in the risk of cardiovascular disease after the ingestion of high amounts of n-6 fatty acids (11%–21% of the total energy intake) during about 10 years (Harris et al. 2009). For quantities usually met in the diet, no major effect on the cardiovascular system has been reported, except a decrease in plasma cholesterol.

Recommended dietary intake of linoleic acid: on the basis of former studies, the FAO/WHO report of 1994 did not rule on the minimum needs of n-6 fatty acids, not more than for n-3 fatty acids, but was primarily interested in the ratio of linoleic acid to linolenic acid within food lipids. Later, on the basis of numerous epidemiological studies, the international experts fixed the minimal physiological requirement in 18:2 n-6 at 2% for total energy intake (ISSFAL 2004, http://www.issfal.org/statements/pufa-recommendations/statement-3). This value is equivalent to 5.5 g/day for an energy intake of 2500 kcal/day, for example, approximately 28 g of rapeseed oil or 9 g of sunflower oil.

In the United States, the standards established in 2005 by the Department of Health and Human Services and the USDA and currently accepted indicate that an intake of n-6 fatty acids from 5% to 10% of the total energy intake (e.g., from 14 to 28 g/day in humans) is compatible with a reduced risk of cardiovascular diseases (http://www.health.gov/dietaryguidelines/dga2005/report/default.htm).

Other recommendations, quite close to the previous ones, have been made by several national and international authorities. Thus, the recommendations of an n-6 fatty acid intake are from 5% to 8% in 2008 by the FAO/WHO, 4% to 8% in 2008 by the European Commission, 6% to 10% in 2008 in the United Kingdom, 4% to 10% in 2008 in Australia and New Zealand, and 3% to 10% in 2007 in Canada.

In France, the ANSES has established in 2010 the ANC (or RDA) of n-6 fatty acids to 4% of the total energy intake, for example, equivalent to a maximum of 11 g/ day for a total energy intake of 2500 kcal/day. This value resulted from a compromise taking into account the usual n-3 fatty acid intake in the French diet and with the aim of respecting the n-6 to n-3 fatty acid ratio whose value must remain lower than 5. These 11 g/day correspond to an ingestion of 55 g of rapeseed oil (4 tablespoons) or 17 g of sunflower oil (1.3 tablespoons).

It is necessary to keep in mind that in humans linoleic acid body reserves are often higher than 1 kg. For the newborn or the infant until 36 months of age, the EFSA panel on dietetic products and nutrition proposed in 2013 a minimum contribution of 4% of total energy intake. Beyond this age, the values established for adults must be used.

Food sources in linoleic acid: few precise data exist on the proportion of n-6 fatty acids in the diets of populations. A study on food consumption carried out in 2009 by L. Elmadfa in 14 countries, including 7 European countries, has shown that the intake of linoleic acid was in the range from 2.7% (India) to 7.2% (Austria) of total energy intakes. In Europe, France had the lowest value (4.2%), as did Finland (3.9%) and Norway (4.3%), these values being lower than those determined in other countries (England 4.8%, Belgium 5.3%, Germany 5.7%, and Austria 7.2%). Linoleic acid accounted for 7.2% of the total energy intake in the American people (Blasbalg et al. 2011). Thus, except in the United States and Austria, the populations of the studied countries have a linoleic acid intake that respects the threshold recommended by international experts (6%). This situation must of course be examined in light of the related n-3 fatty acid contributions.

Linoleic acid is largely present in the diet not only in milk, pork fat, and meats but also in the usually consumed vegetal oils (grape seed, groundnut, sunflower, and soybean oils) (Table 3.11).

Considering the share of various vegetal oils in the French consumption (approximately 46% for rapeseed oil, 19% for soybean oil, and 16% for sunflower oil; data from CETIOM 2008) and their composition, it appears that the linoleic acid intake is supplied at a similar level by rapeseed, soybean, and sunflower oils. The huge increase in the consumption of soybean (more than 1000 times) in the United States during the twentieth century is responsible for the excessive intake of dietary linoleic acid by its populations (Blasbalg et al. 2011).

γ-Linolenic acid (18:3 n-6) is synthesized naturally by cells starting from linoleic acid (18:2 n-6) but does not accumulate. Its formation can be slowed down in certain physiological situations (aging, diabetes, and alcoholism). It is present in some particular oils, which are sometimes regarded as food supplements likely to counteract that inefficient biotransformation. One can thus find γ-linolenic acid concentrations from 18% to 25% in borage seed oil (*Borago officinalis*, Boraginaceae), 16% to 18% in black currant seed oil (*Ribes nigrum*, Grossulariaceae), and 8% to 14% in evening primrose oil (*Oenothera biennis*, Onagraceae). This fatty acid is active at the skin level by improving its barrier function and by limiting the epidermal hyperproliferation characteristic of pathologies such as atopic eczema (Kawamura et al. 2011). Several clinical observations have suggested that this fatty acid can play an important function in the modulation of the inflammatory processes linked to several pathologies (cancer, diabetes, heart disease, arthritis, Alzheimer's disease, and so on) (Kapoor and Huang 2006). As research efforts on the physiology of gamma-linolenic acid are in progress, important therapeutic applications of this fatty acid are expected in a near future.

Arachidonic acid, sometimes considered as the only essential compound of the n-6 fatty acid series, is present in various foods almost exclusively from animal sources. Egg yolk and animal tissues are direct sources. Because it can be metabolized into eicosanoids of the 2 series and leukotrienes of the 4 series, arachidonic

---

TABLE 3.11

Content of Linoleic Acid (18:2 n-6) in Some Food Sources (in Grams per 100 g)

Cow's Milk	Breast Milk	Lard	Sunflower Oil	Groundnut Oil	Rapeseed Oil	Soybean Oil	Grape Seed Oil
0.064	0.43	6–10	62–70	14–41	22	50–62	65–73

acid improves cicatrization and cure of wounds, but it also contributes to the mechanisms of allergic reactions. However, an excess of these eicosanoids may lead to diseases like arthritis, eczema, psoriasis, and several autoimmune reactions.

As arachidonic acid is efficiently synthesized in the adult by a conversion of its precursor linoleic acid, no special recommendation of dietary intake has been promulgated.

In the United States, the average dietary intake of arachidonic acid has been estimated to be about 150 mg/day. In France, the SUVIMAX study has shown that the average intakes are 204 mg/day in men and 152 mg/day in women.

The food sources of arachidonic acid are primarily eggs and all meats, the latter supplying between 30 and 120 mg for every 100 g consumed (Table 3.12).

Supplementation tests (up to seven times the average daily intake in the United States) carried out in humans were not able to detect any undesirable effects after 7 weeks as well at the level of platelets and bleeding time as at the level of blood lipids and immunization responses. It was shown in 1997 by A. Ferretti in the United States that with amounts of arachidonic acid higher than 1.5 g/day the biosynthesis of vasoactive eicosanoids increased significantly. Other investigations have shown that these arachidonic acid intakes had important effects on immunization responses and platelet functions. In any case, these too-elevated intakes remained largely out of the normal nutritional field as they corresponded to the consumption of more than 1 kg/day of meat.

3.2.1.3.4.2 n-3 Fatty Acids   All the essential fatty acids with the *n*-3 structure are biosynthesized by a series of desaturation, elongation, and oxidation (Figure 3.10) beginning with the precursor linolenic acid (Section 2.2.2.3). This fatty acid must be supplied by dietary plants for almost all animals. In the year 1970, linolenic acid was shown to be essential to the vision and cerebral function in the rat, but it was shown to be essential in humans only in 1982 by R.T. Holman.

Linolenic acid (18:3 *n*-3) is first desaturated in stearidonic acid (18:4 *n*-3). These two fatty acids are present in significant amounts only in some vegetal oils. The metabolic pathway of *n*-3 fatty acids will then lead to the formation of EPA (20:5 *n*-3) and DHA (22:6 *n*-3). Those compounds accumulate in the cellular membranes of some

---

TABLE 3.12

Average Contents of Arachidonic Acid in Some Foods (Milligrams per 100 g Fresh Weight)

Hen's egg	100–300	Beef	40–100
Trout	190	Pork	40–80
Salmon	20	Lamb	30–60
Tuna	15–80	Chicken	80–20
Oyster, mussel	40–80	Duck	50–60
Breast milk	15		

Note: Data are from the food composition tables from USDA: http://www.ars.usda.gov/Services/docs
.htm?docid=20957.

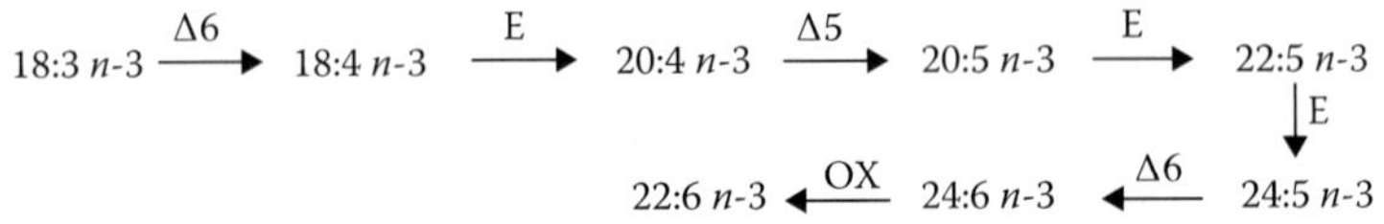

$$18:3\ n\text{-}3 \xrightarrow{\Delta 6} 18:4\ n\text{-}3 \xrightarrow{E} 20:4\ n\text{-}3 \xrightarrow{\Delta 5} 20:5\ n\text{-}3 \xrightarrow{E} 22:5\ n\text{-}3$$

$$22:6\ n\text{-}3 \xleftarrow{OX} 24:6\ n\text{-}3 \xleftarrow{\Delta 6} 24:5\ n\text{-}3$$

FIGURE 3.10 Metabolism of *n*-3 fatty acids. The Δ sign indicates a desaturation step; the figure indicates where the double bond is located by numbering from the carboxyl group. E indicates an elongation step with the addition of two carbon units and OX indicates a β-oxidation step.

TABLE 3.13
Composition of Linoleic and Linolenic Acids in Some Fresh Vegetables (in Milligrams per 100 g)

Vegetable	Linoleic Acid	Linolenic Acid
Cauliflower	23	76
Broccoli	17	21
Brussels sprout	80	170
Cucumber	20	40
Kale	80	100
Green pea	152	35
Lentil	404	109
Green bean	23	36
Spinach	26	138
Water cress	20	100
Potato	32	10
Tomato	80	3
Purslane (*Portulaca oleracea*)	20	120

Note: Data are partially from the USDA food tables (http://www.ars.usda.gov/Services/docs.htm?docid=20957).

animal tissues (brain and retina) and especially in oils of marine animals and unicellular algae. Moreover, both compounds may be metabolized into various oxygenated derivatives playing roles in the regulation of many cellular mechanisms (Section 4.2.4). Thus, EPA can be transformed into thromboxane A3 (weak aggregating agent), prostacyclin PGI3 (vasodilator and inhibitor of platelet aggregation), and leukotriene LTB5 (anti-inflammatory and antichemotactic agent). DHA has protective effects in the brain via specific metabolites such as docosanoids (neuroprotectines) and neuroprostanes.

Linolenic acid is perhaps not the true essential fatty acid of the *n*-3 series, but its metabolic derivatives are more unsaturated and with a longer carbon chain they could be the true essential fatty acids for many cellular functions.

Although the specific physiological properties of linolenic acid, the dietary precursor of all other *n*-3 fatty acids, are not yet completely known, its presence in the diet remains necessary. Rare clinical symptoms, such as dermatitis, can appear after a long linolenic acid deficiency, in spite of the presence of that fatty acid in all vegetables at concentrations from 3 to 200 mg for 100 g of fresh product (Table 3.13) or even more important in some nuts (Table 3.14).

TABLE 3.14
Composition of Linoleic and Linolenic Acids in Some Fresh Fruits and Nuts (in Milligrams per 100 g)

Fruit	Linoleic Acid	Linolenic Acid
Pineapple	23	17
Banana	46	27
Strawberry	182	71
Raspberry	250	126
Orange	31	11
Peach	84	2
Pear	29	0
Apple	43	9
Grape	37	11
Walnut	38,000	8,000
Macadamia nut	2,100	1,380
Chestnut	1,450	170

Note: Data are partially from the USDA food tables (http://www.ars.usda.gov/Services/docs.htm?docid=20957).

Biochemically, linolenic acid is rapidly oxidized into CO_2 (nearly 20%) in tissues (Vermunt et al. 2000) and also transformed in nervous tissues into saturated or mono-unsaturated fatty acids and even into cholesterol (Sinclair et al. 2002). It is moreover a competitive inhibitor of the metabolism of *n*-6 fatty acids contributing to reduce arachidonic acid biosynthesis, thus reducing inflammation. The requirement in *n*-3 fatty acids will thus be dependent on the importance of the *n*-6 fatty acid intake. Some authors have suggested that one could cover the physiological requirement in *n*-3 fatty acids with a 10 times lower intake by strongly limiting the *n*-6 fatty acid intake.

Stearidonic acid (18:4 *n*-3) (Section 2.2.2.3) is the first linolenic acid metabolite. Its rapid transformation into EPA could thus increase the accumulation and amplify the physiological properties of that important long-chain metabolite. Stearidonic acid is present in some vegetal sources used for dietetic applications, some works asserting that it can substitute EPA of animal origin.

Unlike stearidonic acid, the polyunsaturated *n*-3 fatty acids with 20 and 22 carbons (Section 2.2.2.3) EPA and DHA play a significant role at the level of membrane lipids, especially in nervous tissues and retina and also in the cardiovascular system of almost all animals. They are also the precursors of bioactive lipids such as prostaglandins, neuroprostanes, and many hydroxylated and epoxy derivatives. The beneficial health effects of EPA and DHA are being more and more recognized in various fields (Section 4.2.4), research on these effects being under constant development (more than 2300 scientific articles were published in 2013). Their essential character in the development and cellular functions in humans as in all the animal kingdom arises the problem of their metabolism and their dietary supply.

Linolenic acid may be converted, even very moderately, into EPA, 22:5 *n*-3 (timnodonic acid), and DHA, the only *n*-3 fatty acids present in cellular membranes.

As was shown in 1994 by E.A. Emken, this conversion does not exceed 10% in the best case in the presence of normal amounts of n-6 fatty acids.

DHA is a major constituent of membranes of nervous cells, retinal cells, and in general all cells with electric activity (brain and heart). As for its precursor, the 22:5 n-3, its precise function still remains poorly known, but it appears essential for the protection and proper functioning of cells. Taking into account its low level of bio-synthesis from dietary linolenic acid, physiologists have fixed the minimal require-ment in DHA at not less than 250 mg/day for an adult (e.g., about 0.10% of the total energy intake), a value higher than that suggested earlier.

Besides DHA, it is necessary to consider EPA, which appears to hold clini-cal interest especially at the level of nervous system dysfunction (Section 4.2.4.5). Natural fish oil supplements, with a high ratio of EPA to DHA (3:4), could be very useful either mixed with infant formula or as a food supplement in pregnant women. Metabolic research in animals has shown a possible retroconversion of the exogenous DHA into EPA; however, the quantitative aspects of this transformation in humans remain not well defined, especially according to the physiological state of the subject.

Recommended dietary intakes of n-3 fatty acids: for essential fatty acids of the n-3 series (linolenic acid and its polyunsaturated derivatives such as EPA and DHA), recommendations are becoming gradually more and more precise following a great number of investigations concerning various aspects of their physiological roles (Section 4.2.4). Indeed, more than 15,000 works devoted to this subject have been published during the last 30 years. Although various scientific organizations have recommended various levels for EPA and DHA consumption, most experts have adopted a daily recommended allowance of about 500 mg for the sum of both fatty acids. This roughly corresponds to two fish meals (if possible fatty fish) per week, a consumption suitable to profit from the health effects of these lipids (recommenda-tions of the American Heart Association, 2006). That official organism modified that recommendation for a daily intake of 1 g/day of EPA + DHA in patients recovering from heart attack and 2 to 4 g/day in patients suffering from hypertriglyceridemia.

During the global summit on omega-3 held at Bruges, Belgium, in March 2011, international experts vigorously launched six consensus statements to influence policy and grow consumer awareness in alerting the Western population to the urgent necessity of daily ingesting at least 1 g of EPA + DHA (http://www.omega-3summit.org/). This intake must obviously be modulated according to the previous n-3 fatty acid status, defined by the "omega-3 index" (ratio of EPA + DHA to total fatty acids) measured in red cells (optimal value between 8% and 11%) (http://www .omega-3-index.com/fr/index.php). These experts also recommended the reduction of linoleic acid intake while increasing the n-3 fatty acid content in human and animal diets, including linolenic acid. M. Crawford, in the United Kingdom, mili-tated the need of a "Kyoto-type approach in order to tackle the global issue of the long chain omega-3 deficiency which is of similar size and spread for human (men-tal) health as the CO_2 issue may be for the environmental health of our planet." It would thus be desirable that the determination of the omega-3 index becomes a common marker as other biological parameters. Presently, this determination may be carried out using a simple kit obtained from Nutrigenics, Ltd., Berchem, Belgium.

In France, the ANSES and the Nutrition and Health National Plan (PNNS) were assigned to spread this information. Overall, the intake needed to prevent any n-3 fatty acid deficiency should be equal to at least 3% in weight of the lipid ration (approximately 0.8% of the total energy intake). This value corresponds to about 2 g/day of n-3 fatty acids, a quarter of this amount (or 500 mg) being represented by DHA + EPA.

It should be stressed that, because of the competitions at the level of absorption and metabolism, the recommended intake of n-3 fatty acids depends closely on that of n-6 fatty acids. As Western diets have a high content of vegetal oils, the intake of 500 mg/day of EPA and DHA must represent a minimum amount. Except marine fish, this intake may be easily carried out by consuming meat or eggs from animals fed with food enriched with linolenic acid. Generally, the n-3 fatty acids ingested as a supplement are preferred in the form of oil capsules prepared from fish, krill, or algae.

To correct the imbalance currently noted between dietary n-6 and n-3 fatty acids, the contribution of lipids coming from fish should also be supplemented by the consumption of linolenic acid–rich vegetal oil. In the United States, a daily intake of linolenic acid corresponding to 0.6%–1.2% of the total energy intake (e.g., on average 2.5 g/day) was strongly recommended (National Academy of Sciences, IOM, 2002).

It must be recalled that, as the consumption of EPA and DHA is continuously increasing, there is a potential risk of consuming an excess of these fatty acids beyond 3 g/day. Excessive intakes may be associated with adverse effects and, in extreme cases, negative health outcomes.

Recent studies of various infections in animal models demonstrate that EPA and DHA intake may weaken immunity and alter the pathogen clearance resulting in reduced survival (Fenton et al. 2013). Thus, there is a necessity to rapidly study the reliable biomarkers of the effects and risks associated with a supplementation in these fatty acids.

Current situation of n-3 fatty acid intakes: n-3 fatty acids and their properties are objects of continuous research, involving rapid change and frequent revision of the nutritional recommendations on their specific intakes. One of the principal limitations to decide on quantitative recommendations is the disagreement of experts on the conversion rate of the precursor 18:3 n-3, of vegetal origin, in higher unsaturated derivatives as EPA and DHA. According to several specialists, this rate varies from 5% to 10% in the case of EPA and from 2% to 5% for DHA. According to the works of G. Burdge, in 2004, these conversion rates are lower than 1%. Moreover, one possible modulation according to sex and individual physiological status must be considered. The efficiency of this transformation is also influenced by the amounts of ingested linoleic acid. This mechanism has been clearly demonstrated in the case of fetal development (Novak et al. 2012).

It is becoming increasingly obvious that a genetic polymorphism must be taken into account in the determination of nutritional requirements of long-chain n-3 and n-6 fatty acids (Simopoulos 2010). Indeed, it is now well established that the natural activity of Δ-5 and Δ-6 desaturases (or FADS1 and FADS2), enzymes along the metabolic pathways of linoleic and linolenic acids, influences the fatty acid composition of plasma and cellular membranes. Moreover, the hypothesis formulated in 2012 by A. Ameur of a role of these genetic variations (haplotypes A and D) in the distribution of various pathologies is becoming increasingly realistic. The author even proposed the preventive determination of these genotypes to deliver adapted nutritional advice, the subjects belonging to haplotype A having more difficulties than those having haplotype D to convert linoleic acid into arachidonic acid and linolenic acid into EPA and DHA.

The current situation is described here. Several countries, as well as the WHO, have developed recommendations about n-3 fatty acid intake in adults, which may be summarized as follows:

- Linolenic acid (18:3 n-3): from 0.8 to 1.1 g/day
- EPA + DHA: from 300 to 500 mg/day

An international expert committee (ISSFAL) had fixed in 1999 the optimal intake of 18:3 n-3 at 1% of the daily energy intake, for example, approximately 2.2 g/day for a diet providing 2500 kcal. These experts also recommended a minimal intake of 0.3% of the energy intake, for example, 800 mg/day of EPA and DHA for the maintenance of a healthy cardiovascular system (http://www.issfal.org/statements /adequate-intakes-recommendation-table).

These data have not been ratified yet by the American and Canadian authorities. In these two countries, the recommended intake of 18:3 n-3 is determined for the moment at about 0.5% of the total energy intake, for example, 1.1 g/day for women and 1.6 g/day for men. No precise intake has been proposed for the sum EPA + DHA. In the United States, the USDA and the Department of Health and Services (HHS-USA) recommend the population to consume at least 230 g of sea products per week to cover an average intake of 250 mg/day of EPA and DHA. For pregnant or nursing women, it has been recommended to consume from 230 to 340 g of these marine products a week, if possible not contaminated by mercury.

A group of FAO/WO experts on "fats and fatty acids in human nutrition" determined in 2010 the requirements of adults to reach 250 mg/day of EPA and DHA, the value being 300 mg/day, including at least 200 mg/day of DHA, for pregnant or nursing women.

These recommendations are criticized by some experts, who consider them insufficient to allow a reduction of coronary diseases in North America.

At the European level, the EFSA published in 2012, at the request of the European Commission, its own conclusions on the acceptable higher limits of EPA and DHA intakes. Thus, the EFSA considered that an intake of approximately 5 g/day of a mixture of EPA and DHA for a long period does not seem harmful, presenting no risk of bleeding, no disorder of glucose metabolism, and no alteration of immunization functions. The European authority recalls that with respect to cardiovascular risks the intakes recommended for EPA and DHA remain in the range of 250–500 mg/day.

The EFSA decided in 2013 that for the majority of infants and young children (from 0 to < 36 months) the intake of linolenic acid must be 0.5% of the total energy intake. For DHA, an intake of 100 mg/day is recommended from 0 to less than 24 months and 250 mg (with EPA) up to 36 months.

At the French level, the ANSES specified that the values of ANC (or RDA) proposed for adult subjects in terms of prevention of pathologies may also be applied, in the absence of specific experimental data, to pregnant or nursing women. In the newborn baby or infant less than 6 months of age, the ANSES recommended for linolenic acid an ANC (or RDA) not lower than 0.45% of the energy intake (e.g., 1% of the total fatty acids of milk lipids). There is no nutritional interest so that the ANC exceeds the value of 1.5%. In a coherent way with the various current international recommendations, the ANC for DHA is fixed at 0.32% of the total fatty acids. Finally, the EPA intake must be lower than that of DHA. These values can be taken into account also in children up to 3 years old. Beyond that age, the values described for adults are recommended.

The ANSES advised in 2010, from a practical point of view, a dietary intake of linolenic acid (18:3 n-3) in the adult from at least 0.8% of the total energy intake (e.g., 2.2 g/day for 2,500 kcal/day). A nutritional intake of 500 mg/day of EPA + DHA is advised in the adult, as in the child. The minimum intake advised for these two n-3 fatty acids in the infant of more than 6 months and the child up to 3 years is 70 mg, whereas it is 250 mg for children more than 3 years old (AFSSA—saisine N° 2006-SA-0359).

A recent report on the new ANC (or RDA) of these lipids can be consulted in the bulletin of the *Journal of Lipid Nutrition* (April–May 2011): http://www.fncg.fr /fichiers/20110428153752_Lipid_nutri__n9.pdf.

Concerning the requirements of pregnant and nursing women, infants, and young children (6 months to 2 years), it is admitted that mother's milk, except characterized deficiency of the mother, is the most adequate food to ensure a suitable supply of long-chain polyunsaturated fatty acids. The DHA content of the lipids of breast milk, determined in 2006 by R. Yuhas, naturally varied between 0.2% and 1% and that of arachidonic acid varied between 0.3% and 0.7%. If breast-feeding proves to be impossible, the most often employed substitutes are manufactured foods mainly adapted for the first year of life but sometimes until 3 years old. After a first working group organized in 2001 by the Child Health Foundation (http://www.kindergesundheit. de/home.html), the European Commission instructed the research groups PERILIP and EARNEST to establish recommendations specifying the specific lipid requirements of pregnant and nursing women. A large review of the literature and a 2005 consensus conference made it possible to establish conclusions in agreement with the state of knowledge at the time (Koletzko et al. 2007). These conclusions specified that pregnant and nursing women must follow the recommendations already given for other adults, for example, to respect an intake from at least 200 mg/ day of DHA. The experts reminded that the deposit rate of this fatty acid in the fetus is nearly 45 mg/day, especially during the third quarter of gestation, whereas arachidonic acid is deposited in high amounts only after birth. The conference report reminded that this amount of DHA can be supplied by the consumption of one to two fatty fish portions such as herring, mackerel, or salmon. No compensation by an additional n-3 fatty acid ingestion in the form of 18:3 n-3 can be recommended as its transformation into DHA proves to be very reduced. Given the dangers of contamination of fish by heavy metals and pesticides, it is recommended to use as much as possible in the preparations for infants only DHA-enriched oils produced by algae or fungi, or eggs produced by hens fed a diet enriched with linseed oil.

A study published in 2008 by B. Koletzko, Germany, analyzing the world literature and carried out under the auspices of the World Association of Perinatal Medicine (WAPM), has completed and widened the preceding recommendations for the food of young children. The report concluded that it is currently beneficial for the development of the baby's nervous system to supplement food with DHA during the first 6 months to reach a level not less than 0.2% of the total fatty acids (but not more than 0.5%; upper limit recommended because of the lack of clinical data). The present knowledge led the food industry to add also arachidonic acid at the same level as for DHA in infant formula.

These conclusions are largely in line of the recommendations made since 1993 and until 2010 by many qualified national organizations. Among them, we will quote the Food and Drug Administration (FDA), EFSA, FAO, WHO, CAC, ANSES, ADA, Canadian Dieticians, ESPGHAN, Commission of the European Communities, and NAS (the United States).

In France, the ANSES recommends as follows:

- For the infant until 6 months of age, a minimal ANC (or RDA) for linolenic acid (18:3 n-3) of 0.45% of the energy uptake (either 45 mg/kg/day or 1% of the total fatty acids of milk lipids), 1.5% (or 150 mg) representing a maximum value. The ANC (or RDA) for DHA is fixed as 0.32% of the total fatty acids. In addition, the DHA intake must be balanced with the intake of arachidonic acid (0.5% of the total fatty acids) to avoid a deficit in this essential fatty acid in the infant. Lastly, the intake of EPA must be lower than that of DHA.
- For the infant from 6 months to 3 years of age, the linolenic acid values proposed for the infant less than 6 months of age are convenient. An ANC (or RDA) value of 70 mg of DHA per day allows a continuous accumulation of this fatty acid in cerebral membranes.
- For the older child, from 3 to 9 years old, the ANC recommended for the adult applies, for example, 1%. The ANC (or RDA) recommended for DHA is 125 mg/day and 250 mg for EPA + DHA, taking into account a reduced energy intake compared to that of teenagers. For teenagers, the proposed ANC (or RDA) for DHA corresponds to that for adults (250 mg of DHA per day and 500 mg for EPA + DHA) (opinion of the AFSSA as of March 1, 2010; saisine number 2006-SA-0359).

For premature babies, the requirements in n-3 fatty acids have been established by an expert committee appointed by the ESPGHAN in 2007 (Agostini et al. 2010). In spite of the lack of accurate information, the minimum intake in 18:3 n-3 must be 55 mg/kg/day (e.g., 50 mg/100 kcal and 0.9% of the total fatty acids). For DHA, an intake of 12–30 mg/kg/day (e.g., 11–27 mg/100 kcal) is recommended. The intake in EPA should not exceed 30% of that in DHA.

Food sources of n-3 fatty acids: a general inquiry study of consumption in 14 countries, including 7 European countries, has shown that the intake of linolenic acid was lower than 1% of total energy intake everywhere (Simopoulos 2010). This very low level may be regarded as defining a deficiency state, which may worsen if the contribution in EPA and DHA is weak. Moreover, this weak intake does not support the increase in tissue contents in EPA and especially in DHA, considering the very low capacities of biosynthesis of these fatty acids from linolenic acid in humans (Brenna et al. 2009).

In the United States, the intake in linolenic acid was estimated in 2011 by T.L. Blasbalg to 0.72% of the total energy intake, a value two times higher than that estimated 90 years earlier. In this country, soybean oil provides 45% of the linolenic acid intake.

Among the seven European countries in Table 3.15, a great disparity appears, probably as a consequence of the dietary habits specific to each country. Finland, Germany, Belgium, and Austria fill the criteria of a sufficient intake in linolenic acid, but only Finland and Germany have an *n*-6 to *n*-3 fatty acid ratio smaller than 5. France is characterized by a very insufficient intake (about 1 g/day of 18:3 *n*-3), with an elevated *n*-6 to *n*-3 fatty acid ratio. These data confirmed those established by the large SUVIMAX study that had evaluated in 2004 a contribution of linolenic acid for the French population equal to 0.37% of the total energy intake.

All these results would justify a broad diffusion of information aiming at increasing in France the consumption of green plants and vegetal oils rich in linolenic acid (rapeseed, soybean, and walnut) to the detriment of oils rich in linoleic acid (groundnut and sunflower).

The average food supply in EPA and DHA is very disparate among the countries, whatever the continent, in relation to the geography and the dietary habits, as it was shown in 2009 by L. Elmadfa. This consumption study in 12 countries located in four different continents has shown that this intake varied between 0.03 g/day in China and 1.05 g/day in Japan (e.g., lower than 0.4% of the energy intake). Only Japan, South Korea, and Norway exceed the threshold of 400 mg/day of EPA + DHA, value considered to be minimal to prevent any deficiency. Among the studied European countries (Table 3.15), those having the lowest food supplies in EPA + DHA

TABLE 3.15

Average Dietary Intakes of Linolenic Acid (Percentage of Total Energy Intake) and EPA + DHA in Seven European Countries

Country	Linolenic Acid	*n*-6 To *n*-3 Fatty Acid Ratio	EPA + DHA (mg/day)
France	0.35	12	380
Norway	0.55	7.8	830
United Kingdom	0.55	8.7	250
Austria	0.6	12	270
Belgium	0.65	8.1	210
Germany	0.7	4	270
Finland	1	3.9	300

Source: Simopoulos, A.P., *Exp. Biol. Med.*, 235, 785–95, 2010.

(210–270 mg/day) (Belgium, United Kingdom, Austria, and Germany) also have the lowest consumption of marine fish (13.4–24.8 kg a year and per capita) (FAO, 2010). This intake is higher in the countries of northern Europe (300 mg/day in Finland and 830 mg/day in Norway) and intermediate in France (380 mg/day). These relatively higher values are certainly correlated with a greater marine fish consumption (31.9 kg/day in Finland, 35.2 kg/day in France, and 52.3 kg/day in Norway). The very high intake in EPA + DHA among the Japanese (1.05 g/day) seems obviously related to a greater consumption of fish and seafood (61 kg/day) (FAO 2010). On the other hand, in the United States T.L. Blasbalg in 2011 considered the food supply of EPA + DHA at an average of 140 mg/day, half of the contribution estimated 90 years earlier. This fall could be assigned to a major reduction in fish consumption, this being slightly compensated by a rise in poultry and shellfish consumption.

To better inform European consumers about the content of *n*-3 fatty acids of certain commercial products, the European Commission legislated on several nutritional announcements (regulation EC n° 116/2010 of February 9, 2010). Thus, announcement that a product is a source of *n*-3 fatty acids is possible only if it contains at least 0.3 g of 18:3 *n*-3/100 g and 100 kcal, or at least 40 mg of EPA + DHA/100 g and 100 kcal. Announcement that a product is *n*-3 fatty acid rich is possible only if it contains at least 0.6 g of 18:3 *n*-3/100 g and 100 kcal, or at least 80 mg of EPA + DHA/100 g and 100 kcal.

In France, the SUVIMAX investigation in 2004, carried out with approximately 4900 subjects, has shown that adults ingested on average 273 mg/day of DHA for men and 226 mg/day for women. In spite of these figures, it is necessary to note considerable disparities according to individuals, in men the study determined a range from 10 to 1,472 mg/day for the dietary intake of DHA. These variations can result from the age of the questioned people as well as from their geographical location. The same study revealed that the totality of the long-chain *n*-3 fatty acids (20:5, 22:5, and 22:6) was supplied to a mean level of 500 mg/day in men (from 40 to 2600 mg/day) and of 400 mg/day in women (from 11 to 2900 mg/day). Still, a great dispersion of the values was noted there, which suggested the need for information campaigns focused on professionals and mainly on the whole population to try raising the low intake values. The study had also shown that EPA was supplied for 72% by fish and seafood, 22:5 *n*-3 for 55% by meats and eggs and for 32% by fish, and DHA for 65% by fish and seafood and only for 18% by meats and eggs. Another more recent study has shown that, in French, fish and seafood consumed at a rate of approximately 35 kg per capita a year, were in fact the first source of DHA at a rate of 180 mg/day (Bourre and Paquotte 2007). About 70% of this intake came from the consumption of five species: salmon (48 mg/day), sardine (27 mg/day), tuna (20 mg/day), mackerel (17 mg/day), and herring (12 mg/day).

In spite of these reassuring general estimates, the last report by the ANSES organization, based on nutritional inquiries, emphasized the insufficient level of *n*-3 fatty acid intakes in the French population:

> The women are more numerous than the men (82% against 77%) to consume fish. The average levels of consumption are of 26.6 g/day in men and of 26.5 g/day in women, that is to say a little less than 2 portions of fish per week. In a global way, one observes a stagnation

of fish consumption since 1998. The data of the inquiry indicated that only 30% of the adults or the children consumed at least twice fishery products per week, and in a comparable way for the two sexes. It appeared that the intake in linolenic acid was insufficient and the ANC (or RDA) was not reached. Consequently, the AFSSA recommended the installation of a policy aiming at increasing the level of the intakes in n-3 fatty acids in the French population. To this end, the promotion of the consumption of food naturally rich in n-3 fatty acids, like certain fish and certain vegetal oils (rapeseed, nut …) could in particular be considered (AFSSA report on the level of fish consumption in France, June 2006).

After demonstrating the presence of contaminant products (heavy metals, pesticides, dioxans, and polychlorinated biphenyls [PCBs]) in the flesh of carnivorous fish consumed in France, the ANSES recently recommended (opinion of June 14, 2010) to consume two fish servings per week but by associating a fish with elevated amounts of n-3 fatty acids (fatty fish) and a lean fish. It is also necessary to think of varying the species and the sources and of limiting the consumption of fish known to be accumulators of PCB pollutants (eel, common barbel, common bream, carp, catfish, and silurus) or of mercury in carnivorous seawater fish (tuna, shark, bass, and swordfish) (Mahaffey et al. 2011). These recommendations are valid for adults and children from 10 years onward. For 3- to 10-year-old children, the n-3 fatty acids may be supplied by mullet, anchovy, or pilchard (AFSSA—saisine N° 2008-SA-0123). As early as 2004, the authorities of the United Kingdom (Food Standards Agency) have clearly recommended that pregnant women should avoid consuming several marine fish (tuna, shark, marlin, and swordfish) for the same medical reasons (http://cot.food. gov.uk/pdfs/fishreport2004full.pdf). Some works have provided dietetic advice for women in age to procreate and for pregnant women to optimize the intake of long-chain n-3 fatty acids while maintaining at a minimum contamination by mercury (Mahaffey et al. 2011). Information on these parameters in various countries would be desirable so that the population can adapt its consumption of sea products according to the benefit brought by n-3 fatty acids and the risks induced by pollutants.

A report published in 2011 by the FAO and the WHO established that there is at present no obvious cardiovascular risk consecutive to the consumption of methylmercury coming from fish. It would be the same for the dioxans whose risks of cancer are largely compensated by their many health benefits (http://www.fao.org /docrep/014/ba0136e/ba0136e00.pdf).

In all the cases, a food supplement in EPA and DHA coming from purified fish oils (or better from krill), and possibly from marine plants, can be considered especially if intake in the form of fish is considered to be definitely insufficient compared to an excessive n-6 fatty acid intake.

Vegetal sources of n-3 fatty acids: the most important n-3 fatty acids reserve is without any doubt made up by plants. Unfortunately, with some exceptions, only linolenic acid is well represented there, in the green parts or in some oils obtained by seed crushing. For the anecdote, it was determined in 2001 by C. Pereira in Australia that one of the precursors of the linolenic acid in plants, the fatty acid 16:3 n-3, is present in measurable amounts in spinaches (21 mg/100 g fresh weight), cress (45 mg), and parsley (44 mg). Its contribution to the synthesis of other n-3 fatty acids such as linolenic acid has not been explored yet in humans.

Considering the importance of plants in the human diet, they are a considerable source of linolenic acid. The inclusion of a ration of vegetables (approximately 250 g) and of fruits (200 g) in each meal can thus bring nearly 250 mg of linolenic acid to the diet. Some fruits (walnut, macadamia nut, and sweet chestnut) and especially oils (nut and soybean) remain the best sources of linolenic acid (Tables 3.14 and 3.16).

The compositions in linolenic acid (18:3 *n*-3) and linoleic acid (18:2 *n*-6) of some vegetables and fruits usually consumed in France are given in Tables 3.13 and 3.14, respectively.

To these common plants, it is necessary to add several seaweed species containing *n*-3 fatty acids and whose consumption is increasing gradually. They are used in various culinary preparations of fish. Among these seaweeds collected in France and analyzed in 2011 in Netherlands by V. van Ginneken, one finds the following:

- The dulse (*Palmaria palmaria*) containing 9 mg/g dry weight fatty acids (including 92% of EPA)
- The rockweed (*Ascophyllum nodosum*) containing 3.2 mg/g dry weight *n*-3 fatty acids (including 50% of EPA)
- The serrated wrack (*Fucus serratus*) containing 3 mg/g dry weight *n*-3 fatty acids (including 47% of EPA)
- The sea lettuce (*Ulva lactuca*) containing 6.5 mg/g dry weight *n*-3 fatty acids (including 3.5% of EPA)
- The wakame (*Undaria pinnatifida*) containing 6.3 mg/g dry weight *n*-3 fatty acids (including 46% of EPA)
- The tangle or kombu (*Laminaria hyperborea*) containing 7.1 mg/g dry weight *n*-3 fatty acids (including 68% of EPA)

Vegetable oils: vegetable oils rich in *n*-3 fatty acids can be divided into two groups, oils in common domestic use and dietetic (or therapeutic) oils, the latter oils being only dietary supplements. These two categories of oils are in general selected for their high concentrations of linolenic acid (18:3 *n*-3), stearidonic acid (18:4 *n*-3), or DHA. Oils rich in linolenic acid (or α-linolenic acid): all terrestrial crop products, oils included, supply to the diet only the precursor of the *n*-3 fatty acid series, that is, linolenic acid (18:3 *n*-3). Table 3.16 gives the average compositions of oils rich

---

TABLE 3.16

Composition of Linolenic and Linoleic Acids of Some Vegetal Oils Rich in *n*-3 Fatty Acids (Percentage of Total Fatty Acids)

Oil	Linoleic Acid	Linolenic Acid	*n*-6/*n*-3
Rapeseed	22	9	2.4
Walnut	60	12	5
Soybean	51	6,8	7.5
Linseed	22	50	0.44
Hemp	58	20	2.9
Chia	18	72	0.25

in linolenic acid; two of them (rapeseed and soybean oils) are most frequently consumed by the French population. Unfortunately, the walnut oil is poorly consumed (on average 0.1 L per capita per year), except into Dauphiné and Périgord. The other three (flax, hemp, and chia oils) belong to the category of food supplements.

The linolenic acid needs may thus be practically met by consuming regularly, apart from the plants themselves, the following quantities of some common oils (rapeseed, walnut, and soybean) either in cooking or as a form of seasoning:

A total of 2 g of linolenic acid are supplied by

- 15 mL (1 soup spoon) of walnut oil or four to five walnuts (from Dauphiné or Périgord)
- 22 mL (1.5 soup spoons) of rapeseed oil
- 28 mL (2 soup spoons) of soybean oil

Apart from these traditional vegetal sources, the recommended intake of linolenic acid may also be supplied by the intake of modest amounts of oils or seeds belonging to other plant species. Among these, flax (*Linum usitatissimum*), hemp (*Cannabis sativa*), and chia (*Salvia hispanica*) are generally marketed and used as food supplements.

Thus, 2 g of linolenic acid are supplied by

- 4 mL of linseed oil (1 teaspoon) or 15 mL of crushed linseeds or 1.7 kg of flax bread (bread coeur de lin Banette)
- 10 mL (3/4 soup spoon) of hemp oil or 38 mL (2.5 soup spoons) of crushed seeds of hemp
- 3 mL of chia oil or 15 mL (1 soup spoon) of crushed chia seeds

The AFSSA has authorized in 2006, after many countries, the use of linseed oil as a food supplement (saisine 2004-its-0213) and in the most common foodstuffs (saisine 2004-its-0409), either uncooked or mixed with other oils. In 2009, the AFSSA gave a positive opinion on the use of linseed oil in cooking (saisine n° 2008-SA-0392). Hemp and chia oils may also be used without danger as food supplements; they are also known for their cosmetic uses.

Other oils rich in linolenic acid are proposed commercially as dietary or therapeutic products. Thus, one can find the camelina oil (*Camelina sativa*, Brassicaceae) containing 30%–40% of linolenic acid and 12%–19% of linoleic acid and the perilla oil (*Perilla frutescens*, Lamiaceae) containing up to 65% of linolenic acid and 15% of linoleic acid.

Oils rich in stearidonic acid: to face the progressive decrease of the halieutic resources supplying EPA and DHA, it seems probable that clinicians will turn toward stearidonic acid (18:4 *n*-3), an *n*-3 fatty acid of vegetal origin. Indeed, included in foods this fatty acid is more efficient than its precursor, the linolenic acid, in the enrichment of cellular membranes in EPA. It was even proposed in 2012 by W.S. Harris that it could be considered as a "pro-eicosapentaenoic acid" (pro-EPA). Indeed, clinical investigators have verified that only EPA accumulated in red cell membranes in subjects overloaded with 0.75 g/day of stearidonic acid without any slight increase in the concentration of DHA (James et al. 2003). This stearidonic acid effect remained approximately three times lower than that observed with the same amounts of EPA, but it was approximately four times more important than that observed with linolenic acid. Although several works on animals or cultured cells have suggested that stearidonic acid shares many biological effects with EPA and DHA, other clinical investigations will be necessary before promoting this n-3 fatty acid of vegetal origin in the development of foods or supplements intended to efficiently reinforce our intakes of n-3 fatty acids.

Stearidonic acid, also called moroctic acid, is present in various food sources. Except in some cases, fish oils contain only a maximum of 2% of this fatty acid, besides great amounts of EPA and DHA. It is present in marine weeds such as *Undaria pinnatifida*, consumed in Korea, Japan, and now France (wakame), and *Ulva pertusa* (sea lettuce). In spite of the high concentration of this fatty acid in the lipids of these weeds (15%–26% in weight), the practically extractable quantities remained very low (approximately 2 g/kg dry weight).

Several plants of the family Boraginaceae (*Echium* or viper's bugloss) and Grossulariaceae (*Ribes* or black currant) contained interesting amounts of stearidonic acid, especially in oils extracted from their seeds. Thus, *Echium* oil may contain up to 15% of stearidonic acid and black currant oil up to 6% (besides 14% of α-linolenic acid and 16% of γ-linolenic acid).

The European Commission has authorized (decision 2008/558/EC) the marketing of the oil extracted from the seeds of *Echium plantagineum* (purple viper's bugloss) (with at least 10% of stearidonic acid) as a new ingredient for addition in dairy products, cheeses, spreads, sauces, and any dietetic food. The refined *Echium* oil is marketed under the name "BioMega SDA" (www.echiumoil.eu/); it contains 13% of stearidonic acid and 33% of α-linolenic acid.

A mixture of *Echium* oil and marine weeds is marketed under the name NutraVege®; it contains 550 mg of stearidonic acid and 400 mg of DHA in 10 mL of oil (http://www.ascentahealth.com/node/865).

Black currant oil is marketed by Distridiet® in the form of capsules, each one containing 15 mg of stearidonic acid besides 65 mg of α-linolenic acid (http://www.distridiet-nutrition.fr).

Given the increasing demand of this fatty acid, the Monsanto Company has marketed under the name of Soymega™ a soybean oil enriched in stearidonic acid extracted from seedlings modified by genetic engineering (www.soymega.com). These seedlings, variety MY 87769, have been created by the introduction of the gene of Δ6-desaturase of a primrose (*Primula juliae*) and of Δ5-desaturase of a fungus (*Neurospora crassa*). This operation caused a reduction of the quantities of linoleic acid but increased the conversion of linolenic acid into stearidonic acid. This new oil contained 20%–30% of stearidonic acid and 9%–12% of α-linolenic acid, besides 5%–8% of γ-linolenic acid. According to the Monsanto Company, the new plant is intended to be added to various foods (breads, cereals, creams and sauces, prepared vegetables, dairy products, etc.) at a concentration supplying 375 mg of stearidonic acid (e.g., 1.7–2.5 mL of oil) by portion, without exceeding a contribution of 1.5 g/day (e.g., 13–20 mL of oil) of that compound. It could replace fish oil or other products rich in n-3 polyunsaturated fatty acids. This oil obtained the GRAS label from the FDA of the United States. After the deposit of a request at the EFSA, the AFSSA was asked in February 2010 (saisine n° 2010-SA-0046) for an opinion. After authorization and industrial production, it is certain that one will observe a great change in livestock feed not only by using soybean oil but also by using soybean cake. One can also expect a wide use of this genetically modified soybean oil, containing neither EPA nor DHA, but without harmful contaminants and any commercial pressure on the fishing resource.

Oils rich in DHA: the European Commission authorized in 2003 the marketing of an oil extracted from the microalgae *Schizochytrium* sp. as a new food ingredient (decision 2003/427/CE) containing a high percentage of DHA (minimum 32%). In 2009, the commission authorized the marketing of an oil extracted from the microalgae *Ulkenia* sp. as a new food ingredient containing at least 32% of DHA (decision 2009/777/CE). This oil will have to supply a maximum of 60–200 mg of DHA per 100 g of final product (bakery products, cereal bars, and soft drinks).

Food supplemented with *n*-3 fatty acids of vegetal origin: currently, DHA is not added to food of vegetal origin, linolenic acid being the only *n*-3 fatty acid added to some foods usually used by French consumers. Among these foodstuffs, butter and margarines are the most frequently supplemented ones.

- The preparation Saint-Hubert Omega-3 contains 5.3 g% of linolenic acid, with an *n*-6 to *n*-3 fatty acid ratio equal to 1.4. The manufacturer announced that the consumption of two slices of bread with that butter supplied half of the ANC (or RDA) of linolenic acid (2.2 g/day). Calculation showed that the average amount of butter consumed each day by the French (11 g according to the INCA 1998–1999 investigation) supplied only 580 mg of 18:3 *n*-3 with that product.
- The preparation PRO-ACTIV Fruit d'or contains 0.3 g% of linolenic acid, with an *n*-6 to *n*-3 fatty acid ratio equal to 4.8. It is noted that 11 g of this product supply only 33 mg of linolenic acid.
- The preparation Primevère for cooking contains 3.2 g% of linolenic acid, with an *n*-6 to *n*-3 fatty acid ratio equal to 4.2. It is noted that 11 g of this product supply 350 mg of linolenic acid.

Animal sources of *n*-3 fatty acids: whereas linolenic acid is present exclusively in plants, the food sources of polyunsaturated fatty acids (EPA, DHA, 22:5 *n*-3) are in suitable quantities only in fish, some other marine animals (crustaceans and mollusks), and hen eggs. Meats from terrestrial animals have only very moderate contents of these fatty acids (Table 3.17). Many investigations were carried out to enrich chicken and pig meats (INRA studies), but none of these experiments was marketed in France. A similar transformation of beef meat was tried but with little success, a consequence of the biohydrogenation of dietary fatty acids in the rumen.

On the other hand, eggs enriched in DHA and EPA have been produced and marketed for a few years.

Meats: they do not contribute to a consequent *n*-3 fatty acid intake if one considers the amounts consumed by the French.

What is the current situation for fats used for cooking and likely containing *n*-3 fatty acids? Goose fat contains on average 1.4 g% of 18:3 *n*-3 (*n*-6 to *n*-3 fatty acid

TABLE 3.17
Average Contents in Some Meats of *n*-3 Fatty Acids (Milligrams per 100 g)

Meat	22:6 *n*-3	22:5 *n*-3	18:3 *n*-3	*n*-6/*n*-3
Duck magret	30	20	40	13.5
Chicken fillet	16	8	4	6.7
Beefsteak	2	32	18	2
Lamb leg	7	23	38	4.4
Pork rib	4	8	50	18
Horse fillet	0	18	330	1.2
Rabbit leg	3	6	80	6.6

Note: Data for beef, lamb, and horse are from the Information Center of Meats (http://www.lessentieldes-viandes.org/deslipides.php); for pork, rabbit, and chicken from Mourot (2010); and for duck from the USDA. The *n*-6 to *n*-3 fatty acid ratios are calculated from all fatty acids of these two series.

TABLE 3.18
Average Contents of *n*-3 Fatty Acids in Breast Milk and Cow's Milk (Milligrams per 100 mL)

	18:3 *n*-3	20:5 *n*-3	22:5 *n*-3	22:6 *n*-3	*n*-6/*n*-3
Cow's milk	300	0	0	0	10
Breast milk	32	2.5	5	9.2	10

Source: From Lafay, L., and E. Verger, *OCL*, 17, 17–21, 2010; Vaysse et al., Med. Nutr., 47, 5–9, 2011.

ratio = 7), butter contains on average 0.5 g% of 18:3 *n*-3 (*n*-6 to *n*-3 fatty acid ratio = 2.2), and milk cream contains 0.2 g% (*n*-6 to *n*-3 fatty acid ratio = 2).

Fats from milk have a convenient *n*-6 to *n*-3 fatty acid ratio; but they are also rich in saturated fatty acids, a property that limits their recommendation for balancing the sources of unsaturated fatty acids.

Cow's milk: it contributes to about 10% of the fat lipid intake in children from 3 to 17 years of age (Lafay and Verger 2010). Its composition varies primarily according to the animal diet type and the stage of lactation. The mixtures carried out by dairy industry tend to regularize the fatty acid composition of consumed milks. However, the content in *n*-3 fatty acids remains very low, and only in the form of linolenic acid, approximately 10 times less concentrated than linoleic acid (Table 3.18).

Breast milk: it contains an amount of lipids that varies between 2 and 4 g/100 mL, but its fatty acid composition depends partially on the mother's diet and the sampling period, 98% of these fatty acids being in the form of triacylglycerols. Thus, the levels of *n*-3 fatty acids, as the others, vary with the levels of these fatty acids in the maternal diet (Table 3.18). In France, the observed proportions of EPA are from 0.05% to 0.09% of total fatty acids, those of 22:5 *n*-3 from 0.11% to 0.15%, and those of DHA from 0.19% to 0.29% (Vaysse et al. 2011).

TABLE 3.19

Classification of Fish as *n*-3 Fatty Acid Sources

Total Lipid Content	*n*-3 Content (EPA + DHA)	Fish Species
Fatty fish (>2%)	High amount (>1.5 g/100 g)	Salmon, mackerel, herring
	Average amount (0.5–1.5 g/100 g)	Trout, sardine, tuna
Lean fish (<2%)	Low amount (<0.5 g/100 g)	Sole, cod, carp, eel

The data concerning the consumption of polyunsaturated fatty acids collected at the end of the twentieth century in France, as well as the analyses of milk lipids, have shown an imbalance of the *n*-6 to *n*-3 fatty acid ratio within the milk lipids, closer to 15 than 5, the latter value being recommended by nutritionists. It is evident that breast milk contributes to the polyunsaturated fatty acid intake in infants. Milks rich in EPA and DHA may also be used to compensate a possible deficiency in these essential fatty acids.

Fish: in spite of the risks of ingestion of heavy metals and pesticides, fish remain at the present time the most concentrated and most convenient source of long-chain *n*-3 fatty acids, besides being rich in proteins and vitamins. With respect to risks, Sweden has recommended to limit fish consumption during pregnancy to once a month. Among fatty fish, some contain more *n*-3 fatty acids than others and are thus particularly interesting concerning their nutritional value (Table 3.19).

One sees clearly from Table 3.20 that an even modest consumption of marine fish known as fatty fish or salmon or farmed trout is likely to meet the adult's requirements of *n*-3 polyunsaturated fatty acids. A serving from 150 to 200 g of some of these fish (the right-side column) consumed twice per week thus covers the requirement of EPA and DHA in an adult (approximately 500 mg/day).

The composition of farmed fish, estimated to equal 40% of the total fish intake in the world, is still poorly known. It depends closely on the lipids incorporated in the food used in aquaculture installations. Only salmon has been the object of investigations, those showing a profound change in the *n*-3 fatty acids content in animals fed vegetal oils definitely cheaper than fish oils (Seierstad et al. 2005). The salmon from Norway fed with fish oil was twice fattier than the wild animals, the EPA (1.1 g/100 g of flesh compared to 0.6) and the DHA content (2 g/100 compared to 1.1) were about twice higher (Bourre 2005). It would thus be judicious to make available to the consumers the contents of *n*-3 fatty acids in marine fish resulting from aquaculture, those having lost the original qualities of wild fish. Investigations undertaken in farmed fish (mainly salmonids) currently tried as well as possible to improve their lipid composition and to seek effective food components, such as linseed oil, to increase flesh content in *n*-3 polyunsaturated fatty acids.

Eggs: a hen's egg of average weight of 60 g supplies naturally from 30 to 90 mg of DHA (Pieroni and Coste 2010). Thus, this amount constitutes between 10% and 30% of the ANC (or RDA) (about 300 mg).

Given the facility with which the fatty acids of hen food are transferred into eggs, many investigations were carried out to enrich them in polyunsaturated fatty acids.

TABLE 3.20

Composition in Various Fish, Shellfish, and Mollusks of EPA and DHA and Weight of a Serving Supplying 500 mg of These Fatty Acids

Fish	EPA (g/100 g)	DHA (g/100 g)	EPA + DHA (g/100 g)	Weight (g) for 500 mg EPA + DHA
Mackerel	0.90	1.4	2.3	22
Salmon	0.86	1.10	1.96	26
Farmed salmon	1.10	2.0	3.1	16
Herring	0.71	0.86	1.57	32
Sardine	0.47	0.51	0.98	51
White tuna	0.23	0.63	0.86	58
Farmed trout	0.22	0.52	0.74	68
Cod	0.06	0.12	0.18	278
Eel	0.08	0.06	0.14	357
Carp	0.24	0.11	0.35	143
Sole	0.14	0.11	0.25	200
Octopus	0.15	0.16	0.31	161
Mussel	0.19	0.25	0.44	114
Oyster	0.19	0.20	0.39	128
Shrimp	0.3	0.25	0.55	90
Lobster	0.26	0.11	0.37	125

Note: Data are from the USDA food composition tables (http://ndb.nal.usda.gov/ndb/foods/list) and Seierstad and others (2005) for farmed salmon.

For this reason, eggs are practically the only animal products enriched in *n*-3 polyunsaturated fatty acids in the market. The first tests carried out in Canada have shown that an enriched egg may replace nearly 85 g of fish. For a few years, one could find in the French market eggs laid by hens fed a food enriched in linseeds. These eggs have a tenfold content of *n*-3 fatty acids compared with normal eggs and, thus, one egg may cover from 25% to 35% of the daily requirement of *n*-3 fatty acids. The *n*-6 to *n*-3 fatty acid ratio of these eggs is excellent as it varies from 3 to 4. Moreover, the linseed-enriched food contains sufficient vitamin E to protect the *n*-3 fatty acids from oxidation (Bourre 2005).

With respect to DHA, the amounts accumulated in eggs reach a maximum very rapidly according to the amounts of linolenic acid supplied by the diet, in general not more than 2% of the total fatty acid content, for example, 100–120 mg of DHA per egg. Such eggs are marketed in France under the brand Benefic® (http://www.benefic.info/oeuf.html). Amounts of DHA exceeding 300 mg per egg may be obtained by supplementing hen food with fish oil or some powdered algae, the final DHA/EPA ratio reaching values from 5 to 9. The consumption of such DHA-enriched eggs by nursing women ensures an increase in the concentration of that fatty acids in milk and, consequently, in tissues of infants (Jewell et al. 2001).

Food for infants and young children: since 1996, an amendment to the European directive on infant formulas has allowed the marketing of milks enriched in poly-unsaturated fatty acids by specifying the limits to be respected but without fixing the minimum threshold for DHA (and arachidonic acid). For specialists, enrichment similar to those noted in breast milk would seem more convenient.

Many foods (milks) supplemented with polyunsaturated fatty acids for infants or children are already marketed in various countries. Some of them are detailed as follows:

- The product Bon départ by Nestlé contains in 100 mL of preparation 10 mg of DHA (from *Cryptecodinium cohnii*) and 20 mg of 20:4 *n*-6 (from *Mortierella alpina*) and 56 mg of linolenic acid.
- The product Enfamil A+ by Mead Johnson brings in 100 mL of preparation 11.5 mg of DHA.
- The product Candia source d'Omega-3 by Candia is a milk containing 1.7 g of lipids in 100 mL, including 23 mg of EPA and 33 mg of DHA.

Food supplements (oils from marine animals): the ANC (or RDA) for *n*-3 poly-unsaturated fatty acids of 500 g/day, generally in the form of EPA and DHA, may be covered by the daily intake of these fatty acids in the form of capsules (0.5 or 1 g) containing esterified fatty acids (generally ethyl esters) or oils (triacylglycerols) from fish or shellfish (krill). These oils, more or less concentrated in *n*-3 fatty acids, are generally guaranteed to be very low in heavy metals and pesticides. They are in addition enriched with vitamin E to ensure a better conservation of polyunsaturated fatty acids (in general about 10 mg per capsule).

These oils generally come from fish of cool water (anchovy, sardine, cod, and salmon), from shellfish of Antarctic water, and even from marine mammals (seal). Their use in a rough state is very old and began in the nineteenth century with the aim of fighting rickets (Section 3.2.4.2). They are now used after purification, gener-ally by molecular distillation, to eliminate heavy metals and pesticides and after a cold treatment (winterization) to decrease their content in saturated fatty acids. Their *n*-3 fatty acid concentration varies from 30% to more than 90%. Krill oil is charac-terized by its extremely low amount of organic pollutants and heavy metals, justi-fying the intense development of its production. It has been estimated that in 2009 approximately 86,000 t of refined marine fish oils were consumed in the world. The annual production of krill oil would not exceed 6,000 t. Human consumption in the form of food supplements would account for only approximately 10% of the total oil production. The world expenditure on the purchase of food supplements rich in *n*-3 fatty acids has been estimated to be 25.4 billion dollars in 2011, the annual increase in the sales being evaluated at 6.4% (Packaged Facts, Ltd.).

The European Commission gave its authorization in 2009 for the marketing of a lipid extract (50% of phospholipids) of Antarctic krill, *Euphausia superba*, contain-ing at least 15% of EPA and 7% of DHA. This oil is intended to be added to food (dairy products, spreads, cereals, dietetic foods, and food supplements) so as to reach a concentration of DHA + EPA of 200–600 mg per 100 g of final product (decision 2011/513/UE).

Among the oils from marine animals on the market, one can find mainly

- Xtend-Life Omega/DHA Fish Oil (oil from hoki), with 120 mg of EPA, 280 mg of DHA, and 50 mg of 22:5 *n*-3 per one capsule of 1 g.
- Omega-Brite, with 500 mg oil containing 90% of *n*-3 fatty acids in the form of EPA (350 mg) and DHA (50 mg).
- Ultimate Omega, with 1 g of anchovy and sardine oil containing 640 mg of *n*-3 fatty acids in the form of EPA (325 mg) and DHA (225 mg).
- GNC Salmon Oil, with 1 g salmon oil containing 180 mg of EPA and 120 mg of DHA.
- Ysomega, with 1 g fish oil containing 320 mg of EPA and 200 mg of DHA.
- Triglistab, 1010-mg capsules containing 700-mg *n*-3 fatty acid ethyl esters.
- Omacor Caps molle, with 1 g of *n*-3 fatty acid ethyl esters in the form of 460 mg of EPA and 380 mg of DHA.
- ZenixX (IXX Pharma), capsules of 1 g containing 60%–75% of DHA ethyl ester and 15% of EPA ethyl ester.
- Marin EPA Platinum (Nutrogenics), capsules of purified oil containing 764 mg of EPA, 236 mg of DHA, and 300 UI of vitamin D.
- Unicardio (Nutrogenics), capsules containing 460 mg of EPA and 380 mg of DHA.
- Krill NKO, with 300 mg of oil from Antarctic shrimp rich in phospholipids containing 39 mg of EPA and 16.5 mg of DHA. Oemine, the capsule of 500 mg of krill NKO, contains 75 mg of EPA and 45 mg of DHA, plus 0.75 mg of astaxanthine. It seems that the EPA and DHA of this oil has a better bioavailability (30% more) compared with fish oil (Schuchardt et al. 2011). Furthermore, supplements with krill oil may be more effective in raising levels of EPA and DHA that accumulate in human red blood cells (the omega-3 index) than fish oil (Ramprasath et al. 2013).
- Arctic Omega-3 seal oil, with 500 mg of oil containing 120 mg of *n*-3 fatty acids in the form of 36 mg of EPA, 51 mg of DHA, and 31 mg of 22:5 *n*-3.

Ethyl esters of EPA and DHA may be marketed instead of triacylglycerols, sometimes without any label indication by the suppliers. The toxicity of these fatty acid derivatives has often been mentioned but without noticeable effects at reasonable doses. In contrast, the bioavailability of these fatty acid esters is about 20% lower than that of fish oils (Schuchardt et al. 2011). Moreover, the incorporation of EPA and DHA in human erythrocytes as esters would be approximately 15% lower than in the form of triacylglycerols after 3 or 6 months (Neubronner et al. 2011).

It is likely that soon the commercial availability of these oils from marine fish will pose a problem in terms of both their potential toxicity and the ease of obtaining them from a fishery resource known to be in permanent reduction. It seems certain that the industry will use less and less fish and more and more Antarctic krill, a resource considered to be the most abundant and the least contaminated on the planet. The biomass of these animals was estimated in 2007 by C. Watkins, in the United States, to be about 650 million tons. Will the offer follow the demand created by the many recommendations of the medical profession? The results of these incentives are well illustrated by the fact that this market amounted to almost 14 billion dollars (10 billion euros) in the world. In the United States, the percentage of adults who take these supplements has increased from 8% to 17% in 5 years since 2006. Thus, what will be the situation in the world if all populations far from the coasts start following these new recommendations? It is likely that farmed fish and algae cultures will definitely be the preferred sources of EPA and DHA in the near future.

Before considering xenobiotics, it should be noted that fish oils rich in highly unsaturated fatty acids are naturally prone to oxidation. A work performed in Norway in 2011 by B.L. Halvorsen on 34 sources of n-3 fatty acids marketed as food supplements showed that peroxides were present (1–10 meq/kg) and that, besides degradation products, alkenals were found (158–932 nmol/kg). The potential dangers of these oxidation products and the upper limits enabling the definition of a healthy area remain undefined in humans. As the authors pointed out, research in this field should be quickly undertaken. Before any consumption of such supplements, it appears necessary to ask the supplier of the nature and the possible amount of harmful pollutants.

About xenobiotics, the FAO said in 2010:

> If the consumption of seafood has confirmed nutritional and health benefits, the accumulation of contaminants produced by certain fish species can make them toxic. The question that arises is how to take advantage of the consumption of seafood, while minimizing the negative consequences attached to them. It is established that certain fishery products contain methylmercury (the most toxic form of mercury) and dioxins (generic name similar to the dioxin). In general, it is estimated that the amount of marine food contaminants of this type is well below the maximum limits that were determined for ingestion without risk. However, in fish caught in polluted waters or belonging to predatory species with long life, it is possible that contaminant levels exceed the values considered safe for consumers (http://www.fao.org/docrep/013/i1820f/i1820f.pdf).

On the other hand, a report from FAO experts published in 2010 on the "risks and benefits of fish consumption" concluded that there is no convincing evidence of risks from methylmercury and dioxin. The known benefits of eating fish rich in n-3 fatty acids largely outweigh the risks posed by the pollutants that may be present in these fish (http://www.fao.org/docrep/014/ba0136e/ba0136e00.pdf). This work reported many assays that may provide a list of marine animals accumulating the least mercury (≤0.1 µg/g) and dioxin (≤0.5 pg/g): cod, plaice, sole, tilapia, mullet, wild salmon, sardine, squid, clam, and common cockle. In contrast, tuna, sharks, and eels are to be avoided, given their concentrations in these pollutants. A report published in 2012 by M. Wennberg, Sweden, has shown that exposure to methylmercury is associated with an increased risk of myocardial infarction, but this risk is largely compensated for low doses by the intake of n-3 fatty acids. It should be noted that pollutants, especially dioxins, are concentrated in lipids and so in oils prepared from fish or shellfish. It is therefore necessary, besides the financial aspect, to study the conditions of preparation of these oils and if possible to obtain data on pollutants before making this choice. Suppliers' websites are usually quite informative.

A European legislation on the maximum concentration of polychlorinated dibenzo-p-dioxins, polychlorinated dibenzofurans, and PCBs in fish oil was established in 2002 and revised in 2006. It stated that the total amount of these dioxin-like compounds may not exceed 10 pg OMS-TEQ/g in fish oil for human consumption (regulation EC number 199/2006 of February 3, 2006). The upper limit of these pollutants has been set at 8 pg/g fresh weight for fish flesh and fish products (except eel: 12 pg/g).

To avoid these potential sources of pollution, manufacturers endeavor to find other sources of EPA or DHA. The algal cultures are often highlighted because of their richness in polyunsaturated n-3 fatty acids and of the absence of pollutants in culture media subjected to strict controls. It must be noticed that these algal fatty acids are mainly at the origin of those present in animals throughout the marine food chain.

In 2003, Europe authorized the marketing of DHA-enriched oils to supplement a wide variety of foods (DHA-rich oil from OmegaTech) (decision 2003/427/EC). This oil produced from cultures of microalgae of the genus *Schizochytrium* contains 35% DHA and 2.6% EPA and should be very low in contaminants. An oil containing up to 400 mg/g of DHA (DHAid) produced from cultured algae of the genus *Ulkenia* is marketed by Lonza, Ltd.

Importance of the n-3 to n-6 fatty acid ratio: through numerous animal experiments, it has been demonstrated that excessive values of this ratio are detrimental to the synthesis and availability of long-chain fatty acids (EPA and DHA). Indeed, n-3 and n-6 fatty acids compete with several enzymes, including those catalyzing their elongation (elongase) and their desaturation (desaturase). An excess of dietary n-6 fatty acids may therefore prevent the body from making the best use of the various resources of n-3 fatty acids, especially linolenic acid. A too high value of the n-6 to n-3 ratio could indeed increase physiological disturbances and contribute to the occurrence of various diseases (Section 4.2.3.3). Correspondingly, body aging or any disease contributing to slow down fatty acid metabolism (alcoholism, smoking, and stress) accentuates the imbalance of the metabolic ratio by inhibiting the bioconversion of 18:3 n-3 into more unsaturated compounds (EPA and DHA), a bioconversion normally recognized as being inefficient.

If the n-6 to n-3 ratio of lipids in the most commonly consumed meats is less than 8 (Table 3.21), the overall estimate of this ratio for the Western diet is usually between 10 and 30 because of the abundance of vegetal oils in our diet. Epidemiological and clinical studies analyzed by the ANSES indicate that this ratio should not exceed 5. To strengthen these conclusions, we must compare this value to the maximum value of 3, estimated in 2005 by L. Cordain, the United States, for the diet of Paleolithic men. The richness in n-3 fatty acids with respect to n-6 fatty acids in this type of diet is easily explained by an important intake of linolenic acid from plants (leaves, seeds, and fruits) (Tables 3.13 and 3.14) and of EPA and DHA from terrestrial animals (Table 3.17), fish (Table 3.22), and also mollusks and crustaceans. Fish is also the main source of proteins in this type of diet.

The dietetic interest of fish is to lower the value of the n-6 to n-3 ratio of the diet, especially those containing large amounts of long-chain polyunsaturated fatty acids (cod, mackerel, blue whiting, and haddock). Shellfish and crustaceans have little interest in this process because they contain very little amounts of fats, even if the n-6 to n-3 ratio is comparable to that found in fish (0.3 in oyster, 0.66 in mussel, and 0.2 in shrimp).

The knowledge of the n-6 to n-3 fatty acid ratio has less interest today as physiological requirements in 18:2 n-6 and 18:3 n-3 are now better covered than before and the intakes of EPA and DHA are approaching the recommended values. However, it remains useful to have an overall view of the diet. It also retains its importance in the case of imbalance or excess in linoleic acid or deficiency in linolenic acid especially when there is also a deficiency in EPA and DHA, poorly synthesized in humans.

The study of this ratio for other diets in different populations revealed that its value may result from many possible fatty acid levels (Table 3.23). Thus, a comparison between the average American, Japanese, and Mediterranean diets allowed the observation that the low values of the n-6 to n-3 ratio for Japanese and Mediterranean diets resulted from a higher consumption of n-3 fatty acids in the Japanese but a lower intake of linoleic acid in the Mediterranean population. In contrast, the high

TABLE 3.21

Values of the *n*-6 to *n*-3 Fatty Acid Ratio in Some Animal Meats

Duck Magret	Chicken Fillet	Beefsteak	Lamb Leg	Pork Rib	Horse Fillet	Rabbit Leg
13.5	6.7	2	4.4	18	1.2	6.6

TABLE 3.22

Values of the *n*-6 to *n*-3 Fatty Acid Ratio in Some Fish

Species	n-6 + n-3 (g/100 g)	n-6/n-3	Species	n-6 + n-3 (g/100 g)	n-6/n-3
Sea bass	1.7	0.28	Cod	0.20	0.13
Mackerel	3.6	0.12	Haddock	0.18	0.15
Red gurnard	2.5	0.17	Monkfish	0.13	0.15
Salmon	3.6	0.52	Blue whiting	0.21	0.11
Trout	2.8	0.60	Sole	0.16	0.20

Source: From Médale, F., *Cah. Nutr. Diét,* 44, 173–81, 2009.

TABLE 3.23

Comparison of Four Diets (Percentage of the Total Energy Intake)

	Japanese[a]	Mediterranean[a]	American[a]	French[b]
18:3 n-3	0.90	0.50	1	0.36
18:2 n-6	6.3	2.3	6.5	4.1
PUFA n-3	1.3	0.35	0.03	0.2
PUFA n-6	0.08	0.08	0.08	0.08
n-6/n-3 ratio	2.9	2.8	6.4	7.5

Note: PUFA, long-chain polyunsaturated fatty acids.

[a] Data are from the National Institute of Health.

[b] Data are from Astorg et al. (2004).

value of the *n*-6 to *n*-3 ratio in the American diet was primarily due to low amounts of EPA and DHA. The high value of this ratio in the French diet arose obviously from a too low consumption of *n*-3 fatty acids.

Among the most commonly used oils, rapeseed oil is renowned for its *n*-6 to *n*-3 fatty acid ratio similar to that recommended by the ANSES. With a value of 2.4, rapeseed oil is before walnut oil (*n*-6/*n*-3 = 4.2) and soybean oil (*n*-6/*n*-3 = 6.9) but behind linseed oil (*n*-6/*n*-3 = 0.3), the latter being regarded only as a dietary supplement. Mixtures of oils with very low *n*-6 to *n*-3 ratios were marketed in various countries, such as the Nutridan® in Denmark. This product is a mixture of oils from rapeseed, linseed, and camelina (*Camelina sativa*) and contains 18.3% of 18:2 *n*-6 and 28.9% of 18:3 *n*-3 (*n*-6/*n*-3 = 0.63). In France, the ISIO 4® oil from Lesieur has an *n*-6 to *n*-3 ratio of 5. This oil is a mixture of oils from rapeseed, sunflower, and grape seed. This high value of the ratio results from the French legislation allowing a maximum of 5% of linolenic acid in oils used for cooking.

3.2.1.4 Importance of the Triacylglycerol Structure

In rats, F.H. Mattson, United States, was the first to identify in 1962 a greater intestinal absorption of palmitic acid when it was in the sn-2 position of the triacylglycerol molecule than when it was in a free state in intestinal lumen. The knowledge of the laws linking triacylglycerol structure and absorption has been greatly facilitated by the use of a chemical process called transesterification or "randomization." This method, applied for the first time on dietary lipids by the German chemist W. Normann in 1920, is based on the random distribution of the fatty acids on the three stereospecific positions of glycerol in vegetal oils or animal fats (Figure 3.11). This treatment resulted in an increase in the melting point of the fat, comparable to that resulting from the hydrogenation process, but preventing the generation of *trans* fatty acids (Section 3.2.2.1). These physical changes are partly due to the emergence of fully saturated molecular species and also of new crystal structures as those of species such as 1,3-distearoyl,2-oleoyl glycerol.

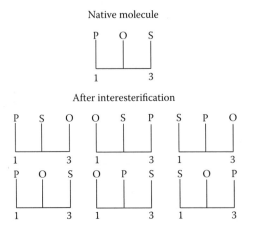

FIGURE 3.11   Structure of a triacylglycerol molecule after interesterification.

The new molecular species are obtained by random redistribution of the fatty acids in each triacylglycerol molecule and also between molecules. If the interesterification is performed in a mixture of dietary lipids, the overall fatty acid composition remains unchanged, but obviously the composition of the molecular species is completely modified. More important for physiology, the method will change the nature of the fatty acids naturally localized at the sn-2 position. Figure 3.11 describes the interesterification of triacylglycerol molecules (POS) containing palmitic acid (P) in the sn-1 position, oleic acid (O) in the sn-2 position, and stearic acid (S) in the sn-3 position. After treatment, these fatty acids are redistributed to give six new molecular species, each fatty acid being able to be acylated in the sn-2 position.

After the first work by F.H. Mattson, several studies using interesterified oils have been carried out in animals, confirming the initial findings. In 1969, the demonstration of a similar effect was done in humans by L.J. Filer, United States. In this study, newborns were fed with lard, native (the dominant palmitic acid in sn-2 position), or interesterified (palmitic acid is then evenly distributed over the three positions of glycerol). After a 3-day treatment, the authors found that palmitic acid of the intact lard is better absorbed than that of the interesterified lard. In the newborn, the lower digestibility of cow milk fats compared to breast milk fats was explained by the lower content of palmitic acid in the sn-2 position in the former than in the latter. These conclusions have been supported by recent findings in adults where it has been shown that postprandial lipemia and apolipoprotein B48 were increasing more slowly when subjects ingested an interesterified sunflower oil (39% of 16:0 at sn-2) compared to normal oil (9% of 16:0 at sn-2) (Sanders et al. 2011). Thus, if all the effects induced by the structure of the triacylglycerol on their metabolism are not yet clarified, it is now clear that the position of 16:0 in the glycerol chain affects the digestion and absorption of triacylglycerol molecules.

Several long-term studies in children and adults are still needed to obtain a clear understanding of the respective influences of the physical structure, quantity, composition and distribution of fatty acids on lipid metabolism in humans. In contrast, many animal studies have shown that enrichment of palmitic acid in the sn-2 position increased the atherogenic power of the ingested triacylglycerols. The easy digestion of cocoa butter is linked to the almost exclusive presence of stearic acid in the sn-1 and sn-3 positions, oleic and linoleic being dominant (90%) in the sn-2 position. This conclusion also applies to palm oil where the saturated fatty acids are found mainly in external positions, the sn-2 position being occupied by oleic and linoleic acids (80%). Thus, this oil does not have the atherogenic effect expected given its strong saturation. The importance of the saturated fatty acids in the changes in blood lipid profile and in the development of atheroma is discussed in Section 4.2.1.1.1, the influence of their distribution in the triacylglycerol molecule being outlined in Section 4.2.6.

What could be the reasons for the importance of the triacylglycerol structure for their metabolism and possibly consumers' health? It has been recognized for many years that the distribution of a fatty acid between the internal position (sn-2) and the outer positions (sn-1 and sn-3) of glycerol influences the further metabolism of that fatty acid (Karupaiah 2007). Indeed, the specificity of pancreatic lipase for the external sn-1 and sn-3 positions facilitates the retention of the fatty acid in the internal *sn*-2 position in the form of a 2-monoacylglycerol that will be converted directly into triacylglycerols in intestinal mucosa before being discharged into lymph and blood. In the postprandial period, the structure of these molecules will be retained in the plasma chylomicrons. Admittedly, in adult men several long-term experiments have suggested that the triacylglycerol structure had little effect on digestibility. It is the same for the effects on postprandial blood lipid profile. As pointed out in England by S.E. Berry in 2009, it is likely that the physical characteristics of ingested lipids are the most important for their digestibility.

It has been shown in animals that in opposition to 18:0 the position of 16:0 at *sn*-2 promoted its atherogenic character. Much of the differential effects observed according to the position seem to be due to the early formation of fatty acid calcium salts (soaps) that are poorly soluble in the intestinal lumen. It has been further shown that the fatty acids in the *sn*-2 position are more easily metabolized in situ than those in other positions. Similarly, *n*-3 fatty acids appear to be protected from an early β-oxidation when acylated in the *sn*-2 position, this situation promoting their metabolism into bioactive derivatives or their transfer to other lipids. This stereo-chemical property is the basis of controversies on the bioavailability of linolenic acid from vegetal oils or polyunsaturated fatty acids (EPA and DHA) from fish oils. Thus, in humans rapeseed oil (58% of 18:3 *n*-3 in the *sn*-2 position) would be more favor-able for the bioavailability than soybean oil (31% of 18:3 *n*-3 in the *sn*-2 position). Similarly, intestinal absorption and lymphatic transport of *n*-3 polyunsaturated fatty acids (EPA and DHA) are at their optimal value when these fatty acids are in the *sn*-2 position, as determined in 1995 by M.S. Christensen in Denmark. This property is also exploited to synthesize structured lipids (molecular species of known structure) for patients with intestinal malabsorption. For this purpose, the essential fatty acids are acylated in the *sn*-2 position and the medium- or short-chain fatty acids are acyl-ated in the outer positions (*sn*-1 and *sn*-3 positions) on the glycerol. Thus, the patient will more easily retain the essential fatty acids for cell membranes and other fatty acids as an energy source.

It is becoming increasingly clear that the inaccuracy of data on the digestion of different triacylglycerols in the diet represents a handicap for understanding their further metabolic fate. Many controversies will be resolved when the differential fatty acid metabolism of the ingested lipids is determined in humans in the most natural conditions.

3.2.2 *TRANS* AND CONJUGATED FATTY ACIDS

These compounds are unsaturated fatty acids, which have at least one double bond in the *trans* configuration as opposed to the *cis* configuration, typical of almost all naturally occurring unsaturated fatty acids. The naturally occurring *trans* fatty acids are monounsaturated (mainly 11-18:1 *trans*, vaccenic acid) (Section 3.2.2.1), di-unsaturated derived from linoleic acid (18:2 *n*-6) (Section 3.2.2.2), or tri-unsatu-rated derived from linolenic acid (18:3 *n*-3). Among the *trans* homologous species of linoleic acid, some have two conjugated double bonds, one being in a *trans* con-figuration. They are grouped under the acronym CLAs (conjugated linoleic acids). Given the differences in structure and properties, the group of CLAs is described separately from other *trans* fatty acids. A general biochemical and physiological study of *trans* fatty acids and CLAs may be found on the AFSSA website (http://www.afssa.fr/Documents/NUT-Ra-AGtrans.pdf).

3.2.2.1   *trans* Fatty Acids

trans Fatty acids have one or more double bonds with the *trans* geometric configuration (Section 2.2.3). From a nutritional point of view, the *trans* monoenes are most important, but the *cis/trans* dienes are also present in small amounts in certain natural products.

The *trans* monounsaturated fatty acids are naturally produced in the rumen of animals (cattle, sheep, and goats) during digestion and transferred into milk and meat. Among the many possibilities, the main isomer formed is vaccenic acid (*trans* 11-18:1 or *trans* 11-octadecenoic acid), which may represent more than 60% of the total *trans* isomers, the elaidic acid (*trans* 9-18:1) amounting to a maximum of 10%. These fatty acids are also formed during technological treatments (partial hydrogenation) of vegetal oils (mainly palm oil) processed to supply fluid or semisolid fats for incorporation into food products such as margarines. In this case, the nature of the *trans* monounsaturated fatty acids is much more varied, the isomerization occurring from carbon 6 to carbon 16, elaidic acid becoming more abundant than vaccenic acid. Discovered in 1897 by the French chemist P. Sabatier (Nobel Prize in 1912), the hydrogenation process was patented for lipids in 1902 by the German chemist W. Normann and applied industrially since 1911 in the United States by Procter & Gamble in the manufacture of the "shortening" Crisco. Originally applied to cottonseed oil, this method was subsequently extended to whale, rapeseed, and soybean oils. In the 1960s, over 60% of the lipids consumed by Americans resulted from such partially hydrogenated oils.

The *trans* fatty acids with two double bonds, especially CLAs, are also present in natural fatty products (meat and dairy). They are described in Section 2.2.4. They may be further formed during thermal or chemical treatments (frying, fat refining, and chemical synthesis). At temperatures exceeding 200°C, the *cis* bonds are easily transformed into *trans* bonds and thus a mixture of fatty acids with one or more *trans* double bonds is obtained. Thus, linoleic acid (*cis*-9,*cis* 12-18:2) may be transformed into two isomers, *cis*-9,*trans*-12 18:2 and *trans*-9,*cis*-12 18:2. The isomerization products may be more complex as a result of processing vegetal oils containing linolenic acid or fish oils containing highly unsaturated fatty acids.

3.2.2.1.1   *Recommended* trans *Fatty Acid Intakes*

Based on previous works, the FAO/WHO 1994 report provided no specific recommendation on the maximum amount of dietary intakes of *trans* fatty acids, except replacing solid fats (butter and margarine) by liquid oils.

After much work, it became clear that the consumption of *trans* fatty acids increased the risk of diabetes and of heart diseases often in the form of sudden death. Thus, in 2002 the experts of the FAO/WHO (TRS 916) recommended that the consumption of *trans* fatty acids should not exceed 1% of the total energy intake. In addition, they advise consumers to make efforts for increasing the proportion of dietary unsaturated fatty acids.

Numerous clinical studies have shown that cholesterol metabolism was altered by *trans* fatty acids when the amount ingested exceeded 4% of the total energy intake (about 7 g per 2500 kcal), but few countries have decided on a limiting threshold. In 2002 the United States (IOM, National Academy of Sciences) fixed no limitation of consumption; but in 2010 the USDA recommended to consume the least possible amount of *trans* fatty acids, especially those of industrial origin (Dietary Guidelines for Americans, 2010, http://www.health.gov/dietaryguidelines/dga2010 /DietaryGuidelines2010.pdf). This position was dictated by the difficulty in industrially eliminating these components and by the risk of introducing some deficiencies. Whereas the American Heart Association wanted the ingestion of *trans* fatty acids to be below 1% of the total energy intake, the majority of other health professional organizations recommended that the levels should be the lowest possible. Whereas Japan, Canada, and Switzerland have recommended no consumption limit, Denmark and the United Kingdom have set a limit of 2% and Netherlands 1%. After reviewing numerous clinical and epidemiological studies (see the report of the EFSA: http:// www.efsa.europa.eu/en/scdocs/doc/81.pdf), it is generally accepted that the *trans* fatty acid intake should not exceed 2% of the fat intake (about 4 g/day).

Denmark was the first country in 2003 to eliminate *trans* fatty acids from industrial sources. A total ban was introduced for all products consumed in restaurants, cafeterias, and hospitals and also for all imported products. In 2009, the European Court of Justice even raised the threat of condemnation with the growing evidence of a health risk. Austria and Switzerland took the same measures in 2009. Since then, the movement has spread in several American states.

The WHO has established that the replacement of *trans* fats by other types of fats would significantly reduce the risk of cardiovascular diseases.

In France, the ANSES also set the limit at 2% and issued its final recommendations in 2009 on the intake of *trans* fatty acids in the French population (http://www .afssa.fr/Documents/NUT2007sa0220.pdf).

In short, the ANSES recommended the following:

- To limit the total *trans* fatty acid intake to less than 2% of the total energy intake
- To reduce in particular the consumption of certain foods of technological origin (viennoiseries, pastry, industrial bread, chocolate bars, and cookies)
- To continue the effort of reduction of the use of *trans* fatty acids of technological origin, as well in human consumption as in animals

Coherent with the drop in consumption of these products, the agency encouraged the manufacturers of margarines and fats intended for the food industry to decrease the contents of *trans* fatty acids in their products.

Finally, and especially for industries, the agency proposed to adopt limiting values of *trans* fatty acid contents in most common foods:

- 1 g/100 g of final products, for example, approximately 0.4% of the total energy intake with a normal serving
- 0.5% for salad oils and 1% for margarines

As emphasized in the AFSSA report in 2005, a lowering of the levels of *trans* fatty acid ingestion of at least 0.15 g/day may be obtained by a reduction of a third of the food consumption based on cookies and pastry or by the replacement of the consumption of ground beef with 15% of fat by another one with 5%.

The labeling of food substances for the *trans* fatty acid contents is currently obligatory only in some countries (United States, Canada, Brazil, and Denmark). In France, no legal provision has still been taken, but the ANSES emphasized that "the obligation of labelling of *trans* fatty acids would be likely to encourage industry to improve the composition of their products since it would apply only in the event of going beyond thresholds corresponding to the standards of the current food on the market." When will such a labeling be obligatory?

3.2.2.1.2 Food Sources of trans Fatty Acids

Industrially hydrogenated lipids are intended to be incorporated into many foods (and ready meals) and thus may supply a *trans* fatty acid amount equivalent to 6% of the daily lipid ration (approximately 7 g/day of *trans* fatty acids). Since the demonstration in the 1980s of their harmful effects on the cardiovascular system (Section 4.2.5), food industry endeavored to improve the manufacturing processes of "vegetal fats" to decrease their *trans* fatty acid content. Concurrently to technical changes at the level of catalytic hydrogenation, the introduction of the interesterification process contributed to gradually lower that amount. Thus, in the United States the *trans* fatty acids consumption per capita decreased from 10 to 4 g/day, respectively, in 1984 and 1995; in Denmark, it decreased from 2 to 0.5 g/day in 1992 and 1999, respectively.

The report launched in April 2007 by the Task Force *trans* Fats Free Americas emanating from the WHO estimated that the *trans* fatty acid consumption was approximately 2% to 3% (4.5–7.2 g/day) of the total energy intake in the United States, 3% (7.2 g/day) in Argentina, 2% (4.5 g/day) in Chile, and 1.1% (2.6 g/day) in Costa Rica (http://www.paho.org/English/DD/PIN/TaskForce_Conclusions-17May07.pdf).

In 1999, an inquiry on the consumption in 14 European countries (the TRANSFAIR study) showed that the *trans* fatty acid ingestion varied according to the countries from 1.2 to 6.7 g/day (average: 2.4 g/day), corresponding to 0.5%–2.1% of the total energy intake (Hulshof et al. 1999). The consumption of these fatty acids thus proved to be lower in Europe than in North America but more important in the northern European countries than in those of the south in connection with a greater consumption of margarine and dairy products.

Between 1996 and 1999, in France (Aquitaine region) a study was carried out on nearly 200 women by analyzing data from food consumption surveys and the composition of fat tissues by the *Laboratoire de Lipochimie Alimentaire de l'Université de Bordeaux* (Combe et al. 2000). The study revealed that *trans* fatty acid consumption in this region was on average 2.7 g/day (approximately 1.3% of the total energy intake). The *trans* fatty acids ingested by the subjects in Aquitaine mainly originated from dairy products (50%) and hydrogenated vegetal fats (40%). A detailed analysis was carried out in 2007 by L. Laloux, the AFSAA, France, starting from the data collected between 1997 and 2002 from several French studies. For this period, the average consumption of *trans* fatty acids of the French people was 2.76 g/day for women and of 3.36 g/day for men, for example, 1.3% of the energy intake in women and 1.2% in men. This consumption placed France in an intermediate situation among the European countries as it was a little higher than that measured in Spain or Greece (1 to 2 g/day) but lower than that measured in Great Britain or the Netherlands (3–5 g/day); it remained largely lower than that measured in the United States or Canada (8–10 g/day). More recently, the average intake in the French population was estimated to be approximately 3 g/day (Léger, 2007).

The determination of the quantity of *trans* fats normally ingested from foods is complicated by the hidden nature of these fats coming mostly from margarine (2 g of *trans* fatty acids per 100 g) used to manufacture industrial foodstuffs (pastry, viennoiseries, and fried foods). The natural products also take part in *trans* fatty acid ingestion. Thus, butter contains from 3 to 7 g of *trans* fats per 100 g, cheeses from 0.3 to 2 g/100 g, whole milks about 0.15 g/100 g, breast milk about 0.05 g/100 g, vegetal oils between 0.5 and 2 g/100 g, and sheep and beef meat from 0.1 to 0.5 g/100g. The contribution of frying oils is far from being negligible as their *trans* fatty acid content can reach 35% according to their origin and utilization period.

The *trans* fatty acid composition of 214 food products collected between 1989 and 1993 on the U.S. market has been published by the USDA: http://www.nal.usda.gov /fnic/foodcomp/Data/Other/trans_fa.pdf. A similar study on products consumed in France would be of great interest.

3.2.2.2   Conjugated Fatty Acids

Conjugated fatty acids are isomers of linoleic acid (*cis* 9, *cis* 12-18:2); they belong to the group of CLAs (Section 2.2.4). They were discovered incidentally in 1978 by M.W. Pariza, in the United States, who investigated mutagen agents formed during the cooking of meat. The most abundant of natural CLAs is rumenic acid (*cis* 9,*trans* 11-18:2).

As for other *trans* fatty acids, rumenic acid is formed in the ruminants during the process of biohydrogenation in the rumen, before being metabolized into monounsaturated fatty acids (like vaccenic acid) and finally transformed into saturated fatty acids (like stearic acid).

In the consumer's body, CLAs are mainly incorporated into neutral lipids, in opposition to the linoleic acid that is especially distributed in phospholipids. Few metabolites of rumenic acid have been detected in humans. Like linoleic acid, this fatty acid may be metabolized into an isomer derivative of arachidonic acid (20:4 *n*-6) by the action of an elongase and two desaturases. On the other hand, compared to linoleic acid rumenic acid seems to be largely directed toward a β-oxidation catabolism as an energy source.

The therapeutic aspects of these compounds are discussed in Section 4.2.5.2.

3.2.2.2.1   Recommended Conjugated Linoleic Acids Dietary Intakes

The low CLA intakes observed in several countries, including France, and the demonstration of their rather positive effects on human health (Section 4.2.5.1) are the root causes of a lack of consumer guidelines on CLAs. Considering the parallelism between the consumption of saturated fatty acids and that of CLAs, nutritional recommendations to lower the consumption of these saturated fatty acids may help to also reduce the CLA intake. Although supplementation with 3.4 g of CLAs per day for 1 year in healthy adults did not lead to any adverse effects, the principle of using CLAs as a dietary supplement imposed some restrictions. The AFSSA specifically has stated in its 2005 report that the consumption of synthetic CLAs should not be allowed, especially for nursing mothers. In fact, the AFSSA opinion was that "studies on the *cis* 9,*trans* 11-18:2 and *trans* 10,*cis* 12-18:2 are insufficient but make it clear that according to the isomer, we may have much to fear from their consumption for public health or nothing to gain in the absence of clearly demonstrated beneficial effects."

3.2.2.2.2 Conjugated Linoleic Acids Dietary Sources

In addition to natural sources of CLAs (milk and meat), these fatty acids may also be derived from the heating of vegetal oils (refining and frying). Their content may then reach 0.5% in waste oils, 50% of these CLAs being di-*trans* fatty acids (*trans* 9,*trans* 11- and *trans* 10,*trans* 12-18:2) (Juanéda et al. 2001). Some, like rumenic acid and *trans* 10,*cis* 12-18:2, are present in commercial vegetal oils but at very low levels (10–70 mg/100 g). Rumenic acid accounts for almost 90% of CLAs present mainly in butter (up to 700 mg/100 g) and cheese (100–250 mg/100 g) (Table 3.24). Breast milk contains the same amount as in cow's milk (10 mg/100 g) (Vaysse et al. 2011).

Meat consumption leads to the ingestion of a not inconsiderable amount of CLAs as beef meat contains up to 120 mg/100 g (ground beef with 15% of fat) and lamb meat approximately 80 mg/100 g. The consumption of products resulting from food industry also contributes to this intake since prepared pastries and cookies may contain up to 0.5 g of CLAs per 100 g of products.

The average CLA intake has been estimated in Germany (0.4 g/day), Sweden (0.16 g/day), and Australia (0.5–1.5 g/day). In France, the average daily intakes of CLAs have been estimated on average as 0.21 g for men and 0.18 g for women, for example, approximately 0.08% of the total energy intake (90% in the form of rumenic acid). The AFSSA 2005 report has also shown that the first contributor is soft butter (35% of the CLA intake), the second cheese (27% of the CLA intake), and the third meat (11% of the CLA intake).

Allegations concerning the ability of CLAs to reduce body fats and develop muscles have led some shops to propose capsules of CLAs as a food supplement "to burn

TABLE 3.24

Average Composition of CLA in Some Dairy Products

Dairy Product	Rumenic Acid (mg/100 g)
Butter	380
Comté cheese	250
30% Milk cream	150
Camembert	100
Cantal	100
Gouda	50
Cow's milk	10

Note: Data are from the 2005 AFSSA report (http://www.afssa.fr/Documents/ NUT-Ra-AGtrans.pdf).

and capture fats." Ingestion of 1–1.5 g of CLAs per capsule, one to three times a day, is often recommended. In general, no detail is given on the composition of the marketed products.

Analyses carried out on these products have shown that they generally contain a mixture of rumenic acid and *trans* 10,*cis* 12-18:2. According to some works, this last isomer could have harmful side effects on health.

CLAs gained GRAS approval in six food categories in 2008 in the United States. This approval authorized its use in milks, yoghurts, meal replacement beverages, nutritional bars, soy beverages, and fruit juices. The market is now open for new products containing CLA mixtures combined with various foodstuffs and also with specific components (protein and lipid). Currently, several products contain a mixture of CLAs (CLARINOL® CLA by STEPAN) approved for reducing body weight.

3.2.3   Cholesterol and Phytosterols

Many works have been devoted to the nutritional aspects of cholesterol and its homologues, the phytosterols. Despite several controversies, it is now admitted that cholesterol is an essential nutrient for the constitution of cellular membranes and the synthesis of molecules as important as steroid hormones, vitamin D, and bile salts. Among the regulating mechanisms of cholesterol homeostasis, one should not neglect the inhibitory role of phytosterols ingested with vegetables on the absorption of cholesterol of animal origin (meat, milk, and egg).

3.2.3.1   Cholesterol

The structure of cholesterol is described in Section 2.6. It has been estimated that the human body contains approximately 100 g of cholesterol and that its daily needs are approximately 1 g. Of this intake, 10%–20% come from the diet (100–200 mg, sometimes up to 500 mg in the Western diet), the remainder coming from endogenous biosynthesis (mainly in liver and intestine). The intestinal absorption of cholesterol is not very efficient (approximately 50%) but variable according to the individual and the diet. Its endogenous metabolism also varies according to the individual and probably according to age. The disturbance of its biosynthesis is at the origin of hypercholesterolemia, either hereditary or acquired.

The regulation of the biosynthesis of cholesterol is carried out, apart from intestinal epithelium, especially at the hepatic level starting from acetyl-coenzyme A by a cascade of enzymatic steps and more particularly via the key enzyme of cholesterol metabolism HMG-CoA reductase. Thus, if more cholesterol is absorbed at the intestinal level its biosynthesis will be reduced at the hepatic level. It results that for the majority of subjects any dietetic modification will have only little impact on cholesterolemia. Many clinical works in healthy people have confirmed this mechanism (Section 4.3.1).

3.2.3.1.1   Recommended Cholesterol Intake

Given the capacity of any tissue to synthesize its own cholesterol and the absence of clear correlation between dietary cholesterol and any pathology, any official recommendation for daily cholesterol intake does not exist. Even if a positive link between the cholesterol food supply and the risk of cardiovascular disease is suspected, a major reduction in its intake does not seem desirable because it would be inevitably accompanied by important changes in dietary habits. These changes could then deteriorate the protein and vitamin intakes, thereby introducing health disorders more serious than those being avoided. Unless a person is diabetic or belongs to a group of people with familial hypercholesterolemia (this condition affects about 1 in 500 people in most countries), it does not seem necessary to limit his or her dietary cholesterol intake.

However, it seems reasonable from a dietetic point of view to not exceed an intake of 300 mg of cholesterol per day (Martin 2001). The cholesterol intake in the French population increased between 1969 and 1980 (+22%) and then regressed until 2001 (−13%) to stabilize to approximately 390 mg for 2500 kcal of food substances (Nichèle et al. 2007). It is recognized that the plasma total cholesterol concentration in the healthy adult must range between 1.8 and 2.5 g/L (e.g., between 4.6 and 6.3 mmol/L). The figure 2 g/L is frequently indicated as the higher limit by analytical laboratories.

The cholesterol intake of the American population has been estimated to be 450 mg/day, whereas the tolerable upper uptake suggested by the American Heart Association is, as in France, 300 mg/day. This threshold was lowered in 2001 by R.M. Krauss to 200 mg/day in subjects at risk, such as those having a high concentration of LDL cholesterol, diabetes, or any cardiovascular pathology. It is also advised by the American authorities that the range of cholesterolemia should be the same as that fixed in France.

The cholesterol dietary supply is close to 280 mg/day in Finland and 360 mg/day in Switzerland.

3.2.3.1.2 Cholesterol Food Sources

The food sources of cholesterol are primarily of animal origin. Table 3.25 gives the cholesterol contents of foods that are the richest in cholesterol (in decreasing order) and the weight of each food supplying approximately 300 mg of cholesterol (in ascending order).

One observes that among these usually consumed foods only five deserve special attention, they are three offal (brain, kidneys, and liver), eggs, and butter. Their high cholesterol content favors the recommendation of a discrete or exceptional consumption of these foods. In healthy adults, the consumption of six eggs per week is apparently associated with no increase in cardiovascular accidents (Djoussé and Gaziano 2008). It is not the same in adults suffering from type 2 diabetes.

For the other foods given in the table, an even important daily ration (between 100 and 200 g) does not constitute a sizeable cholesterol contribution.

3.2.3.2 Phytosterols

The structure of the most abundant phytosterol, β-sitosterol, is discussed in Section 2.6. Since the year 1950, phytosterols are known to efficiently lower cholesterolemia in humans without causing side effects (Section 4.3.2). The interest for these lipids grew in the year 1990 after the discovery of the possibility of adding

TABLE 3.25

Cholesterol Composition of Some Foods

Foodstuff	Cholesterol (mg/100 g)	Weight of Food (g) Supplying 300 mg of Cholesterol
Lamb brain	2220	13
Veal kidney	400	75
Egg	380	80 (2 eggs)
Lamb liver	315	90
Butter	260	115
Shrimp	160	190
Comté	120	250
Camembert	75	400
Lamb leg	70	430
Rabbit	60	500
Chicken fillet	60	500
Trout	57	526
Beefsteak	55	545
Oyster	50	600
Sardine canned	48	625
Mussel	38	790
Pork fillet	33	910
Duck fillet	15	2000

them to various foods in the form of fatty acid esters. Unfortunately, no clinical study was carried out on the direct effects of phytosterol consumption on cardiovascular diseases.

All the plants and some animal products contain phytosterols and, in smaller proportions, their saturated derivatives, stanols, but it is difficult to attribute to these compounds therapeutic effects at the amounts measured in foods.

3.2.3.2.1 Recommended Phytosterol Dietary Intake

In 2000, in the United States the FDA authorized manufacturers to indicate on packings that the consumption of products enriched in phytosterols may contribute to the prevention of coronary diseases by reducing the blood cholesterol level. These foods or these supplements must contain a minimum of 650 mg of phytosterols by portion.

Since May 2010, in Canada the addition of phytosterols is authorized in spreads, margarines, mayonnaise, salad dressings, yoghurts, fruits, and vegetable juices. The recommended daily intake is 2 g (http://www.hc-sc.gc.ca/fn-an/label-etiquet/claims-reclam/assess-evalu/phytosterols-fra.php).

In Europe, a decision by the European Commission on March 31, 2004 authorized the marketing of spreads, seasonings, milk or fermented milk, and drinks containing soya and cheeses, all enriched in phytosterols or phytostanols as new foods or food ingredients (application of the regulation EC n° 258/97).

In 2008, the EFSA recognized that "a reduction of approximately 9% of LDL-cholesterol may be obtained by the daily ingestion of 2 to 2.4 g of phytosterols in a suitable food (plant sterols added to fatty foods, such as margarines, salad sauces, mayonnaises, or to low-fat foods, such as milk, yoghurts, soya drinks). The effect on cholesterolemia can vary according to the type of supplemented food. It is recommended that these products are consumed only by people who want to lower their blood cholesterol level." The European Commission specified in addition that it is preferable to avoid phytosterol intakes higher than 3 g/day and that each product must be clearly labeled by specifying whether it contains plant sterols or stanols. Any limitation of use must also be stipulated, more particularly with respect to pregnant or nursing women and children less than 5 years old. The EFSA approved in April 2012 the fact that ingestion of a mixture of sterol and stanol esters at a rate of 3 g/day lowers the concentration of LDL-cholesterol by 11.3% within a minimum of 2 to 3 weeks.

In France, the AFSSA specified in 2001 to be not favorable to the phytosterol addition in cheeses and cheese-like specialties is not favorable, the consumer risking the intake of higher amounts by ingesting other products enriched with phytosterols (such as low-fat margarines).

In 2002, the AFSSA gave a favorable opinion regarding the addition of phytosterols (in free or esterified form) to margarines but not a favorable one concerning the addition of phytosterols to other foods (seasonings, cereal bars, soft drinks, yoghurts, and modified meats). The AFSSA has indicated that labeling must mention that "it is recommended to consume a significant number of fruits and vegetables to mitigate the drop of the plasma β-carotene level induced by phytostanols."

To date, the AFSSA authorizes sterol enrichment only for margarines and dairy specialties, specifying that "it is not useful to multiply these products, intended exclusively for the people presenting a hypercholesterolemia. Moreover, clinical studies have shown that a daily intake of sterols higher than 3 g did not lower cholesterol anymore."

3.2.3.2.2   Phytosterol Food Sources

The current dietary intake of phytosterols in Western countries varies from 150 to 300 mg/day, vegetarians absorbing even higher amounts, approximately 400 mg/day. Stanols, entirely saturated phytosterols, constitute only 10% of this intake. Given the lack of precise data on the specific effect of each phytosterol, only the value of the total pool of these compounds in various food sources is given.

The richest sources of phytosterols (Table 3.26) are oils (from 100 up to 800 mg/100 g), plants like cabbages (approximately 40 mg/100 g), and fruits like olive (50 mg/100 g). Whole-wheat bread contains about 86 mg of phytosterols per 100 g, whereas white bread contains only half of this amount.

Among seed oils commonly used in human consumption, sunflower seed oil has the highest content of phytosterols, followed by rapeseed oil, groundnut oil, and olive oil. In general, β-sitosterol is the most abundant phytosterol, up to 84% of the total phytosterol amount in almond oil (Table 3.27). The phytosterol concentration of marketed oils depends on the refining method.

It was determined in 2003 by A. de Jong, Netherlands, that phytosterols were poorly absorbed by the intestine in comparison with cholesterol (a maximum of

TABLE 3.26
Total Content of Phytosterols in Some Vegetables, Fruits, and Oils
(Milligrams per 100 g of Fresh Weight)

Vegetable	Sterols (mg/100 g)	Fruit	Sterols (mg/100 g)	Oil	Sterols (mg/100 g)
Brussels sprout	43	Olive (black)	50	Corn	825
Cauliflower	40	Orange	24	Soybean	330
Broccoli	39	Grape	18	Sunflower	280
Mushroom	18	Pineapple	17	Groundnut	235
Carrot	16	Banana	14	Rapeseed	230
Kale	13	Apple	13	Olive	105
Fennel	10	Pear	12		
Onion	8				
Tomato	5				
Potato	4				

Note: Data for vegetables are according to Normen and others (1999) and Marangoni and others (2010) and for oils according to Chan (1994).

0.4%–3.5% instead of 35%–75% for cholesterol). Furthermore, phytosterols enabled a reduction of cholesterol absorption by up to 20%, thus controlling effectively the plasma cholesterol concentration (Section 4.3.2).

Following these observations, several partners of the food industry requested from scientific and medical authorities opinions on the marketing of new products enriched in vegetable sterols. These products were prepared from various vegetal oils (soybean, rapeseed, and sunflower) or from tall oil (a by-product of wood pulp manufacture). They are sometimes hydrogenated to obtain stanols or acylated by fatty acids to make them more lipophilic (http://www.fao.org/ag/agn/agns/jecfa /cta/69/Phytosterols_CTA_69.pdf). The sterol esters are generally prepared by using an enzymatic process containing fungi lipases (Novozym and Lipozyme).

From 1954 and until 1982, a sitostanol-rich specialty, Cytellin (Eli Lilly), was marketed in the United States for the treatment of hypercholesterolemia. The most common doses employed were from 6 to 18 g/day, amounts up to 45 g/day having no side effects. After forgetting these prescriptions for a long period, phytosterols were rediscovered and reappeared in 1995 on the market in Finland in the form of foods enriched especially in phytostanol esters (Benecol margarines).

In 2008, 17 suppliers of these products were present in the world. One thus finds on the worldwide market various products enriched in phytosterols, the most abundant being β-sitosterol: spreads, yoghurts, cheeses, orange juice, cereal bars, chocolate bars, soya drinks, and so on.

Among the most recent products, Diminicol by Teriakia, Ltd., obtained a marketing authorization in Europe in 2004 (decision 2004/336/EC). Its use is reserved for dietetic foods (spreads and dairy products) for adults as well as young children. The Cognis Deutschland company from Dusseldorf, Germany, one of the most important

TABLE 3.27

Concentration of the Most Important Phytosterols in Some Oils (Milligrams per 100 g of Oil)

Oil	β-Sitosterol	Stigmasterol	Campesterol
Almond	229	19	9
Groundnut	184	24	44
Macadamia	104	1	9
Hazelnut	134	2	10
Walnut	225	35	19
Rapeseed	220	—	136
Sunflower	190	36	34
Olive	52	1	2

Source: From Alasalvar, C., and E. Pelvan, *Eur. J. Lipid Sci. Technol.*, 113, 943–9, 2011.

European phytosterol suppliers, obtained in July 2004 the approval of the European Commission for phytosterol ester supplementation in various types of foods. Since then, many other producers have obtained the same authorization.

Several milk-derived products enriched with phytosterols are present on the French market:

- Danacol, a yoghurt with phytosterols by Danone. The product, authorized in July 2004 by the European Commission, is elaborated like a yoghurt with skimmed milk and lactic acid bacteria, and added with plant sterols. There exists two types, one to eat with a spoon, containing 0.8 g of sterols for a yoghurt pot of 125 g, and the other to drink, containing 1.6 g of sterols for a 100-g bottle.
- Ilô of St. Hubert, a yoghourt enriched in stanol (0.8 g per yoghurt pot of 125 g).
- Pro-Active Fruit d'Or, a spread in two versions (35% or 62% fat content) containing 0.75 g of plant sterols for 10 g.
- Isio ActiStérol, a salad sauce by Lesieur. Sauce containing rapeseed oil (naturally rich in *n*-3 fatty acids) enriched with 1.6 g of plant sterols and presented in personal doses (20 mL).

Phytosterols are also offered on the Internet in the form of supplements (capsules), some products containing free phytosterols and others having esterified forms.

3.2.4 Fat-Soluble Vitamins and Related Compounds

The so-called fat-soluble vitamins, which are also regarded as lipids, are vitamins A, D, E, and K. National authorities have established several ANC (or RDA) values, published in France by the ANSES. These ANC values are only benchmarks for defined groups of subjects. In many countries, foods such as margarines, dairy products, cereals, and dietetic products are fortified with vitamins A, D, and E, baby foods often being supplemented with vitamin K. The French regulation is more restrictive, recognizing only the possibility of vitamin supplementation to compensate for the losses during manufacture and food storage. This concept has led to the emergence of food "with guaranteed content of vitamins." The total content obtained is then a value ranging between 80% and 200% of the amount of that vitamin naturally present in the raw materials. Any enrichment in vitamins for the purpose of preventing or correcting a deficiency shall not exceed 40% of the recommended daily allowance (RDA, equivalent to the average values of ANC for various categories of individuals).

3.2.4.1   Vitamin A and Carotenoids

3.2.4.1.1   Discovery

In 1913, the American biochemist E.V. McCollum discovered, thanks to his works on nutrition and the growth of rat, a lipid-soluble factor essential for growth and survival. He named it "liposoluble factor A," thus making a difference with other accessory factors called "water-soluble factors B." In 1931, the famous Swiss chemist P. Karrer (Nobel Prize in 1937) established the chemical structure of retinol, following that of carotene in the previous year (Section 2.7.1). Also in 1913, the relationships between serious damages to the cornea, loss of night vision, and retinal purple pigment were established, but the visual function of vitamin A was described only in 1968 by the American biochemist C. Wald. He was awarded the Nobel Prize in 1967 "for the discoveries concerning the fundamental processes of the physiology and the chemistry of the vision by the eye." On the metabolic level, in 1930 the biosynthesis of vitamin A from carotene was shown in vitamin A–deficient rats but receiving food rich in β-carotene. The mechanisms of this transformation were elucidated only in 1965 by the American biochemist D.S. Goodman. Thus, he discovered that the cleavage of the central double bond of carotene gave rise to two retinal molecules, which were later reduced in retinol. The discovery in 1987 in the laboratory of Prof. P. Chambon in Strasbourg, France, of the nuclear receptor of retinoic acid has widened the field of research on vitamin A in the field of cellular differentiation and carcinogenesis. This discovery made it possible to clearly regard vitamin A as a hormonal compound.

Except β-carotene, many carotenoids of the xanthophyll group, originating initially from bacteria or plants, have increasingly obvious biological functions, some of which are now being taken into account in human treatment. Among the most important carotenoids that are objects of advanced research, one can quote lutein and zeaxanthin, both involved in the protection of the retina. Lycopene is also a powerful antioxidant and a potential anticancer molecule, and astaxanthin has multiple therapeutic activities.

3.2.4.1.2   Vitamin A and Provitamins A

3.2.4.1.2.1   General Functions   Compounds having a vitamin A activity (retinol, retinol esters, retinal, and retinoic acid), called retinoids, are present in animal tissues and result from the enzymatic conversion of carotenoids in the intestinal wall and the liver. If many carotenoids are potentially provitamins A (the all-*trans* forms being the most active), only three are present in food in significant quantities: β-carotene (Figure 2.27) giving two molecules of retinol, α-carotene and β-cryptoxanthin

(hydroxylated derivative of the previous one), each one giving only one molecule of retinol (Section 2.7.1).

Vitamin A is involved in many physiological processes (Section 4.4.1), such as vision, development of bone, mucus secretion, spermatogenesis, and control of gene expression. It also has various functions in the physiology of skin, development, and cellular differentiation. Retinal, an aldehyde derivative of retinol, is known to be involved in the maintenance of normal vision and exerts specific effects in the development of adipose tissue and regulation of glucido-lipid homeostasis. The acidic derivative of retinol, retinoic acid, is a very powerful lipid mediator involved in gene transcription. Physiologically, β-carotene has the same properties as vitamin A but may also act as an antioxidant (destruction of free radicals).

3.2.4.1.2.2 Units Amounts of vitamin A are expressed in milligrams or in retinol equivalents (mg RE), which makes it possible to convert all the sources of vitamin A into only one expression. Thus, given the yield of their enzymatic conversion and intestinal absorption, it is admitted that 1 RE is equivalent to 1 μg of retinol, 1.78 μg of retinyl palmitate, 6 μg of β-carotene, or 12 μg of other carotenoids (primarily cryptoxanthin) (these values are adopted by the FAO). The International Unit (IU) is also used; it corresponds to 0.3 μg of retinol. These RE values are still discussed as the Food and Nutrition Board of the IOM, Washington, DC, estimated in 2001 that 6 β-carotene molecules diluted in oil are necessary to generate 1 molecule of retinol, but 12 molecules are necessary when they are included in complex foodstuffs. Recent studies in developing countries have demonstrated an even lower biological efficiency as under the nutritional conditions present in these countries one would need practically 21 β-carotene molecules instead of 6 (West et al. 2002). Considering this figure and the dietary habits found in Asia, Africa and South America, it clearly appears that the populations of these countries are confronting a real state of vitamin A deficiency.

3.2.4.1.2.3 Recommended Vitamin A and Carotenoid Intakes The WHO has recommended a daily intake of 600 μg RE in men and of 500 μg RE in women. In the United States, the FDA has recommended an intake of 900 μg RE in men and 700 μg RE in women but 770 μg RE in pregnant women and 1300 μg RE in nursing women. In the United Kingdom, health authorities have recommended an intake of 700 μg ER in men and 600 μg ER in women.

In 2013, the EFSA scientific panel considered that the observed mean vitamin A intake in Europe from breast milk of 350 μg RE/day is adequate for the majority of infants in the first year of life. A value of 400 μg RE/day is adequate for children from 12 to 36 months.

In France, the AFSSA has fixed the following recommended daily intakes in vitamin A: for adults (from 16 years) at 800 μg RE in men and 600 μg RE in women, 400 μg RE in very young children (1–3 years), and 450–700 μg RE in children from 4 to 15 years.

Simulation studies have indicated that a supplementation equivalent to 15% of the recommended daily intake (e.g., approximately 120 μg RE) is likely to double the percentage of the general population exceeding the safety threshold. Consequently, the AFSSA indicated that enrichment does not appear to be desirable. With respect to food supplements, the maximum daily amount of 3000 μg RE is maintained.

The status of vitamin A may be established by the analysis of retinol in the blood, the average plasma concentration in retinol (retinolemia) being about 1–1.2 μmol/L (286–340 μg/L) in the French. The retinolemia in several populations has been determined to be normally between 2.5 and 4 μmol/L (700–1150 μg/L or 2300–3800 IU/L) in the serum (Grzelinska et al. 2007). Actually, this concentration may undergo momentary increases, according to food ingestions. It should be noticed that the plasma concentration is a faithful consequence of the food supply in vitamin A and provitamins A but not of those in the hepatic reserves.

Although a state of deficiency in vitamin A in the developed countries was observed in some subjects having difficulties in intestinal absorption (Crohn's disease and celiac disease) or chronic hepatic pathologies (alcoholism type), reports of partial deficiency have been established in several countries. Indeed, the WHO has estimated at several million the number of vitamin A deficient children in the world. Vitamin A deficiency led each year to blindness nearly 500,000 children in the developing countries, mainly in South Asia and Africa. Thanks to the efforts of the WHO and its partners, it is estimated that more than one million deaths could be avoided since 1998 in 40 countries by the installation of policies organized against vitamin A deficiency. A database concerning the world state of deficiency in vitamin A (xerophthalmia and plasma retinol) in each country has been published by the WHO (http://www.who.int/vmnis/vitamina/data/database/countries/en/index.html).

The populations of developed countries are not safe from the deficiency of vitamin A. In Quebec, 28% of men had an intake lower than the average requirement in vitamin A estimated as 625 μg/day in this country (National Institute of Public Health).

It may also be considered that the French population is currently in a deficient state as shown in a study of food supplies (the SUVIMAX study, 1997–2003). Approximately 72% of the population tested (from 50 to 68 years) had vitamin A intakes lower than that advised and 45% of the population had intakes lower than two-thirds of the advised intake.

Besides the direct nutritional problems, physiological studies have shown that the deficiency of vitamin A can have other origins, the most frequent being an iron deficiency and an excess of alcohol consumption.

The National Institute of Health has established that 3 mg/day for long periods is the tolerance limit for a supplementation in vitamin A in a healthy adult population. Since 1978, the American authorities have required milk producers to maintain a final concentration of 600 μg/L (2000 UI/L) of vitamin A, to compensate the losses induced by skimming.

A hypervitaminosis by too elevated intakes of vitamin A is not recommended because it can interfere with the status of vitamin C and vitamin K. For significant amounts, abdominal pains, anorexia, vomiting, and headaches have been observed.

In 2012, the EFSA stipulated that the addition of β-carotene to food did not pose any problem if the ingested amount did not exceed the amount naturally ingested in our food (5–10 mg/day), the total being not higher than 15 mg/day. The AFSSA has drawn attention to the multiplicity of the intakes of β-carotene and has recommended great caution in the supplementation of food substances, taking into account the risk for people exposed to potent carcinogens from the environment (tobacco and asbestos).

The whole status of carotenoids may be established by a blood analysis defining a health index for these compounds that has been correlated with the risks of cancer (Section 4.4.1.4) and cardiovascular disease (Section 4.4.1.5). This approach would require to be widened to various pathologies before this index can become a faithful reference to estimate the status of these compounds.

To maintain without supplementation an intake of vitamin A in conformity with the standards accepted by medical authorities, it is recommended to have a diet diversified with several fruits and vegetables. The latter will preferably be consumed cooked in the presence of a little vegetal oil to facilitate the absorption of carotenoids. The consumption of liver, dairy products, and cheeses is also recommended. When a deficiency is suspected, retinolemia may be determined and, followed if necessary by a daily supplementation from approximately 1 mg of vitamin A, without exceeding 3 mg/day.

3.2.4.1.2.4 Food Sources of Vitamin A The sources of vitamin A, primarily in the form of retinol, are livers of mammals and poultry, fatty fish, milk, and butter, but it also occurs in the form of carotenoids, fruits, and vegetables (Table 3.28). It has been estimated that a maximum of 34% of the ingested vitamin A in the developed countries are in the form of carotenoids, this proportion being approximately 80% in Africa and Asia. Thanks to the EPIC study, A. Jenab, France, estimated in 2009 that

TABLE 3.28

Food Sources of Vitamin A (Expressed in Micrograms of Retinol Equivalent per 100 g Fresh Weight)

Animal Product	µg ER/100 g	Vegetal Product	µg ER/100 g
Liver (mammals, poultry)	10,000–15,000	Dry apricot	5,800
Butter	800	Melon	3,600
Egg (whole)	180	Raw carrot	2,000
Roquefort, goat cheese	400	Cooked carrot	1,400
Camembert, brie, comté, gouda	200–300	Spinach, pumpkin, corn salad	700–800
Tuna	450	Fennel, sweet pepper	500–700
Mackerel	100	Grapefruit	390
Salmon	41	Fresh apricot	280
Oyster	93	Celery, lettuce, tomato sauce	200–300
Mussel	54	Tomato, asparagus, margarine	80–170
Herring	38	Plum	65
Skimmed milk	20	Peach	16
Cod	6.5		

Note: Data are from the CIQUAL 2008 database and the food composition and nutrition tables, S.W. Souci et al., CRC Press, 2000.

the average dietary intake of retinol was 835 µg across 10 countries, France having an intake of 578–728 µg according to the areas.

The β-carotene intake observed across the 10 European countries (EPIC study) is on average 2.8 mg/day. The dietary supply of carotenoids among French people is the highest among these countries, which has been estimated in 2001 by M.E. O' Neill, England, to 10–22 mg/day, including 3.8–8 mg/day for β-carotene (median value: 5.8 mg/day). This inquiry study has determined that β-carotene originates mainly from spinach and carrot consumption.

A table of the β-carotene contents of the most common nutrients available on the U.S. market has been published by the USDA (http://www.nal.usda.gov/fnic/food-comp/Data/SR21/nutrlist/sr21a321.pdf).

The contents of several different carotenoids in a large number of plants are reported in a food composition table by the USDA Nutrient Data Laboratory (http://www.nal.usda.gov/fnic/foodcomp/Data/car98/car98.html) and in a CIQUAL table published on the website of the ANSES (http://www.afssa.fr/TableCIQUAL/).

3.2.4.1.3   Lutein and Zeaxanthin

Lutein and zeaxanthin are the main xanthophylls of green plants. They are very interesting as they are concentrated in the macula of the human retina, in protecting the eyes against oxidative stress (ultraviolet [UV] radiation and cigarette smoke) and thus avoiding retinal degeneration.

Lutein (Figure 2.28) and its isomer zeaxanthin are two carotenoids related to car-otene (Section 2.7.1) without being provitamins A. They are present in practically all plants. The lutein intake required to cover human needs has not yet been determined.

The main food sources of lutein are vegetables with green leaves: (in the order of decreasing concentration) cabbage, cress, spinach, chard, garden peas, broccoli, and lettuce (Table 3.29). These sources may supply about 1–15 mg/100 g. Fruits such as oranges and peaches are also potent sources of zeaxanthin. On the whole, the average of lutein to zeaxanthin concentration ratio in the Western diet is close to 5. Furthermore, all studies indicate that these two carotenoids have the same physiological properties.

TABLE 3.29
Lutein Content of Some of the Most Common Plants (Grams per 100 g Fresh Weight)

Species	Lutein (µg/100 g)	Species	Lutein (µg/100 g)
Cabbage	15,000	Brussels sprout	670
Cress	10,700	Pumpkin	630
Spinach	6,260	Cucumber	570
Chard	1,960	Green peas	490
Garden peas	1,840	Sweet pepper (red)	290
Broccoli	1,600	Artichoke	275
Lettuce	1,250	Orange	92
Celery	860	Peach	60
Zucchini	750		

Apart from plants, the egg is the richest food in lutein and zeaxanthin compared to other carotenoids (nearly 90% of the total carotenoid amount). However, the contents of lutein and zeaxanthin are relatively variable across eggs (from 50 to 150 µg). Eggs rich in vitamins and n-3 fatty acids (Benefic eggs) containing up to 1 mg of these carotenoids are available in the market (Bourre 2005). In Belgium, eggs containing 1.5 mg of lutein are produced by the company Belovo (Columbus egg, http://www.columbus-egg.be/).

The intakes observed in various populations of industrialized countries are on average from 1.6 to 3.2 mg/day. The dietary intake of lutein + zeaxanthine among the French people has been estimated to be 1.7–3.9 mg/day (median value: 2.5 mg/day), these carotenoids originating mainly from lettuce and spinach consumption (O'Neill et al. 2001).

To increase the bioavailability of these carotenoids, foods must be steamed, cooked, and finally prepared with a little amount of oil to improve their intestinal absorption. It has been shown that egg yolk lipids represent the best mixture allowing an optimal absorption of lutein, either included in vegetables or as supplement in capsules (Chung et al. 2004).

A carotenoid fortification could be necessary in patients with pancreatic or intestinal deficiency (Crohn's disease and celiac disease), gall bladder or liver disease, tobacco addiction, or alcoholism. It must be recalled that the consumption of high amounts of these carotenoids on the long run has not been the object of clinical investigation.

In 2012, an evaluation made by the EFSA showed that the total ingested amount (natural content and added quantity) should not exceed on average 1 mg of carotenoids per kilogram of body weight per day (e.g., approximately 75 mg/day for an adult). In 2004, the European Commission gave its consent to the marketing of food supplements enriched with astaxanthin in the form of capsules containing an oleoresin extracted from the microalga *Haematococcus pluvialis* (4 mg/capsule) provided by Herbal Sciences International, Ltd., Loughton, United Kingdom.

In France, the AFSSA has estimated that as yet there is no scientific element that makes it possible to justify a supplementation of lutein and zeaxanthin in a healthy human having a varied diet (saisine No 2009-SA-0287, March 2010). However, the AFSSA has considered that there is no scientific element against an additional intake of 6 mg/day of lutein, alone or associated with zeaxanthin. In spite of this opinion, health professionals (mainly ophthalmologists) recommend their patients to take additional food supplements present on the market and containing a mixture of lutein and zeaxanthin (sometimes with astaxanthin). An amount of 6 mg appears to be the usually recommended value, and higher amounts (10–20 mg) are prescribed for the prevention of cataract and macular degeneration in patients at risk.

3.2.4.1.4 Lycopene

Lycopene (Section 2.7.1 and Figure 2.30) is abundant in the human body; it is a very potent antioxidant mainly with respect to the oxygen singlet. For several years, attention was focused on the protective action of lycopene against prostate cancer (Section 4.4.1.4). Lycopene is abundant mainly in tomato (2.7 mg/100 g), watermelon (3.5 mg/100 g), and also pink grapefruit (3.4 mg/100 g) and dry apricot (0.86 mg/100 g). As for all carotenoids, cooking increases lycopene's bioavailability. Thus, tomato sauce contains four times more bioavailable lycopene than fresh tomato. A table of lycopene contents of the most common nutrients available on the U.S. market has been published by the USDA (http://www.nal.usda.gov/fnic/foodcomp /Data/SR21/nutrlist/sr21a337.pdf).

The dietary intakes of lycopene observed in various populations of industrialized countries were in the range 1.6–5 mg/day. The dietary intake among the French people has been estimated at 2.1–8.3 mg/day (median value: 4.75 mg/day), the supply originating mainly from tomatoes and pizza pie (O'Neill et al. 2001).

In 2006, the European Commission gave a marketing authorization for the lycopene produced from a Zygomycetes fungus (*Blakeslea trispora*) (Vitatene Antibiots S.A.U) (decision 2006/721/EC). In 2009, the EC authorized the marketing of lycopene extracted from tomatoes (Berry Ottaway & Associates, Ltd.) (decision 2009/355/EC).

3.2.4.1.5 Astaxanthin

The distribution of astaxanthin (Figure 2.29) in foodstuffs is very selective. It is present in many marine organisms whose flesh is orange in color, such as fish, mollusks, and especially shellfish. Its concentration in the flesh of animals is for some of them due to the ingestion of algae synthesizing that carotenoid and for the others due to the consumption of the previous ones, humans being included in the last group. Astaxanthin is an antioxidant that is at least as powerful as lycopene.

Many studies have focused their interest on this compound for the prevention and treatment of several diseases. Thus, pathologies as varied as cancer, immunization diseases, cardiovascular diseases, diabetes, gastrointestinal diseases, and neurodegenerative pathologies were the objects of conclusions very favorable to a role of astaxanthine. Research on these effects is very recent and was primarily carried out *in vitro*, but a growing number is now devoted to studies on humans. The most important aspects are evoked in Section 4.4.1.

Marine fish may contain up to 15 mg/kg of astaxanthin, but the shrimps living in cool waters may contain up to 1200 mg/kg. Cultured Atlantic salmon contains a maximum of 10 mg/kg and trout a maximum of 4 mg/kg. The Pacific salmon contains up to 40 mg/kg. The astaxanthin contained in the flesh of fishes raised in specific farms originates from a fortification of their food with synthetic astaxanthin or rough extracts from shellfish carapace (by-product of the marketing of shrimps and crabs), these carapaces being able to contain from 5 to 150 mg/100 g.

According to the WHO, the average daily intake in the world population is between 0.34 and 0.85 mg.

In 1987, the FDA authorized the use of astaxanthin as a dye in the manufacture of foods for fish and other livestock for human consumption. Its use as a food supplement in human nutrition is legalized in the United States, United Kingdom, Finland, and Norway. The European Commission indexed it only as a food coloring (E161j number).

In 2005, the EFSA indicated that the astaxanthin supply in human foods originating from fortified animals should not exceed 2 mg/day, and 4 mg/day for the largest consumers (http://www.efsa.europa.eu/fr/efsajournal/doc/291.pdf).

In France, the AFSSA emphasized that, in addition to children and pregnant women, patients taking drugs should avoid foods supplemented with astaxanthin as *in vitro* studies have indicated possible drug interactions linked to the stimulation by that carotenoid of the P450 cytochromes involved in the metabolism of some drugs.

Due to the lack of information on long-term toxicity and metabolism of astaxanthin in humans, the AFSSA gave an unfavorable opinion on the addition of this carotenoid to common foods (milk, fruit juice, etc.) (saisine N° 2009-SA-0026).

Since 2004, several companies have obtained from the European Commission authorization for marketing oleoresin capsules enriched with astaxanthin, extracted from *Haematococcus pluvialis* (maximum of 4 mg astaxanthin/capsule) (Herbal Sciences International, United Kingdom; AstaReal AB, Sweden; Cyanotech Corporation, the United States; and Alga Technologies, Israel).

Other xanthophylls have been the object of biological investigations likely to emerge rapidly toward clinical applications. Among these compounds, fucoxanthin and neoxanthin, present in green and brown algae, have been recognized as inhibitors of carcinogenesis in animals and cultured cells. Fucoxanthin has also been determined to be potentially active against allergy and obesity. Unfortunately, no clinical study was still devoted to both xanthophylls.

3.2.4.2 Vitamin D

3.2.4.2.1 Discovery

The vitamin D complex is formed of several compounds with antirickets properties deriving first from sterols by UV irradiation (Section 2.7.2). The discovery of this vitamin is historically related to the treatment of rickets and the close discovery of vitamin A (Section 3.2.4.1). Since 1872, cod-liver oil has been used in the United States for the treatment of rickets, in 1922 the American chemist E.V. McCollum discovered in that oil an antiricket factor improving the calcium deposit, he named this factor vitamin D. The demonstration of a beneficial effect of UV radiation in the treatment of rickets justified many investigations that led to knowledge of the structure of vitamin D_2 (or ergocalciferol) derived from vegetal ergosterol and of vitamin D_3 (or cholecalciferol) (Figure 2.31) biosynthesized in humans from cholesterol. For his whole work carried out in the field of vitamins and cholesterol, in 1928 the famous German chemist A. Windaus (1876–1959) was awarded the Nobel Prize in chemistry. The circulating form of vitamin D (calcidiol) (Section 2.7.4) was described in 1969 by H.F. DeLucas, in the United States, and the true active form (1.25-dihydroxy-vitamin D_3 or calcitriol) was discovered in 1971 by the English chemist D.E. Lawson.

3.2.4.2.2 General Functions

Vitamin D is often compared to a hormone ("sun hormone") because its active form is synthesized in the kidney and then transported via the blood toward intestinal mucosa where it controls the absorption of calcium and phosphorus, and also toward any cell equipped with specific receptors where it achieves multiple functions (Section 4.4.2). Since 2000, it is known that the production of calcitriol is also carried out in the skin under the action of UV radiation.

Units: the amounts of vitamin D are expressed in micrograms (or nanograms) of cholecalciferol (or ergocalciferol). IU is also used, knowing that 40 IU are equivalent to 1 µg of vitamin D. The blood concentration is expressed in micrograms per liter or nanomoles per liter (1 µg/L = 2.6 nmol/L).

3.2.4.2.3   Recommended Vitamin D Intake

Vitamin D has an endogenous origin by synthesis in the skin and an exogenous one from food containing preformed vitamin D, the amounts required to maintain a normal calcification status have primarily been defined from studies of plasma-level variations (Calvo and Whiting 2005). It has been recognized in 2010 by Mr. Ashwell, England, that with a moderate but regular sun exposure, the vitamin D dietary sources may be reduced. For the majority of the populations of many countries, the dietary supply of vitamin D does not exceed 5 μg/day, the amount corresponding to only approximately 30% of the required intake.

In November 2010, the IOM of the United States recommended that the daily intake of vitamin D should be tripled compared to the preceding announcements. Thus, these daily values must reach 15 μg (600 IU) from 1–70 years old and 20 μg (800 IU) beyond that age. In addition, the IOM has increased the maximum daily intake to 100 μg (4000 IU) for adults (http://www.iom.edu/Reports/2010/Dietary-Reference-Intakes-for-Calcium-and-Vitamin-D.aspx). These recommendations were accepted by the American and Canadian medical authorities.

Better than using a value of a daily intake, which remains difficult to establish, the human deficiency or insufficiency status of vitamin D is appreciated by measuring the plasma level of calcidiol (25-hydroxycholecalciferol), which gives a clear picture of the vitamin D general status resulting from sun exposure and food supply. The recommended concentration in the majority of European countries is at least 30 μg/L (75 nmol/L). This threshold has been defined by obtaining a plateau of calcium absorption for that value (Heaney 2004). Moreover, it has been functionally defined, as it corresponds to an important fall in the secretion of parathyroid hormone, a vitamin D antagonist with respect to calcemia. In 2011, the IOM fixed the values that enable the estimation of vitamin D status by defining a state of deficiency for a concentration less than 12 μg/L (30 nmol/L), a state of sufficiency for a concentration ranging between 12 and 20 μg/L (30 and 50 nmol/L), and a satisfactory concentration beyond 20 μg/L (50 nmol/L). In Canada, medical authorities have accepted the same values. For several specialists, these values appear much too low.

It is recommended to not exceed a concentration of 50 μg/L (125 nmol/L), which is capable of increasing the risk of undesirable events. These effects are poorly known, but a recent investigation has shown that a vitamin D concentration of 50–60 nmol/L was associated with the lowest mortality in a large Danish population (Durp et al. 2012); with a higher concentration (140 nmol/L), the mortality rose by more than 42%, but it was lower (2.13 times) than that observed for very low concentrations (10 nmol/L). Despite these observations, causal relationships have not been elucidated.

It is generally accepted that a concentration ranging between 20 and 30 μg/L (50 and 75 nmol/L) defines the insufficiency status, a physiological situation that would affect more than one billion people in the world and more half of all postmenopausal women (Holick 2007). The deficiency status has been established for a level lower than 20 μg/L (50 nmol/L), a value typically accompanied in children by rickets and in adults by osteomalacia (osseous decalcification induced by a defect of mineralization). In the Netherlands, the health authorities have fixed a lower threshold of 12 μg/L (30 nmol/L) in women less than 50 years of age and men less than 75 years and of 20 μg/L (50 nmol/L) in older people.

In France, clinicians have considered that the desirable values for the plasma concentrations of vitamin D_3 must be between 30 and 80 µg/L (75 and 200 nmol/L). The higher limit has been arbitrarily selected to be sufficiently far away from potentially toxic values. The deficiency status has been associated with values lower than 10 µg/L (25 nmol/L) (Souberbielle et al. 2009).

The International Association for Research on Cancer has established that serum concentrations in vitamin D lower than 16 µg/L (40 nmol/L) were associated with an increasing risk of colorectal cancer (Vitamin D and Cancer. Working Group Report 2008 vol. 5. Lyon: International Agency for Research on Cancer, http://www.iarc.fr/en/publications/pdfs-online/wrk/wrk5/index.php).

In 2009, a group of experts from 12 countries made recommendations on the requirements of vitamin D for people presenting a risk for cardiovascular diseases, autoimmune diseases, and cancer in a meeting in Paris, France (Souberbielle et al. 2010). A basal plasma level of 30 to 40 µg/L was considered as normal, if lower the level must be checked 3 months after a supplementation. People with black skin or those wearing a veil and the elderly people will have to be regularly supplemented with 20 µg/day (800 IU/day). The expert group agreed to evaluate at 100 µg/L the higher limit of the plasma concentration of vitamin D.

In 2012, the French Academy of Medicine communicated its recommendations on vitamin D status, the extra-bone roles and daily requirements of vitamin D (http://www.academie-medecine.fr/publication100036502/).

In short, on the one hand, the French Academy of Medicine recalled the following:

- The vitamin D status of a subject is defined by the determination of calcidiol concentration in serum, in agreement with the whole international scientific community.
- The serum level of calcidiol must be higher than 30 µg/L (or 75 nmol/L) to define a normal vitamin D status but must not exceed 80 µg/L (or 200 nmol/L), according to the majority of the authors.

On the other hand, the academy also recommended the following:

- That the vitamin D status of the French population must be examined with great attention
- That the determination of the serum calcidiol concentration must be carried out more frequently during bone, intestinal, and renal diseases and is necessarily associated with the determination of the serum parathyroid hormone concentration
- That the daily vitamin D intake must be estimated taking into account the age and sex whatever the season
- That the vitamin D deficit must be corrected only by oral supplementation without leading the concerned subjects to increase their sun exposure or to use a sunshine cabin, which is known to induce skin cancers

When necessary, supplementation may be done by daily ingestions or by intermittent administration of calculated doses monthly or every 2 months.

In short, the recommended daily intakes are from 600 to 800 IU from 1 to 8 years, 800 to 1000 IU from 9 to 50 years, 1000 to 1500 IU from 51 to 70 years, and greater than 1500 IU beyond 70 years.

Given the high prevalence of vitamin D deficiency, especially in late winter, the Nutrition Committee of the French Pediatric Society has also suggested in 2012 to reestimate upward the recommendations of vitamin D supplementation in pregnant women, children, and teenagers. This supplementation must be 80,000–100,000 IU in pregnant women (at the beginning of the seventh month of pregnancy); from 1,000 to 1,200 IU/day in nursed infants and in children less than 18 months of age fed with cow's milk; and from 80,000 to 100,000 IU in children of 18 months to 18 years (two doses in November and February).

Considering the recommendations promulgated in various countries and the new updates taking into account the scientific advances in the field of vitamin D physiology, it is important to adopt the latest rules issued by the French Academy of Medicine, namely,

- Maintain a serum vitamin D concentration equal to or higher than 30 µg/L (75 nmol/L), without exceeding 80 µg/L (200 nmol/L).
- Check this concentration especially in patients with bone, digestive, cardiac, and mental diseases and among the elderly.
- Correct any deficit by an adapted oral supplementation and, if possible, by direct and regular sun exposure (at least 15 min/day).

3.2.4.2.3.1 Contribution of the Sun   Short solar exposures may cover our needs for vitamin D, with approximately 15 minutes of exposure (hands, forearm, and face) two to three times a week between 9:00 and 16:00, from spring to autumn, under average latitudes. Thus, in France only from April to September the organism can synthesize and store enough vitamin D for the winter, however, by supplementing that skin supply by foods or dietary complements rich in vitamin D. It should be specified that the synthesis of vitamin D cannot be carried out behind a pane, or under a layer of sun lotion. It indeed seems that the instructions of protection against the risks of skin melanoma following prolonged sun exposure could contribute to a vitamin D deficiency status, especially in children and the elderly. It must be noticed that not all dermatologists share that fear. It remains accepted that subjects with dark skin must expose themselves to the sun longer than the others because melanin limits the skin penetration of UV-B radiation. For this reason, Afro-Americans have a plasma concentration of vitamin D that is approximately half of that measured in white subjects in the same locality. This difference has also been observed at birth in the blood of the umbilical cord (Hanson et al. 2011). The physiological consequences of these observations have been largely analyzed (Dawson-Hughes 2004).

The mutation that has generated a loss of pigmentation in our African black ancestors makes it possible for humans to survive in not very sunny regions by synthesizing sufficient vitamin D (Jablonski and Chaplin 2012). Indeed, it is probably this genetic change that allowed the first great migrations of *Homo erectus* from Africa toward Asia, 2 million years ago, and toward Europe, a million years later.

Curiously, there is a well known north–south gradient of vitamin D status in the world, but the highest concentrations are observed in the northern countries of Europe and America (van Schoor and Lips 2011). Thus, in Europe higher vitamin D levels have been described in Norway or Sweden compared to Spain, Greece, or Italy. Nutrition and behaviors with respect to sun exposure explain these differences apparently not connected to the geographical location.

The effect of sun underexposure of people with a light skin is particularly measurable among Japanese women, as determined in 1983 by T. Kobayashi. These women who do not have the direct radiations of the sun have a vitamin D plasma level 36% lower than that measured in men (19.3 µg/L compared to 26.2 mg/L). Even in the sunniest countries, like Saudi Arabia and Australia, up to 50% of subjects have insufficient vitamin D blood levels, a situation related to lifestyles and local vestimentary habits. The geographical location also has a great importance even in European countries such as the United Kingdom where the dietary intake of vitamin D covers less than 20% of the needs. Clear differences in plasma concentrations of vitamin D have thus been measured in 2011 by H.M. Macdonald in the United Kingdom: he observed that women had significantly lower concentrations (from 20% to 30% according to the season) in the north (Scotland) compared to those living in the south (Surrey) of the country. The situation is even more critical for the populations living in the Artic regions, where sunshine is extremely restricted most of the year and where vitamin D intake is less and less provided by the traditional diet (Sharma et al. 2011). It seems that part of the French population is also affected by a lack of sun exposure as the vitamin D blood levels are much lower among townsmen of the north (29% of hypovitaminosis) compared to those living in the south (6% to 7% of hypovitaminosis) (Chapuy et al. 1997). At the latitude of France, the sun exposure is too reduced from November to February to produce sufficient skin synthesis of vitamin D. Moreover, a cloud cover divided the energy of UV radiations by two and shade reduced it by 60%.

On the nutritional level, the capacity of an organism to absorb or synthesize vitamin D is known to decrease with aging, which could be at the origin of a deficiency status in one adult on seven. This modification would have an important effect on the incidence of osteoporosis in elderly people. For this reason, it seems beneficial to supplement the majority of old subjects of more than 60 years with at least 50 µg/day (2000 IU/day) of vitamin D_3 (Heaney 2006). A recent intervention study on the long run in elderly people living by 58° of northern latitude (Gothenburg, Sweden) has indeed shown that a daily supplementation of 40 µg of vitamin D was necessary to maintain a serum level equal to or higher than 19 µg/L (50 nmol/L) (Toss et al. 2012).

It should be specified that a UV overexposure leading to an erythema does not increase any more the synthesis of vitamin D but produces inactive metabolites. There is thus no risk of hypervitaminosis D except from a food supplementation, the sun thus participating in the regulation of the body's vitamin D supply. It must be emphasized that estimation of the time of a sun exposure necessary for a known supply of vitamin D is possible with an automatic calculator included in a website (the Norwegian Institute of Research on the Air): http://nadir.nilu.no/~olaeng/fastrt/VitD-ez_quartMEDandMED.html. Thus, after entering some geographical parameters, one can calculate that on June 21 around midday in Paris an exposure of the face, forearms, and hands for 4–12 minutes (according to the skin color) allows the biosynthesis of 25 μg of vitamin D. In a subject with black skin, 20 minutes will be necessary to get the same result. According to the standards established by the FAO/ WHO in 2002, under latitudes ranging between 42°N and 42°S, a sun exposure of the face and arms for 30 min/day is enough to maintain an adequate level of vitamin D. Apart from these latitudes, this minimum exposure is efficient only in summer. For example, even on a sunny day, the cutaneous synthesis of vitamin D is worthless from November to February among inhabitants of European cities like Vigo, Barcelona, Ajaccio, Rome, Istanbul, or Tbilissi, all located at a latitude of approximately 42°N. With a latitude of approximately 52°N, as in London, Rotterdam, Berlin, Warsaw, or Edmonton, this period is extended from October to March.

It should be specified that in spite of some advertisements edited in the press, the use of bronzing tanning bed is not at all adapted to the skin biosynthesis of vitamin D. Within the framework of the French Cancer Plan 2009–2013, the INCA published a report on UV radiation (artificial and solar), vitamin D, and noncutaneous cancers. The INCA recalled that "the practice of bronzing with artificial UV is not at all recommended, because of the well known carcinogenic potential of artificial UV. The use of tanning bed as source of vitamin D for the general population cannot be justified as there is an easy possibility of vitamin D oral intake (food supplies, consumption of food enriched in vitamin D and/or ingestion of vitamin D supplements)."

3.2.4.2.3.2 Contribution of Foods The nutritional contributions have been defined as a complement to the endogenous production of vitamin D in the skin during sun exposure. It is admitted that, in general, food does not bring more than 5 μg of vitamin D per day, that is to say, approximately 20% of the human needs. The contribution of the diet has a major prevalence during winter at latitudes higher than 40°N, where solar radiation is inefficient for cutaneous vitamin D synthesis. The contribution of food to the vitamin D supply has been appreciated at 38% of the total intake as this proportion is the value of the difference estimated in winter in England between meat eaters and vegan subjects (Crowe et al. 2010). Moreover, one considers that very young children, pregnant women, obese, and elderly may have increased needs as a consequence of their usually weak sun exposure. With equal sunning, a 70-year-old person synthesizes nearly four times less vitamin D than a 25-year-old person (Holick et al. 1989). Thus, a slowed down biosynthesis of vitamin D and a weak sun exposure in elderly people justify an oral supplementation in vitamin D estimated at 65 μg/day (2600 IU/day) to avoid a deficiency that is detrimental to health (Heaney 2006).

To better appreciate the effect of a supplementation, it is necessary to consider that taking 1 μg (40 IU) of vitamin D increased its plasma content of approximately 1.2 nmol/L (Black et al. 2012). Thus, to maintain a level higher than 75 nmol/L in people already having a level of 30 nmol/L, they must ingest each day about 40 μg of vitamin D. In November 2010, the Canadian and American authorities recommended a daily dietary intake in vitamin D not lower than 10 μg (400 IU) from 0 to 1 year, 15 μg (600 IU) from 1 to 70 years, and 20 μg (800 IU) beyond 70 years (Ross et al. 2011). The tolerable maximum intakes have been established at 50 μg/ day

(2000 IU/day) for infants up to 6 months and at 100 μg/day (4000 IU/day) for the remaining population. The Canadian Cancer Society has recommended a supplementation of 25 μg (1000 IU) per day of vitamin D in autumn and winter. In Europe, the official recommendations vary according to countries between 0 and 15 μg/day (Doets et al. 2008).

In Europe, the scientific panel of the EFSA considered in 2013 that a vitamin D intake of 10 μg (or 400 IU) per day is adequate for the majority of young children (from 0 to 36 months) having minimal sun exposure.

For people suffering from osteoporosis, an intake from 20 to 50 μg (800–2000 IU) per day has been recommended. Some Canadian experts recommend a daily intake of 50–100 μg (2000–4000 IU) of vitamin D, except in summer, but with regular sun exposure when possible. In pregnant women, a food supply in vitamin D of at least 5 μg (200 IU) per day is recommended in the United States (http://www.nrv.gov.au/nutrients/vitamin%20d.htm), Australia, New Zealand, and Canada, but a twofold value (10 μg or 400 IU/day) is recommended in the United Kingdom (http://www.healthystart.nhs.uk/en/fe/vitamin_supplement_recommendations.html). These recommendations are generally considered too low by specialists and are prone to controversy considering the uncertainty that still remains about the concept of sufficiency.

In France, the AFSSA has recommended a daily ANC (or RDA) of 5 μg (200 IU) of vitamin D in teenagers and adults and of 10 μg (400 IU) in children less than 4 years, pregnant women, and elderly people. Compared to the recommendations of other countries, these ANC values for the French population, except perhaps young children, appear too low. Indeed, recent studies have suggested that higher intakes would be necessary: 25–50 μg/day to reach a plasma level of 25-hydroxyvitamin D of 30 μg/L (approximately 80 nmol/L). A daily supplementation of at least 10 μg (or 400 UI) has been proved to be necessary for an elderly person even in good health. It is the same for a person with dark or black skin or a person practicing low sun exposure during autumn and winter, unless he consumes plenty of fatty fish.

In Italy, the authorities decided in December 2012 to increase from 10 to 25 μg the daily recommended intake of vitamin D.

A supplementation must also be used to correct an insufficiency or a deficiency of vitamin D. These are two types of intervention that require the use of different amounts of supplementation. First, it is necessary to determine the circulating level of 25-hydroxyvitamin D before deciding on the administration of an adequate amount of vitamin D. The prescribed vitamin D is generally in the form of a drinkable solution, generally taken weekly or monthly, but can also be in the form of a daily food supplement, enclosed in capsules from 25 to 50 μg (1000–2000 IU). The prescription of calcium does not seem essential to reinforce the beneficial effect of a vitamin D supplementation, even in elderly people (Dinizulu et al. 2011).

As a convincing example, the recent experimentation in the scientific base of McMurdo in the Antarctic has shown that a supplementation of 50 μg/day (2000 IU/day) of vitamin D efficiently increased by 40% the serum level of that vitamin after a 6-month winter, without any UV exposure (Zwart et al. 2011). On the other hand, a supplementation of 10 μg/day (400 IU/day) among members of the crew of the International Space Station did not preserve after 6 months the serum level measured before their departure (Smith et al. 2005). The nutritionists and clinicians of Scandinavian countries have well integrated these results as they strongly recommend a vitamin D supplementation for all populations living close to the poles to maintain a suitable circulating level of that vitamin.

It should not be forgotten that people suffering from intestinal malabsorption (biliary diseases, cystic fibrosis, celiac disease, and Crohn's disease) must receive a supplementation in vitamin D. Other causes of malabsorption induced by surgical or pharmacological treatments (Orlistat, Olestra, and sequestering of biliary acids) also require a supplementation in vitamin D to avoid any deficiency state.

It is also necessary to draw the attention of vegans and people consuming voluntarily phytosterol supplements since the recent demonstration of a competitive interaction in the intestine between the absorption of vitamin D and that of phytosterols (Goncalves et al. 2011). Other particular risks may justify a supplementation all year long or with more important amounts: dermatological affections or wearing covering clothing preventing sun exposure, cholestase, nephrotic syndrome, renal insufficiency, obesity, and some pharmacological treatments (rifampicine, phenobarbital, and phenytoine).

In general, investigations on the impact of relatively high amounts of vitamin D are only at their beginning. Several decades ago, nutritionists proposed a daily amount of 10 μg (400 IU). Currently, the tendency consists of prescribing higher amounts (up to 2.5 mg or 100,000 IU/month, e.g., 3,333 IU/day); this approach will certainly evolve as the scientists better understand the metabolism and multiple functions of vitamin D. Now, long-term studies tend to prove that taking up to 250 μg (10,000 IU) of vitamin D per day would perfectly fulfill its physiological role without any detectable side effect. In 2012, considering the effects on calcemia, the EFSA recommended as a tolerable maximum the value of 100 μg/day (4,000 IU/day) for subjects of 11 years and of 50 μg/day (2,000 IU/day) for children from 1 to 10 years.

The state of deficiency is widespread in the countries situated in South Asia and the Middle East, in relation to age, sex, latitude, skin color, and clothing habits (Mithal et al. 2009). A recent study in schoolboys living in Mongolia has shown that the average concentration of their serum vitamin D was 8 μg/L (20 nmol/L) and that 98% of these children have a concentration lower than 20 μg/L (50 nmol/L) (Rich-Edwards et al. 2011). The use of milk enriched with vitamin D has been used but with only little effect.

The developed countries are also not spared from this medical problem as 7% of the children (8–11 years) in Quebec are in a state of deficiency (Mark et al. 2011) and a quarter of the white population (61% in colored people) does not reach the level of 20 μg/L (50 nmol/L), the level recommended in that country.

In the United States, it was reported by A.A. Ginde in 2009 that nearly 77% of the population had a serum level of vitamin D equal to or lower than 30 μg/L (75 nmol/L), a value considered in the country as an insufficiency threshold, and 36% had a level lower than 20 μg/L (50 nmol/L), a value close to the deficiency threshold. More worrying for the American population, its vitamin D status was largely degraded since the 1988–1994 survey. The degradation was probably linked to several factors among which the most important could be the increased incidence of overweight and obesity, a reduction in milk consumption, and the increasing use of sunscreens (Ganji et al. 2012).

Studies carried out in several European countries have shown that a great proportion of the inhabitants are vitamin D deficient (Lips et al. 2006). A third of the Spanish population may be regarded as deficient, half of the population having a serum vitamin D level equal to or lower than 22.5 µg/L (or 56 nmol/L). This situation is definitely more catastrophic in elderly people. In the United Kingdom, approximately 26% of the population between 19 and 24 years are deficient (in that country, the deficiency threshold is officially 10 µg/L, or 25 nmol/L) and 37% of the people of more than 65 years and in those in institution or home for the elderly are also deficient.

In France, the INCA2 study of 2009 has revealed that between 18 and 74 years more than 4% of the population were deficient (plasma concentration < 10 µg/L [or 25 nmol/L]) and nearly 37% of the population have a clear insufficiency (between 10 and 20 µg/L, or between 25 and 50 nmol/L). The study has confirmed an insufficiency state of vitamin D dietary intake as it was only 2.6 µg/day in adults, an amount twice weaker than that recommended by the AFSSA, a value now generally recognized as largely being too weak. The study of the qualitative and quantitative statistical data of fish consumed in France has shown that the average dietary intake in vitamin D is 2.2 µg/day, a value close to the previous one (Bourre and Paquotte 2006). Among fish, the first six suppliers were salmon (0.57 µg/day), tuna (0.24 µg/day), herring (0.23 µg/day), sardine (0.22 µg/day), mackerel (0.12 µg/day), and trout (0.08 µg/day).

In 2012, the INVS published the detailed status of vitamin D in the French adult population (www.invs.sante.fr/content/download/35388/.../beh_16_17_2012.pdf). The average serum concentration of vitamin D was 23 µg/L; 80% of the adults presented an insufficiency (<30 µg/L), 42.5% a moderated to severe deficit (10–20 µg/L), and 4.8% a severe deficit (<10 µg/L). The risk of moderate to severe deficit has been associated with being born out of Europe, not leaving on vacation, having a low level of physical activity, being sedentary, and living in an area of low sunlight. The risk of severe deficit has been associated with being born out of Europe, living alone, and not leaving on vacation but was independent of the level of physical activity and sedentarity.

The INVS emphasized the need for a public action aiming at reinforcing the vitamin D supplies by sun exposure and also by dietary fortification.

3.2.4.2.4   Vitamin D Dietary Sources

In fish, the vitamin D contents are related to its fat content. The foods that are the richest in vitamin D are fatty sea fish (5–20 µg/100 g) and eggs, with smaller amounts of vitamin D being present in milk, butter, and some meats (Table 3.30). In milk and butter, one-third of the vitamin D amount is in the form of ergocalciferol. The highest contents are 2.50 mg/g in cod-liver oil and especially 60 mg/g in halibut-liver oil.

It is thus seen that, except for fatty fish, great amounts (from 200 to 5000 g) of the foodstuffs quoted in Table 3.30 should be consumed each day to cover the recommended intakes in the absence of skin synthesis. Consequently, vegans, who do not consume meat, fish, egg, or dairy products, have marked risks of vitamin D deficiency, even with a moderate sun exposure.

TABLE 3.30
Vitamin D Content of Some of the Most Common Foods
(Micrograms per 100 g)

Foodstuff	Vitamin D	Foodstuff	Vitamin D
Sardine in oil	11	Tuna (canned)	2.3
Salmon	9	Egg	1.6
Mackerel	7.5	Butter	1.3
Oyster	5	Camembert	0.8
Smoked Salmon	3.7	Calf liver	0.3
Trout	2.5	Chicken	0.2

Note: Data are from the CIQUAL 2008 table.

Plants, fungi, and yeast contain only very low amounts of vitamin D_2 originating from the natural irradiation of ergosterol (<1 µg/100 g fresh weight), except someone after being subjected to a UV radiation under well-defined conditions. Thus, a Californian society (Monterey Mushrooms, Inc.) launched on the market cultivated mushroom (*Agaricus bisporus*) with an increased content of vitamin D_2. The mushroom is simply exposed, for a few minutes, to a lamp emitting UV-B radiations to increase its vitamin D_2 content (410 µg/100 g of dry weight). The Dole Company has just marketed that mushroom in the form of a dry powder (Dole® Portobello), allowing the supply of approximately 400 IU of vitamin D per coffee spoon (http://www.dole.com/mushroompowder/#/home). Another dried mushroom of Chinese origin, the shiitake (or black or scented mushroom), can also be used as a source of vitamin D if it has been sun dried. Its content of vitamin D is unfortunately variable and is often badly indicated.

Recently (January 2014), in Europe the EFSA considered that UV-treated baker's yeast exhibiting an enhanced content of vitamin D_2 was safe under the intended use for food supplements and for fortification of yeast-leavened bread, rolls, and fine pastry at maximum concentrations of 5 µg of vitamin D_2 per 100 g of these foods.

In the United States and Canada, legislations have authorized the supplementation of vitamin D in milk, fruit juices, cereals, and margarine. Following the initiatives taken in the year 1930, breast milk substitutes (infantile milks) sold in North America were regularly supplemented so as to contain approximately 400 IU (10 µg) of vitamin D per liter. In Canada, the Ministry of Health decided in 2007 the obligation to add vitamin D to milk (11 µg/L) and margarine (132 µg/kg). The supplementation of other food products (soya milk, orange juice, cheese, and yoghurt) was also allowed. The decision has risen from the absence of sun in the majority of the country, from October to March, and for one period even longer in the northern part of the country.

The British division of Kellogg's announced that it was going to add vitamin D in its cereals for children by the end of 2012. The brands Cornflakes, Special K, and Mini Max and the cereals Choco Pops already contain vitamin D, but the group decided to extend this supplementation to Rice Krispies, Ricicles, Miel Pops, and the other variations of Choco Pops.

It is the same in Finland where a government decree of February 2003 has authorized and recommended the supplementation of dairy products at a level of 5 μg/L and of spreads at a level of 10 μg/100 g. These decisions would have a detectable impact on the vitamin D status in the Finnish population 1 year later (Lanham-New et al. 2011).

In France, an investigation organized by the INSERM from 1988 to 1990 showed that rickets accounted for 0.2% of the hospitalizations of children of less than 2 years. This observation led the health authorities to require in February 1992 manufacturers to add 400–600 IU of vitamin D per liter of infantile milk, in accordance with European standards. Since June 2001, France has also authorized the sale of milk enriched with vitamin D: "The AFSSA gives a favorable opinion to that project as it adopts again the enrichment levels suggested by the CSHPF, e.g., 20% of the recommended daily intake (1 μg) for 100 ml of milk and 25% of the recommended daily intake (1.25 μg) for 100 g of fresh dairy products." One thus finds on the market dairy products enriched in vitamin D such as the milk Lactel (7.5 μg/L), milk Growth of Candia (14 μg/L), and yoghurts Gervais (1.3 μg/100 g) and Yoplait (Calin+, 4 μg/100 g). One also finds on the market a vegetal oil supplemented with 25 μg/100 mL of vitamin D (oil Isio4 by Lesieur).

Within the framework of free sale of supplemented food, the decree of May 9, 2006 published in the *Official Journal of the French Republic* has indicated that the use of vitamin D in the form of cholecalciferol or ergocalciferol is allowed but not intakes beyond a daily intake of 5 μg, taking into account the consumed daily portion of the product recommended by manufacturers on the foodstuff labeling.

It should be stressed that many works have proved that the enrichment of some foodstuffs with vitamin D was a simple and effective way to improve the status of this vitamin in consumers (O'Mahony et al. 2011). It is practically certain that this supplementation has generally only one reduced impact on health as only low doses are authorized. Other investigations will be necessary to ensure a greater bioavailability of vitamin D by establishing precise measurements and by choosing the types of the best adapted foodstuffs and also the foodstuffs most frequently consumed by adults or young children (dairy products, cereals, bread, and fruit juice).

The French Ministry of Economy, Finances and Industry, in its technical specification n° E4-05 of March 31, 2005 concerning vegetal oils, declared that the use of products enriched with vitamin D (milks, fresh dairy products, and some oils) may mitigate the lack of fish consumption (it is necessary to consume fish two times per week), but it does not replace solar exposure (http://www2.economie.gouv.fr/directions_services/daj/marches_publics/oeap/gem/huiles/huiles.pdf).

Apart from medical products (in general containing high doses) whose sale is subjected to medical prescription, several vitamin complements enriched with vitamin D are freely at the disposal of consumers in the form of polyvitamin formulas or better fish oil. Capsules containing 500 mg or 1 g of these oils are sold without a prescription, the vitamin D contents being usually 2.5 μg/g for cod-liver oil and from 10 to 60 μg/g for halibut-liver oil. Recently, it became possible to find a formulation rich in vitamin D_2 (ergocalciferol) generated from UV irradiation of brewers' yeast (*Saccharomyces cerevisiae*). The preparation Oemine D^2 contains 5 μg of vitamin D^2 in each capsule. The company Lallemand Health Ingredients has marketed (July 2011) a product (Lalmin® Vit D) containing 200 μg of vitamin D^2 per gram of dry product. The consumer of these food supplements must be informed that the physiological efficiency of vitamin D_2 is less than one-third of that of vitamin D_3, as shown in 2004 by L.A. Houghton, Canada.

3.2.4.3   Vitamin E

3.2.4.3.1   Discovery

The discovery of Vitamin E goes back to 1922 when in the United States Prof. H.M. Evans discovered the existence of a liposoluble factor (factor X) that improved animal fertility. Later, B. Sour confirmed these first results and called this substance vitamin E (the vitamins A, B, C, and D being already named). H.M. Evans invented the term tocopherol in 1936. In 1938, the structure of α-tocopherol was elucidated, and its synthesis was carried out in the same year by the famous Swiss chemist P. Karrer. Tocotrienols were described only in 1963 by P. Schudel and were synthesized in 1976 by J.W. Scott. Details on the structure of the biologically active molecules (vitamers) are described in Section 2.7.3.

3.2.4.3.2   General Functions

Since 1937, besides being involved in a still quite mysterious way in the mechanisms of animal reproduction, vitamin E is especially regarded as a group of powerful antioxidants protecting lipoproteins and membrane lipids. This last characteristic is probably at the base of the numerous physiological properties described for this vitamin (Section 4.4.3). This antioxidant property is far from being homogeneous as it varies according to the form and also the nature of the oxidizing agent. Thus, in a traditional test (generation of conjugated dienes) α-tocopherol is approximately twice more active than β- and γ-tocopherol, but three times more active than δ-tocopherol. Many studies in various fields of medicine tended to prove the existence of specific actions for α- and γ-tocopherol. These differences are more and more often taken into account in new clinical studies (Section 4.4.3). The tocotrienols are less known but are the object of intensive investigations tending to prove that, besides having antioxidant properties, they are also efficient neuroprotective, anticancer, and especially hypocholesterolemic compounds.

Some cases of vitamin E deficiency, apart from the presence of a lipid malabsorption, have been described in humans in 1981 by U. Burck. This deficiency was not characterized, as in animals, by an anemia but by a peripheral neuropathy.

3.2.4.3.3   Units

The amounts of vitamin E are now expressed in milligrams of α-tocopherol equivalent, and those of the other vitamers are classically expressed after a correction according to a specific coefficient (0.56 for β-tocopherol, 0.16 for γ-tocopherol, 0.5 for δ-tocopherol, and 0.16 for α-tocotrienol). These correction coefficients are very criticizable as they are historically founded on the restoration of fertility in deficient rats. The IU is practically not used anymore; an IU corresponds to 0.67 mg of natural α-tocopherol (RRR-α-tocopherol) or 0.91 of synthetic α-tocopherol (all-rac-α-tocopherol).

3.2.4.3.4   Recommended Intake of Vitamin E

The recommended intakes of vitamin E are rather close from one country to another, but the continuous revisions by the national health organizations tend to gradually increase these values. In 2000, in the United States the Food and Nutrition Board of the IOM proposed to increase the daily intake from 10 to 15 mg in subjects from 14 years to adulthood. The intakes have been set to 4 to 5 mg from 0 to 1 year, 6 mg from 1 to 3 years, 7 mg from 4 to 8 years, and 11 mg from 9 to 13 years. These values were ratified by the Canadian authorities.

In Europe, the scientific panel on dietetic products and nutrition considered in 2013 that the observed mean vitamin E intakes of 3 mg of α-tocopherol equivalent per day from breast milk are adequate for the majority of infants in the first year of life. An intake of 6 mg/day is adequate for the majority of young children (12–36 months).

In France, the recommended daily intakes of vitamin E (α-tocopherol equivalent) vary from 6 to 9 mg for the young child (from 1 to 9 years) and 12 mg from 10 years until adulthood, these values being supposed to take into account the absorption efficiency. After 75 years, the recommended intake is more important (from 20 to 50 mg/day). Obviously, these recommendations do not take into account the individual factors that could increase the antioxidant needs. Among these factors, one can retain a diet rich in unsaturated fatty acids, any disease, physical effort, tobacco addiction, and pollution. It should not be forgotten that people suffering from intestinal malabsorption (biliary disease, cystic fibrosis, celiac disease, and Crohn's disease) must receive a supplementation of vitamin E or, better, must take a vitamin complex containing vitamin E. Other causes of malabsorption induced by pharmacological treatments (Orlistat, Olestra, and sequestering agents as biliary acids) require a vitamin E supplementation to avoid any deficiency.

It should be stressed that all these values are still based more on the prevention of deficiency symptoms than on the prevention of chronic diseases concerning the cardiovascular system, brain, or immunostimulation (Section 4.4.3). The Linus Pauling Institute (Oregon State University) has declared that there existed credible evidence that a supplement from 135 to 270 mg/day of α-tocopherol of a natural source (RRR-α-tocopherol) taken during a meal can help to protect adults against these chronic diseases. We hope that investigators will soon bring additional evidence of the interest of a supplementation in one or the other of various forms of vitamin E (tocopherols and tocotrienols). It should be recalled that the maximum tolerable amount of α-tocopherol has been fixed at 1 g/day by the American medical authorities. The subjects having an anticoagulant treatment or having a vitamin K deficiency should take such supplements only under strict medical supervision. The relationships between vitamin E and blood coagulation still remain poorly known.

The daily dietary intake of vitamin E must be at least 12 mg, the value being able to reach 50 mg in elderly people or subjects suffering from intestinal malabsorption. Any supplement must be avoided in patients treated with anticoagulants. A balanced and diversified diet including fruits and vegetables allows a sufficient intake of vitamin E, and the occasional seed and nut consumption (almonds and hazel nuts) is recommended to allow an optimum intake.

3.2.4.3.5 Vitamin E Food Supplies

The survey carried out in the United States in 2004 by J. Ahuja has shown that the average food supply is 7.8 mg/day in adult men, for example, approximately half the recommended intake. Consequently, 90% of the American population have intakes lower than the appreciated needs (12 mg/J in that country).

On a more global level, several studies also showed that a great proportion of the European population did not receive the recommended intake in vitamin E; however, it is deliberately fixed at a low level.

In France, the great study on the situation and evolution of food supplies of the population (SUVIMAX) revealed that the percentages of men and women having vitamin E intakes lower than the ANC were 41% and 63%, respectively, in 1997, but increased to 65% and 82%, respectively, in 2002. The percentages of men and women whom one can regard as being deprived of vitamin E (contributions lower than two-thirds of the ANC) were 10% and 20%, respectively; they doubled between 1997 and 2002 for the two sexes. The INCA2 study (2006–2007) of the AFSSA revealed that the average food supply was 11.4 mg/day in adults (close to the average intake in the United States). The study of the proportions of consumed foods made it possible to determine that this contribution in vitamin E was ensured for 34% by oils and fats; 15% by fruits and vegetables; and 12% by bread, pastry makings, and cookies.

3.2.4.3.6 Vitamin E Food Sources

One finds in the literature quantities of data more or less reliable on the contents of various components of the vitamin E complex. The published values varied according to technological developments and also the period of collection or the mode of conservation of plants before analyses. The food sources are especially of vegetable origin (portions of plants and oils) and incidentally of animal origin.

The products of vegetable origin are the main sources of vitamin E for humans, the richest elements being oils resulting from seeds (Table 3.31) and cereals (Table 3.32). The distribution of various compounds of vitamin E varies according to the source: wheat, corn, soya, and groundnut contain especially tocopherols, whereas palm oil, rice, oats, and barley contain especially tocotrienols. The content of vitamin E equivalents is very weak in fruits but elevated in seeds (Table 3.33) and that of vegetables varies according to the consumed part, while being higher in the green parts with maturity (Table 3.34). A table of the α-tocopherol contents of the principal nutrients available on the U.S. market was published by the Ministry for Agriculture of the United States (USDA) (http://www.nal.usda.gov/fnic/foodcomp/Data/SR21/nutrlist/sr21a323.pdf).

Values can also be obtained from the CIQUAL composition tables of food of the ANSES.

The animal products contain especially α-tocopherol, but in small quantities (between 1 and 7 mg/kg) (Table 3.35). It is the same for cow's milk, but its content of vitamin E depends much on the diet of the animal (about 1.6 mg of vitamin E per kilogram of semiskimmed milk) and butter contains approximately 24 mg/kg. Fish and mollusks contain moderate quantities of vitamin E (2–15 mg/kg).

After this review, one realizes that the recommended intake in France (12 mg/ day) may be partially covered by the consumption in moderate quantities of the vegetables

TABLE 3.31
Contents of Tocopherols and Tocotrienols of Some Oils

	α-T	γ-T	δ-T	α-T3	γ-T3	δ-T3
Palm (mg/L)	198	—	11	210	408	87
Sunflower (mg/L)	765	—	—	—	—	—
Corn (mg/L)	260	1360	88	—	—	—
Wheat germ (mg/L)	2,570	257	257	—	—	—
Almond (mg/L)	257	18	—	—	—	—
Hazelnut (mg/L)	425	68	17	—	—	—
Walnut (mg/L)	12	517	61	—	—	—
Olive (mg/L)	80	9	3	—	—	—
Groundnut (mg/L)	43	55	19	—	—	—
Rapeseed (mg/L)	29	49	—	—	—	—
Soybean (mg/L)	150	940	360	—	—	—
Macadamia (mg/L)	—	—	—	29	17	15
Cocoa butter (mg/L)	14	225	37	9	—	—

Source: From Alasalvar, C., and E. Pelvan, *Eur. J. Lipid Sci. Technol.*, 113, 943–9, 2011; Cuvelier et al., *Ann. Med. Vet.*, 147, 315–24, 2003.

Note: T, tocopherols; T3, tocotrienols.

TABLE 3.32
Contents of Tocopherols and Tocotrienols of Some Cereals

	α-T	γ-T	δ-T	α-T3	γ-T3	δ-T3
Wheat (mg/kg)	9	8	0.3	4	42	—
Barley (mg/kg)	6	4	0.3	19,5	17	—
Oat (mg/kg)	3,4	2	0.4	8	2	—
Corn (mg/kg)	14	46	—	—	—	—
Soya (mg/kg)	4	19.5	5	0.3	0.8	—

Source: From Cuvelier et al., *Ann. Med. Vet.*, 147, 315–24, 2003.

Note: T, tocopherols; T3, tocotrienols.

richest in vitamin E (spinach, cabbage, sweet pepper, tomato, and pumpkin) and of meats or seafood. A complement in oil, if possible corn oil (seasoning), and in seeds (almond, hazel nut, and groundnut), all in small amounts, makes it possible to reach values close to those recommended. For the intake of therapeutic or even preventive amounts of vitamin E (up to 200 mg), the vegetable sources (mainly oils and seeds) are not appropriate anymore as too large quantities of such sources should be ingested. It then becomes necessary to use concentrated vitamin supplements prepared either from a chemical synthesis or from by-products of oil refining (palm oil). The bioavailability of natural vitamers are not well known but is likely about two times more important than that of the products of synthesis.

TABLE 3.33

Contents in Tocopherol Equivalent of Some Fruits and Seeds

(Milligrams per 100 g)

Apple	0.490	Blueberry	2.1
Apricot	0.500	Avocado	1.3
Peach	0.965	Cashew nut	5.7
Plum	0.862	Groundnut	11
Grape	0.666	Hazelnut	26
Blackberry	0.720	Almond	26
Black currant	1.900	Walnut	6
Chestnut	100		

TABLE 3.34

Contents in Tocopherol Equivalent of Some Vegetables

(Milligrams per 100 g)

Potato	0.053	Lettuce	0.60
Carrot	0.465	Spinach	1.40
Celeriac	0.539	Green pea	0.13
Cauliflower	0.089	Pumpkin	1.10
Broccoli	0.621	Tomato	0.81
Cabbage	1.700	Sweet pepper	2.60

Source: From Souci et al., *Food Composition and Nutrition Tables*, Medpharm, Stuttgart, Germany, 2000.

TABLE 3.35

Contents of α-Tocopherol in Some Common Animal Products

(Milligrams per 100 g)

Beef (muscle)	0.40	Cod	0.23
Calf liver	0.33	Herring	1.07
Pork (muscle)	0.08	Mackerel	1.52
Lamb (muscle)	0.62	Mussel	0.74
Hen's egg	0.70	Oyster	0.85
Chicken (muscle)	0.29		

Source: From McLaughlin, P.J., and J.L. Weihrauch, *J. Am. Diet. Assoc.*, 75, 647–65, 1979.

3.2.4.3.7 Legislation

Vitamin E is often used by the pharmaceutical industry in the preparation of additional food supplements and by food industry in the preparation of products prone to oxidation before consumption. A specialty rich in tocopherols and tocotrienols (Tocomin®, Carotech, Inc.), purified from crude palm oil, recently obtained in April

2010 a GRAS approval from the FDA of the United States for drink and food supplementation. This product is, in addition, rich in other micronutrients such as squalene, phytosterols, coenzyme Q10, and carotenoids.

French legislation recognizes four types of tocopherols whose European denominations are from E 306 to E 309. The E 306 (tocopherol-rich extract of natural origin) was allowed as an additive in margarines and dietary fats at a maximum concentration of 500 mg/kg of enriched product. E 307 (synthetic α-tocopherol) was allowed for the same uses and with the same amounts, and similarly for E 308 (synthetic γ-tocopherol) and E 309 (synthetic δ-tocopherol). However, the last two are not frequently used.

The decree of May 9, 2006 published in the *Official Journal of the French Republic* indicated that the use of vitamin E in the form of D-α-tocopherol, DL-α-tocopherol (racemic mixture), D-α-tocopherol acetate, DL-α-tocopherol acetate, or D-α-tocopherol acid succinate was allowed but should not go beyond the daily amount of 30 mg of α-tocopherol equivalent, taking into account the daily portion of product recommended by the manufacturer such as it is indicated on the label.

Many preparations with a composition close to that prepared from palm oil are proposed in the market, with claims of having antioxidant, hypocholesterolemic, and antithrombotic effects.

3.2.4.4 Vitamin K

3.2.4.4.1 Discovery

Vitamin K was discovered in 1929 by H. Dam, Denmark, by observing the appearance of hemorrhages in chickens fed a delipidated food. In 1939, a group under E. Doisy, the United States, isolated the vitamin and defined its structure. For these discoveries, C. Prejudice and E. Doisy were awarded the Nobel Prize in 1943.

3.2.4.4.2 General Functions

Historically, vitamin K has an antihemorrhagic function, which rises from its implication in the biosynthesis of prothrombin (factor II) and of three other blood coagulation factors (factors VII, IX, and X). It also improves the uptake of calcium on the protein matrix of bone via a specific protein, osteocalcine, and the formation of cellular growth factors (Section 4.4.4). It is now well proved that this vitamin intervenes as a cofactor in a metabolic cycle including mainly a specific carboxylase of glutamic acid.

Vitamin K is a set of related molecules of varied origin but with similar activities. One distinguishes the vitamin K_1, or phylloquinone (Figure 2.36), found in plants and the vitamin K_2 (Figure 2.37) formed by a set of similar molecules (vitamers) called menaquinones, of bacterial origin (Section 2.7.4). It is generally recognized that in humans vitamin K_1 is the main actor in charge of the natural antihemorrhagic activity. The menaquinones play a role that is still little known but likely in the field of maintaining the quality of skeletal bones and in the calcification of arterial walls (Section 4.4.4.2). This ignorance undoubtedly comes from the small amounts present in foodstuffs (a maximum of 10%–20% of the total vitamin K complex) and the uncertainty on the importance of intestinal flora contribution. It should be stressed that the organism may also synthesize vitamin K starting from menadione (vitamin K_3), a synthetic precursor, by the addition of an isoprenoid side chain. This water-soluble precursor is used in humans only in certain countries but is especially used in the manufacture of animal foods.

The employed unit for vitamin K is microgram.

3.2.4.4.3 Vitamin K Food Supplies

The vitamin K supplies were badly defined as the chemical analyses were difficult and inaccurate until the last years. Furthermore, the estimation of menaquinones in foods and bacteria remains poorly appreciated. In the United States, the average intake of vitamin K has been estimated to be 90 μg/day for women and 120 μg/day for men. These values were adopted by the IOM for the recommended intake of vitamin K. This intake has been estimated to be 230 μg/day in Japan and 250 μg/day in the Netherlands and the north of China. Lower values were measured in Finland (120 μg/day), England (103 μg/day), and Scotland (70 μg/day) (Booth 2012). No French study (including the SUVIMAX study) was carried out in this field.

As the FAO did in 2001, the majority of the European food agencies recommended an intake of about 1 μg of vitamin K by kilogram of body weight per day. The American or Japanese nutritional recommendations are a little higher, about 1.5–1.7 μg/kg. These figures resulted in appreciating the needs at 110–130 μg for vitamin K per day for an individual of 75 kg. In France, the medical authorities considered that the requirements of vitamin K are little documented and they have fixed the recommended daily intake to 45 μg in men, 35 μg in women, and from 10 to 35 μg according to age in children (Martin 2001). In Europe, the scientific panel on dietetic products and nutrition considered in 2013 that for the majority of infants a vitamin K intake of 5 μg/day is adequate in the first half-year of life, 8.5 μg from 6 to 12 months, and 12 μg up to 36 months.

These recommendations were defined by taking into account only the optimal synthesis of the coagulation factors by the liver. Physiologists have determined that the plasma level of vitamin K_1 did not increase anymore when the daily intake was higher than 200 μg. On the other hand, it was suggested that a higher intake (up to 350 μg of vitamin K_1) was necessary to satisfy the needs at the bone level. Furthermore, it appeared that a specific intake of menaquinones was beneficial for the maintenance of the bone status in postmenopausal women (Section 4.4.4.2). For the moment, in the absence of work contradicting these results, no qualified organization has recommended such high intakes, but this position is certainly prone to modification in the near future. It would be indeed important to make consumption studies of vitamin K (vitamins K_1 and K_2) in various populations and to follow their evolution. Indeed, a decrease of more than 38% of vitamin K intake was shown in young children between 1950 and 1990. This fall could be due to a change in food habits characterized by a preference to meat and vegetal oils instead of crop products, the main sources of vitamin K supply.

People suffering from intestinal malabsorption (biliary diseases, cystic fibrosis, celiac disease, and Crohn's disease) must have a beneficial supplementation of vitamin K or, better, a regular dose of a vitamin complex containing vitamin K.

Other causes of malabsorption induced by pharmacological treatments (cephalosporines, certain anticonvulsivants, Orlistat, Olestra, and bile acid sequestrants) require a supplementation of vitamin K to avoid any deficiency. Conversely, the high concentration of vitamin K in some plants (Brassicaceae: cabbage, cauliflower, Brussels sprout, and broccoli) led to their removal from the diet of patients treated by an antivitamin K (warfarin) used to fight against venous thromboses. A limited consumption of rapeseed and soybean oils may also be recommended. It was indeed proved that a dietetic monitoring of vegetal sources rich in vitamin K allowed a better adjustment of anticoagulant treatments on the long term (De Assis et al. 2009).

The recommended daily intake of vitamin K is badly defined but is appreciated in France at 35–45 µg. This supply is normally ensured by a balanced diet rich in green vegetables, primarily cabbages, broccoli, spinaches, and salads. If necessary, it is possible to compensate an insufficiency by a supplementation in vitamin K with a maximum of 25 µg/day. Obviously, a medical opinion is recommended in people suffering from intestinal malabsorption. People under anticoagulant treatment must avoid consuming green vegetables rich in vitamin K.

3.2.4.4.4 Vitamin K Food Sources

Vitamin K is largely present in green plants (cabbages, salads, and spinaches) at a level between 100 and 1000 µg/100 g (Table 3.36). Other vegetables and fruits contain much less (<10 µg/100 g), the majority containing only traces (<1 µg/100 g).

TABLE 3.36

Contents of Vitamin K_1 in Some Common Foodstuffs

<1 µg/100 g	1–10 µg/100 g	10–100 µg/100 g	100–1000 µg/100 g
Avocado	Eggplant	Button mushroom	Broccoli
Banana	Butter	Cauliflower	Brussels sprout
Mushroom	Carrot	Red cabbage	Kale
Wheat flour	Zucchini	Sauerkraut	Water cress
Milk	Date	Cucumber	Spinach
Corn	Veal liver	Green pea	Lettuce
Watermelon	Cheese	Kiwi	Parsley
Egg	Whole-grain bread	Garden pea	
Orange	Pear	Plum	
White bread	Leek		
Grapefruit	Apple		
Fish	Grape		
Potato	Tomato		
Rice			
Meat			

Source: From Shearer et al., *J. Nutr.*, 12c6, 1181S–6S, 1996.

Vegetable oils are a considerable source of vitamin K_1, the richest being soybean oil and rapeseed oil (Table 3.37). It should be specified that hydrogenated oils, leading to the manufacture of margarines, contain only one inactive derivative of vitamin K_1, that is, 2′,3′-dihydrophylloquinone.

Fish and meats practically do not contain vitamin K_1, except the liver of ruminants, which may contain up to approximately 7 µg/100 g vitamin K. The vitamin K content of many foodstuffs present on the U.S. market may be consulted on the database established by the USDA (http://ndb.nal.usda.gov/ndb/foods/list). Some values are given for oils and milk and its derivatives in the CIQUAL table published by the ANSES. As yet, there exist only few reliable results for the vitamin K_2 content in foodstuffs. It is only known that menaquinone-4 is especially present in cheeses (up to 80 µg/100 g), eggs (25 µg/100 g), meats (1–30 µg/100 g), and ruminant liver (10–20 µg/100 g), other menaquinones with longer carbon chains being present in cheeses and other milk products (up to 80 µg/100 g). Menaquinone-7 is concentrated (about 900 µg/100 g) in natto, a fermented soybean product in Japan. Menaquinones present in derived products of milk, meat, and eggs originated probably from the metabolism of menadione used in the manufacture of animal feed.

The bioavailability of vitamin K_1 in foods is little known in humans, but it appears to be low when the source is a green plant where it is strongly linked to the chloroplast membranes. Experimentation in humans has shown that its bioavailability is about six times lower than that of the free form (Garber et al. 1999). As with other liposoluble vitamins, its absorption is facilitated by baking and is highly dependent on the presence of fats and oils. These considerations are taken into account for vitamin K from plant sources. Unlike menaquinones supplied by some foods, those present in intestinal flora seemed to have a low bioavailability and therefore would be a slow and low intake of vitamin K_2 for the organisms (Conly et al. 1994).

3.2.4.4.5 Legislation

In the United Kingdom, a vitamin K supplement is admitted at a daily dose of 45 µg/ day (in the form of vitamin K_1 or K_2).

In the decree of May 9, 2006, the *Official Journal of the French Republic* stated that "it is possible to use vitamin K as phylloquinone in the manufacture of food supplements, but this addition must not result in a dose exceeding the daily dose of 25 µg given the daily ration of the product as recommended by the manufacturer and indicated on the label."

TABLE 3.37
Vitamin K Content (µg/100 g) of Some Vegetal Oils

Groundnut	0.65	Walnut	15
Corn	2.9	Olive	55
Almond	6.7	Rapeseed	141
Sunflower	9	Soybean	193

Source: From Ferland, G., and J.A. Sadowski, *J. Agric. Food Chem.*, 40, 1869–73, 1992.

On July 3, 2009, the AFSSA considered that the existence of many subjects following anticoagulant therapy with vitamin K antagonist (VKA) justified extreme caution regarding the enrichment of common foods with vitamin K. The AFSSA therefore rejected the possibility of enrichment of common foods because it posed a difficulty to control risk in individuals under VKA. Moreover, the AFSSA maintained the maximum dose of 25 µg/day for dietary supplements, a dose appearing not to cause risk to the population under VKA, and maintained the rejection of the use of higher doses, even with proper labeling. However, no research could show toxic risks with elevated intakes of vitamin K. The vitamin K_2 extracted from natto, used in the Japanese cuisine, is now available in France and Belgium as a dietary supplement. The company J-Oil Mills (Chuo-ku, Japan) obtained in October 2010 the EFSA agreement for marketing in Europe the vitamin K_2 (menaquinone-7) produced from natto, as a dietary supplement [Novel Foods Regulation (258/97/EC)]. Other suppliers have since obtained a similar license (BIORESCO, Basel, Switzerland). The society NattoPharma ASA, Oslo, Norway, provides a synthetic menaquinone-7 (MenaQ7) in the powder form or as an oil. According to suppliers, this dietary supplement is available in capsules containing up to 100 µg of vitamin K_2. Its use should be considered after checking detailed information about potential risks. The company Nutrogenics supplies capsules containing 45 µg of menaquinone-7 (isolated from natto) mixed with n-3 fatty acids, vitamin D_3, and coenzyme Q10.

3.2.5 PHOSPHOLIPIDS

Phospholipids are present in our foods, whether of vegetable or animal origin. It is possible to estimate our average dietary phospholipids to about 2 g/day. Except for a few sources, they form only a small fraction of ingested lipids. Indeed, the richest animal sources are egg yolk (11%), sheep brain (5.4%), liver (2.2%), calf kidney (1.4%), and beef meat (1%). In plants, soybean (1.5%–3%) and groundnut (1.1%) are the richest in phospholipids; they are prepared industrially as a mixture called "lecithin." In addition to phospholipids, unrefined lecithin also contains glycolipids and triacylglycerols. Phospholipids, almost exclusively in the form of soybean lecithin, are used as a food additive (E322) in a large quantity of manufactured products to modify their texture. Lecithin is incorporated into margarines, chocolate, confectionery, instant powders (cocoa and coffee), pastries, cheeses, and many other products. Legally, the label must always mention its presence.

Among the phospholipids, phosphatidylcholine, phosphatidylserine, and sphingomyelin were the subject of numerous studies, not for their intrinsic value but for the food intake of choline or serine, primarily components involved in the normal function of the nervous system. Sphingomyelin is a source of both choline and ceramide, as glycolipids.

Synthetic lecithins are banned in France, but they are allowed in other countries like the United States and Great Britain.

3.2.5.1 Phosphatidylcholine

Phosphatidylcholine (Section 2.5 and Figure 2.20) represents only a small proportion of dietary lipids. No evidence has been provided on the necessity of this lipid for growth and maintenance of a healthy body. The situation is probably different for pathological conditions, but the literature is sparse on this topic. This phospholipid forms, respectively, over 35% and 54% of phospholipids of muscle and beef liver, and more in egg yolk (66%). Phosphatidylcholine is mainly a dietary source of choline. The knowledge in this area results mainly from nutritional investigations with soybean lecithin. Moreover, this product is also a source of vitamins A and E and promotes intestinal absorption of all fat-soluble vitamins. Marketed lecithin may also be prepared from rapeseed and also from chicken eggs.

Phosphatidylcholine is a good supplier of unsaturated fatty acids; it seemed to be even better than triacylglycerols (Cansell 2010). It is possible to find on the market food supplements rich in lecithin from marine crustaceans (krill) presented as providing an excellent source of DHA and EPA (Oemine Phytobiolab of NKO Neptune Krill Oil). This product is within the scope of an n-3 fatty acid supplementation but not choline (Section 3.2.1.3.2). Vegetable lecithin also contains linolenic acid (18:3 n-3) but in a small proportion (6% soybean lecithin and 11% for rapeseed). Without any supplementation, dietary intake of fatty acids from phospholipids is negligible compared to that resulting from the digestion of triacylglycerols (oils and fats) found in abundance in the diet.

Interest on phosphatidylcholine, and to a lesser degree sphingomyelin, was renewed after many studies showing that choline must be considered an essential substance for the human body, especially when there is an inadequate intake of methionine and folate (vitamin B9). Moreover, it appears that despite the existence of an endogenous biosynthesis of choline dietary intake is very often necessary. Via S-adenosylmethionine, choline participates in the biosynthesis of membrane phospholipids and acetylcholine and in the transfer of methyl groups for many methylation reactions involving DNA and proteins. It is within this framework that a biochemical phosphatidylcholine supplementation has been proposed to primarily improve cognitive performance in deficient individuals (Section 4.5.1.2). Several studies have shown that phosphatidylcholine is a better source of choline than choline chloride, the water-soluble form generally used as a food supplement. Choline chloride is often regarded as a water-soluble vitamin B complex (sometimes called B4 and B7). Choline is becoming a key factor in the normal development and health of animals. In 1998, the IOM officially pronounced itself in favor of its essential character and recommended a daily consumption. In 2001, the FDA asked manufacturers to inform consumers about products that could be considered "excellent sources of choline" if they contain not more than 110 mg per serving, or "good sources of choline" containing between 55 and 110 mg per serving.

3.2.5.1.1   Recommended Dietary Choline Intake

Although uncertainty exists about the needs of humans for choline, the authorities have set not an RDA but an adequate supply. The value of this intake, based on mean intakes of healthy adults, was estimated at 550 mg of choline per day, equivalent to about 4 g of phosphatidylcholine. The daily maximum allowable value was 1 g of choline for children and 3.5 g for adults. The FDA has also authorized a choline supplementation of processed foods for children.

As the actual total dietary choline for the French population was estimated at 220 mg, one must conclude from the American standards that this amount is inadequate. This status could justify the advice of a dietary supplementation rich in lecithin. It is unfortunate that a lack of reliable studies on the levels of consumption and absorption of lecithin and choline and its derivatives precludes any nutritional recommendation for the French population in this area. Determining a too low choline intake through dietary surveys should lead dieticians to advise an increase in the consumption of choline-rich foods (eggs and veal liver) or a supplementation with lecithin (1–3 g/day). Supplementation with products containing choline seems to be advisable in patients who seek to limit their dietary cholesterol level through a reduction in their consumption of meat and eggs, the main sources of phosphatidylcholine (Ishinaga et al. 2005).

3.2.5.1.2 Dietary Phosphatidylcholine Intake

It is very difficult to quantify the average dietary intake of phospholipids as it varies according to the food habits of the consumer. Despite these reservations, an estimation may be made by taking into account the data of the CNAM 1997–2003 study on the average dietary intakes of French and the values published on the phospholipid compositions of some foodstuffs. Thus, it may be calculated that dairy products provided about 250 mg of phospholipids per day (80 mg of phosphatidylcholine and 80 mg of sphingomyelin), whereas fish and meat provided 800 mg of phospholipids per day (280 mg phosphatidylcholine and 60 mg of sphingomyelin). The contribution of plants is less clear, given the small number of analyzes and the wide variety of diets. Using CNAM data on the consumption of French people and some published analyses, we can estimate that plants contributed on average 310 mg of phospholipids per day per capita, including 150 mg of phosphatidylcholine per day per capita. The total dietary intake (vegetable, meat, and dairy) of phospholipids may be estimated at 1360 mg/day per capita, including 650 mg of choline phospholipids (phosphatidylcholine and sphingomyelin).

3.2.5.1.3 Choline Intake

The previous data allowed the estimation that phospholipids contributed to a dietary choline intake of about 90 mg/day per capita. It should be noted that the consumed foodstuffs contain, besides the choline included in the phosphatidylcholine and sphingomyelin, water-soluble metabolites such as free choline, phosphorylcholine, and glycerophosphocholine. Data on the food consumption of French people (INCA2 study, 2006–2007) and the total choline contents for various foods published in 2003 by S.H. Zeisel allowed to assess the total daily intake of choline to about 220 mg per adult. This value is much lower than that calculated for American people (537 mg for adult men and women) but similar to that determined for the Japanese (214 mg) (Ishinaga et al. 2005). Yet, nothing can explain this difference: is it the nature of the food, diet composition, or accuracy of the surveys? Only more detailed studies could clarify this aspect of choline intake in all its forms (phospatidylcholine, sphingomyelin, free choline, and phosphocholine).

The choline intake in infants is seemingly simpler as the diet begins for half of them with breast milk, 42% consuming that milk even after 8 weeks, at least in France. For other infants, the first milk is replaced by infant formulas adapted from cow's milk. In both cases, a daily intake of about 500 mL of milk provided the infant approximately

20 mg of choline, 80% being in the form of lipids in the breast milk and 70% with the cow's milk (Zeisel et al. 1986). Thus, on the basis of body weight, choline intake is about two times greater in infants than in adults. In all cases, the efficiency of intestinal choline absorption in its different molecular forms remains unknown.

3.2.5.1.4   Phosphatidylcholine Food Sources

Besides the natural egg representing a phospholipid concentrate (approximately 1.7 g per egg), phosphatidylcholine is present in a large number of foods, particularly meats, fish, seeds (soybean, peanut, and wheat), breast milk, and cow's milk (Table 3.38). Fruits and vegetables have relatively very low levels of phospholipids (Table 3.38). It is evident that the fatty acid composition of this phospholipid varies greatly depending on its origin.

The data in these tables show that, except eggs, fatty fish, and some meats, traditional foods are generally low in phosphatidylcholine, a major source of choline (14% of the phospholipid molecule) for the body. The richest sources of total choline are natural calf liver, which supplies 246 mg/100 g; hen's egg, 180 mg in an egg; and meats, from 60 to 110 mg/100 g.

Apart from natural foods, the phospholipids are prepared from cheap products and are rich in fats: soybean, sunflower, and rapeseed, and also chicken eggs and, more recently, krill oil (crustacean from Antarctic seas). Supplements of phosphatidylcholine prepared from these products are commercially available in the form of capsules, tablets, or granules.

TABLE 3.38

Average Levels of Phosphatidylcholine in Common Foods (Milligrams per 100 g Fresh Weight)

Animal Product		Vegetal Product	
Hen's egg	1,870	Broccoli	150
Calf liver	1,760	Spinach	130
Beef meat	440	Avocado	16
Pork meat	500	Apple	22
Calf meat	250	Peach	26.5
Chicken	310	Grape	1.5
Salmon	340	Raspberry	30
Herring	1,400	Potato	20
Tuna	640	Green pea	170
Shrimp	410	Oatmeal	144
Cow's milk	4.3	Groundnut	230
Breast milk	6	Soybean	670
Cheese	52	Wheat seed	470
Yoghurt	7.4		
Krill oil	29,400		

Note: Recalculated data are mainly from Zeisel et al., *J. Nutr.*, 133, 1302–7, 2003.

The product most commonly marketed under the name of lecithin is soybean powder made after solvent extraction, which contains 25%–30% of phosphatidylcholine (16% in the liquid form), the amount reaching 40% for rapeseed. Of course, other phospholipids are present, among them phosphatidylethanolamine and its derivatives (27%), phosphatidylinositol (17%), and phosphatidic acid (6%) (Malton 2002). Soybean lecithin may contain up to about 4% of choline, the lecithin prepared from chicken egg containing about 80% of phosphatidylcholine, or about 6% of choline.

The production of soybean lecithin is generally parallel to that of soybean oil itself. According to the U.S. Census Bureau Service, the U.S. production of soybean lecithin was 74,000 t in 2009. In the absence of other published data from other countries, we can estimate the world production to be about 150,000 t. Soybean is gradually replaced by rapeseed or sunflower as industrial sources of lecithin. This change is likely motivated by the reduced risk of allergy related to soybean and the elimination of all consumption of genetically modified organism products.

The production of egg lecithin is more limited and poorly quantified. Krill oil is rich in phospholipids (about 32%), consisting almost entirely of phosphatidylcholine (29%). Current scarcity and price make it a dietary supplement more of unsaturated *n*-3 fatty acids than of choline. Indeed, the phospholipids of that oil contain 15% of DHA and 35% of EPA.

Lecithin (E322), mostly extracted from soybean, sunflower, and to a much lesser degree hen's egg, is used as an emulsifier and stabilizer in the food industry at a maximum level of 1% of the final preparation (chocolate, sauces, pastry, pastries, cheese, margarine, and food for children) (Chanussot 2008). In addition to its physical action during manufacturing, this product saves eggs and fat in making pastries and cakes. Apart from this main use, lecithin is given as a dietary supplement to restore or enhance certain physiological functions (Section 4.5.1). Lecithins are the subject of many other nonfood uses in various industrial sectors (pharmaceuticals, paints and coatings, and plastics).

3.2.5.1.5   Legislation

Since 1995, at the European level soybean lecithin is authorized as a food additive (E322) (directive 95/2/EC); egg lecithin (purity above 85%) was authorized in 2000 (decision 2000/195/CE) and krill oil (Neptune krill oil) in 2009 (EFSA-Q- 2008-027).

The French company ASL (Health Applications of Fat) was authorized by the AFSSA to put on the market egg yolk phospholipids naturally rich in DHA. The maximum permitted amounts of lecithins are 30 g/L in oils and fats of animal or vegetable origin, 1 g/L in food preparations for infants and toddlers, and 10 g/kg for products based on cereals (cookies, crackers, etc.); no limit was specified for other foods.

Synthetic lecithins are banned in France but are allowed in the United States and Great Britain; they are mainly used in the manufacture of chocolate.

3.2.5.2   Sphingomyelin

In the majority of tissues and foods, sphingomyelin is 10 times less abundant than phosphatidylcholine. This lipid (Section 2.4 and Figure 2.21), present almost exclusively in foods of animal origin, is hydrolyzed in the intestine, giving rise to choline and ceramide; the latter is optionally converted into a fatty acid and a long-chain amine (sphingosine like). In contrast to phosphatidylcholine, sphingomyelin contains only a long-chain fatty acid that may be saturated and sometimes hydroxylated.

Sphingomyelin is a major component of the nervous system, especially the nerves, but to a lesser degree all animal products commonly consumed (meat and dairy) contain this complex lipid (Table 3.39). This sphingolipid is considered mainly as a source of ceramide; it may thus be regarded as a bioactive lipid (Section 4.6). The ceramide intake will be discussed with the glycolipids as they have a common structure (Section 3.2.6).

3.2.5.3 Phosphatidylserine

In mammals, phosphatidylserine (Section 2.4 and Figure 2.22) is the only phospholipid containing an amino acid, serine. Except in the brain and erythrocytes, this phospholipid represents no more than 8% of the total tissue phospholipids. Its biosynthesis is largely provided in all tissues by an exchange between serine and choline in a phosphatidylcholine molecule or by reaction of serine with CDP-diacylglycerol. It is therefore not surprising that no investigation has been made on the importance of its dietary intake.

3.2.5.3.1 Phosphatidylserine Dietary Sources

Given the biological importance of this phospholipid, it has nevertheless been established that the dietary intake of phosphatidylserine in Western countries is around 130 mg/day. All foodstuffs contain this phospholipid but to varying levels (Table 3.40). The brain contains a higher amount, about 10 times more than red

TABLE 3.39
Average Contents of Sphingomyelin and Ceramide Equivalents of Some Animal Products

Animal Product	Sphingomyelin (mg/100 g Fresh Weight)	Ceramide Equivalents (mg/100 g Fresh Weight)
Hen's egg	120	94
Calf liver	190	148
Beef meat	37	29
Pork	46	36
Chicken	80	62
Salmon	13	10
Herring	115	90
Tuna	150	117
Cow's milk	9	7
Breast milk	8.3	6.5
Cheese	32	25
Yoghurt	7.7	6
Butter	29	22.6

Note: Recalculated mainly from Zeisel et al., *J. Nutr.,* 133, 1302–7, 2003 and Jensen, R.G., *Prog. Lipid Res.,* 35, 53–92, 1996.

TABLE 3.40

Phosphatidylserine Content of Some Foods

Animal Product	mg/100 g Fresh Weight	Vegetal Product	mg/100 g Fresh Weight
Brain	710	White bean	110
Mackerel	480	Soybean lecithin	10–40
Herring	360	Carrot	2
Tuna	190	Potato	1
Chicken fillet	85–130		
Beef meat	70		
Pork meat	60		
Trout	14		

Source: From Souci et al., *Food Composition and Nutrition Tables*, Medpharm, Stuttgart, Germany, 2000.

meat; fish is also a good source. Except white beans and soybean lecithin, vegetables are very poor phosphatidylserine suppliers.

For experimental studies in animals or humans, nutritionists used powdered lipid concentrates obtained from soybean lecithin and containing up to 40% phosphatidylserine. Almost no product on the market is derived from bovine brain, as a result of the risk of encephalopathy associated with that animal. Despite the lack of scientific results, the doses proposed for consumption by retailers of dietary supplements are ranging from 100 to 300 mg of phosphatidylserine per day.

Supplementation of phosphatidylserine was proposed to improve cognitive performance, primarily in mentally retarded subjects (Section 4.5.2.1).

The European Commission has given authorization to put on the market an extract of soybean phospholipids containing 73% of phosphatidylserine (Enzymotec, Ltd.) (decision 2011/513/EU).

3.2.6 GLYCOLIPIDS

Glycolipids, limited here to glycosphingolipids (ceramide glycolipids) (Section 2.5), do not represent an essential lipid fraction in foodstuffs. Although a wide range of compositions is observed in animal foods, most glycosphingolipids present in plants are glucosylcerebrosides. The majority of plant glycolipids are not glycosylated ceramides but glycosylated diacylglycerols.

There is no evidence that these lipids are essential for growth and health of the human body. It may be different in some pathological conditions, but the literature is quite limited on this topic (Section 4.6). Similar to sphingomyelin, all glycosphingolipids may be considered to be dietary sources of ceramides, which give after digestion a fatty acid and a sphingosine base, phosphorylated or not. These bases are well known to be bioactive lipids.

Research carried out in rats have shown that sphingolipids (sphingomyelin and glucosylceramide) undergo very gradual degradation in the intestine and colon, but only 75% are digested. The hydrolysis products (ceramide, sphingosine, fatty acid, and phosphorylcholine) are absorbed through the intestinal epithelium and are mostly converted into various sphingolipids, phospholipids, and triacylglycerols; some remain in the intestine before evacuation. Some studies have suggested that in humans sphingolipid catabolism is not different from that described in the rat. It is unfortunate that few studies are devoted to these mechanisms, whereas many animal experiments have shown a clear influence of ingested sphingolipids on the hydrolysis of triacylglycerols, cholesterol absorption (Section 4.6.4), and the growth and health of the intestine itself (Nilsson and Duan 2006).

3.2.6.1 Glycolipids Dietary Intake

It is difficult to quantify the average dietary intake of ceramide from glycolipids (and sphingomyelin) because little reliable quantitative data exist for animals and plants commonly ingested by consumers. It should be noted that the great diversity of this group of polar lipids does not simplify the biochemist's task. However, taking into account the overall concentrations of sphingolipids in main consumed foodstuffs (Vesper et al. 1999) and the dietary intakes of the French people determined by the CNAM 1997–2003 study, it is possible to estimate the intake of ceramide to about 130 mg/day per capita. Animal products contribute to 42%, 24% as eggs and dairy products. Nearly 30% of the total comes from fruits and vegetables and 28% from cereals (bread and pasta).

The situation in infants is special as they are under the age of 4 months and exclusively fed with breast milk or formula. In both cases, a daily intake of about 500 mL of milk provides about 45 mg of ceramide to the infant (Jensen 1996). Thus, on the basis of body weight this intake is nearly eight times higher in infants than in adults. The nutritional status has been linked to specific properties of the milk diet for brain development and prophylactic activities against bacterial toxins and cancer (Section 4.6.3). The same daily intake was estimated among U.S. adults of about 320 mg of ceramide (Vesper et al. 1999).

It is obvious that the actual absorption of sphingolipids, by both the French and the Americans, is largely dependent on the type of food consumed, large consumers of eggs, dairy, and cereals having an intake far more higher than the aforementioned estimates.

3.2.6.2 Glycolipid Food Sources

The most abundant sources of glycolipids in nutritional practice are eggs and dairy products, cereals, and soybean (Table 3.41). The soybean contains 60–200 mg of glycolipids per kilogram dry weight and lecithin, prepared from that seed, may contain up to 16% of glycolipids on a dry weight basis. The ceramide included in these glycolipids is usually dihydrosphingosine (Figure 2.23), sometimes sphingosine, phytosphingosine (4-hydroxy-sphingosine), or sphingadienine, each being linked to a fatty acid, normal or hydroxylated with a 16- to 26-carbon chain.

Some foods have been the subject of joint analysis of sphingomyelin and glyco-sphingolipids, but unfortunately the work published in 1999 by H. Vesper, United States, did not make a distinction as to the biochemical source of ceramides. Table 3.41 summarizes the main foods providing ceramides in the human diet.

TABLE 3.41
Contents in Ceramide Equivalents (Sum of Sphingomyelin and Glycosphingolipids) of Some Common Foods

Foodstuff	Ceramide (mg/100 g Fresh Weight)	Foodstuff	Ceramide (mg/100 g Fresh Weight)
Hen's egg	135	Potato	4
Beef meat	23	Spinach	4
Chicken meat	32	Lettuce	3
Fish	8	Apple	4
Cow's milk	8.6	Banana	1
Breast milk	8.8	Wheat flour	34
Cheese	196		

Note: Data are recalculated from Jensen, R.G., *Prog. Lipid Res.*, 35, 53–92, 1996, Vesper et al., *J. Nutr.*, 129, 1239–50, and Newburg, D.S. and Chaturvedi, P., *Lipids* 27, 923–7, 1992.

3.2.6.3 Dietary Supplements

Soybean lecithin may provide significant amounts of ceramide, as phytoglycolipids (Section 2.5), but nutritional supplements based on ceramides extracted from wheat germ are also available on American, Japanese, and French markets. They are sold with the allegation of a possible preservation and improvement in skin health. Despite the lack of accurate scientific data, a consumption of 1.2–4.8 g/day of lecithin enriched with ceramide is proposed by resellers.

Tests performed in the rat in 1997 by T. Kobayashi have shown that the administration of food highly enriched with sphingolipids (1% in the diet) did not induce any disease.

REFERENCES

Agostini, C., Buonocore, G., et al., 2010. Enteral nutrient supply for preterm infants: Commentary from the European Society of Paediatric Gastroenterology, Hepatology and Nutrition Committee on Nutrition. *J. Pediatr. Gastroenterol. Nutr.* 50:85–91.

Alasalvar, C., Pelvan, E., 2011. Fat-soluble bioactives in nuts. *Eur. J. Lipid Sci. Technol.* 113:943–9.

Astorg, P., Arnault, N., et al., 2004. Dietary intakes and food sources of n-6 and n-3 PUFA in French adult men and women. *Lipids* 39:527–35.

Black, L.J., Seamans, K.M., et al., 2012. An updated systematic review and meta-analysis of the efficacy of vitamin D food fortification. *J. Nutr.* 142:1102–8.

Blasbalg, T.L., Hibbeln, J.R., et al., 2011. Changes in consumption of omega-3 and omega-6 fatty acids in the United States during the 20th century. *Am. J. Clin. Nutr.* 93:950–62.

Bonthuis, M., Hughes, M.C., et al., 2010. Dairy consumption and patterns of mortality of Australian adults. *Eur. J. Clin. Nutr.* 64:569–77.

Booth, S.L., 2012. Vitamin K: Food composition and dietary intakes. *Food Nutr. Res.* 56:5505.

Bourre, J.M., 2005. L'oeuf naturel multi-enrichi: des apports élevés en nutriments, notammant acides gras oméga-3, en vitamines, minéraux et caroténoïdes. *Méd. Nutr.* 41:116–34.

Bourre, J.M., Oaland, O., et al., 2006. Les teneurs en acides gras oméga-3 des saumons atlantique sauvages (d'Ecosse, Irlande et Norvège) comme références pour ceux d'élevage. *Méd. Nutr.* 42:36–49.

Bourre, J.M., Paquotte, P., 2006. Contribution de chaque produit de la pêche ou de l'aquaculture aux apports en DHA, iode, sélénium, vitamines D et B12. *Méd. Nutr.* 42:113–27.

Bourre, J.M., Paquotte, P., 2007. Apports en DHA (acide gras omega-3) par les poissons et les fruits de mer consommés en France. *OCL* 14:44–50.

Brenna, J.T., Salem, N., et al., 2009. α-Linolenic acid supplementation and conversion to *n*-3 long-chain polyunsaturated fatty acids in humans. *Prostaglandins Leukotr. Essent. Fatty acids* 80:85–91.

Calvo, M.S., Whiting, S.J., 2005. Overview of the proceedings from experimental biology 2004 symposium: Vitamin D insufficiency: A significant risk factor in chronic diseases and potential disease-specific biomarkers of vitamin D sufficiency. *J. Nutr.* 135:301–3.

Cansell, M., 2010. Marine phospholipids as dietary carriers of long-chain polyunsaturated fatty acids. *Lipid Technol.* 22:223–6.

Chan, W., 1994. *Miscellaneous Foods*, Royal Society of Chemistry, Cambridge, United Kingdom.

Chanussot, F., 2008. *Lécithine, métabolisme et nutrition*, Tec & Doc, Lavoisier, Paris, France.

Chapuy, M.C., Preziosi, P., et al., 1997. Prevalence of vitamin D insufficiency in an adult normal population. *Osteoporos. Int.* 7:439–43.

Chung, H.Y., Rasmussen, H.M., et al., 2004. Lutein bioavailability is higher from lutein-enriched eggs than from supplements and spinach in men. *J. Nutr.* 134:1887–93.

Combe, N., Boue, C., et al., 2000. Aspects nutritonnels des acides gras trans—Consommation en acides gras trans et risque cardio-vasculaire: Étude Aquitaine. *OCL* 7:30–4.

Conly, J.M., Stein, K., et al., 1994. The contribution of vitamin K2 (menaquinones) produced by the intestinal microflora to human nutritional requirements for vitamin K. *Am. J. Gastroenterol.* 89:915–23.

Crowe, F.L., Steur, M., et al., 2010. Plasma concentrations of 25-hydroxyvitamin D in meat eaters, fish eaters, vegetarians and vegans: Results from the EPIC-Oxford study. *Public Health Nutr.* 14:340–6.

Cunnane, S.C., 2003. Problems with essential fatty acids: Time for a new paradigm. *Prog. Lipid Res.* 42:544–68.

Cuvelier,C., Dotreppe, O., et al., 2003. Chimie, sources alimentaires et dosage de la vitamine E. *Ann. Med. Vet.* 147:315–24.

Dawson-Hughes, B., 2004. Racial/ethnic considerations in making recommendations for vitamin D for adult and elderly men and women. *Am. J. Clin. Nutr.* 80:1763S–6S.

De Assis, M.C., Rabelo, E.R., et al., 2009. Improved oral anticoagulation after a dietary vitamin K-guided strategy. *Circulation* 120:1115–22.

Dinizulu, T., Griffin, D., et al., 2011. Vitamin D supplementation versus combined calcium and vitamin D in older female patients—An observational study. *J. Nutr.* 15:605–8.

Djoussé, L., Gaziano, J.M., 2008. Egg consumption and cardiovascular disease and mortality. The physicians' health study. *Am. J. Clin. Nutr.* 87:964–9.

Doets, E.L., de Wit, L.S., et al., 2008. Current micronutrient recommendations in Europe: Towards understanding their differences and similarities. *Eur. J. Nutr.* 47:17S–40S.

Durp, D., Jørgensen, H.L., et al., 2012. A reverse J-shaped association of all-cause mortality with serum 25-hydroxyvitamin D in general practice: The CopD study. *J. Clin. Endocrinol. Metabol.* 97:2644–52.

Eaton, S.B., Eaton, III S.B., et al., 1997. Paleolithic nutrition revisited: A twelve-year retrospective on its nature and implications. *Eur. J. Clin. Nutr.* 51:207–16.

FAO, 2010. *World apparent consumption statistics*. Rome. Accessed at ftp://ftp.fao.org/FI/CDrom/CD_yearbook_2008/root/food_balance/yearbook_food_balance.pdf

Fenton, J.I., Hord, N.G., et al., 2013. Immunomodulation by dietary long chain omega-3 fatty acids and the potential for adverse health outcomes. *Prostaglandins Leukotr. Essent. Fatty acids* 89:379–90.

Ferland, G., Sadowski, J.A., 1992. Vitamin K1 (phylloquinone) content of edible oils: Effects of heating and light exposure. *J. Agric. Food Chem.* 40:1869–73.

Gaillard, D., Laugerette, F., et al., 2008. The gustatory pathway is involved in CD36-mediated orosensory perception of long-chain fatty acids in the mouse. *FASEB J.* 22:1458–68.

Ganji, V., Zhang, X., et al., 2012. Serum 25-hydroxyvitamin D concentrations and prevalence estimates of hypovitaminose D in the US population based on assay-adjusted data. *J. Nutr.* 142:498–507.

Garber, A.K., Binkley, N.C., et al., 1999. Comparison of phylloquinone bioavailability from food sources or a supplement in human subjects. *J. Nutr.* 129:1201–3.

Goncalves, A., Gleize, B., et al., 2011. Phytosterols can impair vitamin D intestinal absorption *in vitro* and in mice. *Mol. Nutr. Food Res.* 55 (suppl 2):S303–11.

Grotto, D., Zied, E., 2010. The standard American Diet and its relationship to the health status of Americans. *Nutr. Clin. Pract.* 25:603–12.

Grzelinska, Z., Gromadzinska, J., et al., 2007. Plasma concentrations of vitamin E, vitamin A and b–carotene in healthy men. *Polish J. Environ. Stud.* 16:209–13.

Hanson, C., Armas, L., et al., 2011. Vitamin D status and associations in newborn formula-fed infants during initial hospitalization. *J. Am. Diet. Assoc.* 111:1836–43.

Harris, W.S., Mozaffarian, D., et al., 2009. Omega-6 fatty acids and risk for cardiovascular disease. *Circulation* 119:902–7.

Heaney, R.P., 2004. Functional indices of vitamin D status and ramifications of vitamin D deficiency. *Am. J. Clin. Nutr.* 80:1706S–9S.

Heaney, R.P., 2006. Barriers to optimizing vitamin D_3 intake for the elderly. *J. Nutr.* 136:1123–5.

Henry, C.J., 2012. How much food does man require ? *Nutr. Bull.* 37:241–6.

Holick, M.F., 2007. Vitamin D deficiency. *New Engl. J. Med.* 357:266–81.

Holick, M.F., Matsuoka, L.Y., et al., 1989. Age, vitamin D, and solar ultraviolet. *Lancet* 2(8671):1104–5.

Hu, F.B., Stampfer, M.J., et al., 1997. Dietary fat intake and the risk of coronary heart disease in women. *N. Engl. J. Med.* 337:1491–9.

Hudgins, L.C., 2000. Effect of high-carbohydrate feeding on triglyceride and saturated fatty acid synthesis. *Proc. Soc. Exp. Biol.* 225:178–83.

Hulshof, K.F., van Erp-Baart, M.A., et al., 1999. Intake of fatty acids in western Europe with emphasis on trans fatty acids: The TRANSFAIR Study. *Eur. J. Clin. Nutr.* 53:143–57.

Ishinaga, M., Ueda, A., et al., 2005. Cholesterol intake is associated with lecithin intake in Japanese people. *J. Nutr.* 135:1451–5.

Jablonski, N.G., Chaplin, G., 2012. Human skin pigmentation, migration and disease susceptibility. *Phil. Trans. R. Soc.* B 367:785–92.

James, M.J., Ursin, V.M., et al., 2003. Metabolism of stearidonic acid in human subjects: Comparison with the metabolism of other *n*-3 fatty acids. *Am. J. Clin. Nutr.* 77:1140–5.

Jensen, R.G., 1996. The lipids in human milk. *Prog. Lipid Res.* 35:53–92.

Jewell, V.C., Northrop-Clewes, C.A., et al., 2001. Nutritional factors and visual function in premature infants. *Proc. Nutr. Soc.* 60:171–8.

Juanéda, P., Cordier, O., et al., 2001. Conjugated linoleic acid (CLA) isomers in heat-treated vegetable oils. *OCL* 8:94–7.

Kapoor, R., Huang, Y.S., 2006. Gamma linolenic acid: An anti-inflammatory omega-6 fatty acid. *Curr. Pharm. Biotechnol.* 7:531–4.

Karupaiah, T., 2007. Effects of stereospecific positioning of fatty acids in triacylglycerol structures in native and randomized fats: A review of their nutritional implications. *Nutr. Metabol.* 4:16.

Kawamura, A., Ooyama, K., et al., 2011. Dietary supplementation of gamma-linolenic acid improves skin parameters in subjects with dry skin and mild atopic dermatitis. *J. Oleo Sci.* 60:597–607.

Koletzko, B., Cetin, I., et al., 2007. Dietary fat intakes for pregnant and lactating women. *Brit. J. Nutr.* 98, 873–7.

Lafay, L., Verger, E., 2010. Les apports en lipides d'origine animale de la population française: résultats de l'étude INCA2. *OCL* 17:17–21.

Lanham-New, S.A., Buttriss, J.L., et al. 2011. Proceedings of the rank forum on vitamin D. *Br. J. Nutr.* 105:144–56.

Léger, C.L., 2007. Health risks and benefits of trans fatty acids including conjugated fatty acids in food—synopsis of the AFSSA report and recommendations. *Eur. J. Lipid Sci. Technol.* 109:887–90.

Legrand, P., Rioux, V., 2010. The complex and important cellular and metabolic functions of saturated fatty acids. *Lipids* 45:941–6.

Lips, P., Hosking, D., et al., 2006. The prevalence of vitamin D inadequacy amongst women with osteoporosis: An international epidemiological investigation. *J. Int. Med.* 260:245–54.

Mahaffey, K.R., Sunderland, E.M., et al., 2011. Balancing the benefits of n-3 polyunsaturated fatty acids and the risks of methylmercury exposure from fish consumption. *Nutr. Rev.* 69:493–508.

Malton, J.L., 2002. *Additifs et auxiliaires de fabrication dans les industries agroalimentaires.* Tec & Doc, Lavoisier, Paris.

Marangoni, F., Poli, A., 2010. Phytosterols and cardiovascular health. *Pharmacol. Res.* 61:193–9.

Mark, S., Lambert, M., et al., 2011. Higher vitamin D intake is needed to achieve serum 27(OH)D levels greater than 50 nmol/L in Québec youth at high risk of obesity. *Eur. J Clin. Nutr.* 65:486–92.

Martin, A., 2001. *Apports nutritionnels conseillés de la population française.* Tech&Doc, Lavoisier, Paris.

McLaughlin, P.J., Weihrauch, J.L., 1979. Vitamin E content of foods. *J. Am. Diet. Assoc.* 75:647–65.

Médale, F., 2009. Teneur en lipides et composition en acides gras de la chair de poisson issus de la pêche et de l'élevage. *Cah. Nutr. Diét.* 44:173–181.

Mithal, A., Wahl, D.A., et al., 2009. Global vitamin D status and determinants of hypovitaminosis D. *Osteoporos. Int.* 20:1807–20.

Mourot, J., 2010. Que peut-on attendre des pratiques d'élevage pour la viande de porcs et autres monogastriques ? *OCL* 17:37–42.

Neubronner, J., Schuchardt, J.P., et al., 2011. Enhanced increase of omega-3 index in response to long-term n-3 fatty acid supplementation from triacylglycerides versus ethyl esters. *Eur. J. Clin. Nutr.* 65:247–54.

Newburg, D.S., Chaturvedi, P., 1992. Neutral glycolipids of human and bovine milk. *Lipids* 27:923–7.

Nichèle, V., Andrieu, E., et al., 2007. *La consommation d'aliments et de nutriments en France. Evolution 1969–2001 et déterminants socio-économiques des comportements.* INRA, working paper 05-07, 133p., http://pmb.santenpdc.org/doc_num.php?explnum_id=908

Nilsson, A., Duan, R.D., 2006. Absorption and lipoprotein transport of sphingomyelin. *J. Lipid Res.* 47:154–71.

Normen, L., Johnsson, M., et al., 1999. Plant sterols in vegetables and fruits commonly consumed in Sweden. *Eur. J. Nutr.* 38:84–9.

Novak, E.M., King, D.J., et al., 2012. Low linoleic acid may facilitate delta 6 desaturase activity and docosahexaenoic acid accretion in human fetal development. *Prostaglandins Leukotr. Essent. Fatty acids* 86:93–8.

O'Mahony, L., Stepien, M., et al., 2011. The potential role of vitamin D enhanced foods in improving vitamin D status. *Nutrients* 3:1023–41.

O'Neill, M.E., Carroll, Y., et al., 2001. A European carotenoid database to assess carotenoid intakes and its use in a five-country comparative study. *Brit. J. Nutr.* 85:499–507.

Pegorier, J.P., Le May, C., et al., 2004. Control of gene expression by fatty acids. *J. Nutr.* 134:2444S–9S.

Pereira-Lancha, L.O., Coelho, D.F., et al., 2010. Body fat regulation: Is it a result of a simple energy balance or a high fat intake? *J. Am. Coll. Nutr.* 29:343–51.

Pieroni, G., Coste, T.C., 2010. Composition en acides gras des oeufs. Interêt nutritionnel et valeur santé. *Cah. Nutr. Diét.* 45:261–6.

Ramprasath, V.R., Eyal, I., et al., 2013. Enhanced increase of omega-3 index in healthy individuals with response to 4-week *n*-3 fatty acid supplementation from krill oil versus fish oil. *Lipids Health Dis.* 12:178.

Rich-Edwards, J.W., Ganmaa, D., et al., 2011. Randomized trial of fortified milk and supplements to raise 25-hydroxyvitamin D concentrations in schoolchildren in Mongolia. *Am. J. Clin. Nutr.* 94:578–84.

Ross, A.C., Manson, J.E., et al., 2011. The 2011 dietary reference intakes for calcium and vitamin D: What dietetics practitioners need to know. *J. Am. Diet. Assoc.* 111:524–7.

Sanders, T.A., Filippou, A., et al., 2011. Palmitic acid in the sn-2 position of triacylglycerols acutely influences postprandial lipid metabolism. *Am. J. Clin. Nutr.* 94:1433–41.

Schuchardt, J.P., Schneider, I., et al., 2011. Incorporation of EPA and DHA into plasma phospholipids in response to different omega-3 fatty acid formulations—a comparative bioavailability study of fish oil vs. krill oil. *Lipids Health Dis.* 10:145.

Seierstad, S.L., Seljeflot, I., et al., 2005. Dietary intake of differently fed salmon; the influence on markers of human atherosclerosis. *Eur. J. Clin. Invest.* 35:52–9.

Sharma, S., Barr, A.B., et al., 2011. Vitamin D deficiency and disease risk among aboriginal Arctic populations. *Nutr. Rev.* 69:468–78.

Shearer, M.J., Bach, A., et al., 1996. Chemistry, nutritional sources, tissue distribution and metabolism of vitamin K with special reference to bone health. *J. Nutr.* 126:1181S-6S.

Simopoulos, A.P., 2010. Genetic variants in the metabolism of omega-6 and omega-3 fatty acids: Their role in the determination of nutritional requirements and chronic disease risk. *Exp. Biol. Med.* 235:785–95.

Sinclair, A.J., Attar-Bashia, N.M., et al., 2002. What is the role of alpha-linolenic acid for mammals. *Lipids* 37:1113–23.

Smith, S.M., Zwart, S.R., et al., 2005. The nutritional status of astronauts is altered after long-term space flight aboard the International Space Station. *J. Nutr.* 135:437–43.

Souberbielle, J.C., Body, J.J., et al., 2010. Vitamin D and musculoskeletal health, cardiovascular disease, autoimmunity and cancer: Recommendations for clinical practice. *Autoimmunity Rev.* 9:709–15.

Souberbielle, J.C., Prié, D., et al., 2009. Actualité sur les effets de la vitamine D et l'évaluation du statut vitaminique D. *Rev. Fr. Lab.* 414:31–9.

Souci S.W., Fachmann W., et al., 2000. *Food Composition and Nutrition Tables,* Medpharm, Stuttgart, Germany.

St.-Onge, M.P., Jones, P.J.H., 2002. Physiological effects of medium-chain triglycerides: Potential agents in the prevention of obesity. *J. Nutr.* 132:329–32.

Tiffin, R., Arnoult, M., 2011. The public health impacts of a fat tax. *Eur. J. Clin. Nutr.* 65:427–33.

Toss, G., Magnusson P., et al., 2012. Is a daily supplementation with 40 microgram vitamin D(3) sufficient? A randomised controlled trial. *Eur. J. Nutr.* 51:939–45.

U.S. Senate Select Committee on Nutrition and Human Needs, 1977. *Dietary oals for the United States.* Government Printing Office, Washington, DC.

Valenzuela, A., Delplanque, B., et al., 2011. Stearic acid: A possible substitute for trans fatty acids from industrial origin. *Grasas y Aceites* 62:131–8.

van Schoor, N.M., Lips, P., 2011. Worldwide vitamin D status. *Best Pract. Res. Clin. Endocrinol. Metab.* 25:671–80.

Vaysse, C., Billeaud, C., et al., 2011. Teneurs en acides gras polyinsaturés essentiels du lait maternel en France: évolution des teneurs en acides linoléique et alpha-linolénique. *Med. Nutr.* 47:5–9.

Vermunt, S.H., Mensink, R.P., et al., 2000. Effects of dietary alpha-linolenic acid on the conversion and oxidation of 13C-alpha-linolenic acid. *Lipids* 35:137–42.

Vesper, H., Schmelz, E.M., et al., 1999. Sphingolipids in food and the emerging importance of sphingolipids to nutrition. *J. Nutr.* 129:1239–50.

West, C.E., Eilander, A., et al., 2002. Consequences of revised estimates of carotenoid bioefficiency for dietary control of vitamin A deficiency in developing countries. *J. Nutr.* 132:2920–6.

Zeisel, S.H., Char, D., et al., 1986. Choline, phosphatidylcholine and sphingomyelin in human and bovine milk and infant formulas. *J. Nutr.* 116:50–8.

Zeisel, S.H., Mar, M.H., et al., 2003. Concentrations of choline-containing compounds and betaine in common foods. *J. Nutr.* 133:1302–7.

Zwart, S.R., Mehta, S.K., et al., 2011. Response to vitamin D supplementation during antarctic winter is related to BMI, and supplementation can mitigate Epstein-Barr virus reactivation. *J. Nutr.* 141:692–7.

4 Lipids and Health

4.1 INTRODUCTION: IMPORTANCE OF LIPID INTAKE

Although a relationship between high-fat food and atherosclerosis was earlier suspected, an interest in the impact of lipids on health appeared only in the 1950s when dietary lipids were implicated in the genesis of cardiac atherosclerosis. The vast epidemiological study by the American chemist A. Keys (1953) established for the first time relationships between the level of dietary lipids and mortality from cardiovascular events. In the field of overweight and obesity studies, it appeared that in patients maintained with a normocaloric diet an increase in the proportion of fat in the diet did not necessarily lead to an increased adiposity; it may even be accompanied by weight loss (Shikany et al. 2010). These conclusions were made in 2010 by J.M. Shikany, the United States, after an analysis of 13 highly controlled studies that were achieved between 1976 and 2005. On the reverse, under conditions of low energy intake a decrease in the lipid ration induced a weight increase. Thus, in the United States, following the recommendations of nutritionists to the population, there has been a decrease in fat intake, which was nevertheless accompanied by increasingly common obesity. It is obvious that the increase in body weight is better correlated to excessive energy intake than to excessive fat intake. Lipids may obviously contribute to obesity, but they are only one factor in a complex and multifactorial process. It is therefore easy to understand why fighting obesity cannot be reduced to limiting fat intakes. It must be necessary first to restrict the energy sources contained in the plate and to increase energy expenditure by an appropriate physical activity.

Nevertheless, obesity and diabetes, its usual companion, are two diseases characterized by what is now called the metabolic syndrome. It is well known from the work carried out in 1999 by C. Magnan, France, that a fat-rich diet has a lipotoxic effect inducing dysregulation of insulin function and weight control. The deleterious effect of a high-fat diet also finds its origins in an ongoing inflammatory condition induced by microbial endotoxins absorbed from the intestine. These endotoxins appearing in the circulation would cause insulin resistance, but they also cause vascular disorders leading to atherosclerosis and hypertension. The demonstration of these effects in humans shortly after the ingestion of a meal enriched in lipids (50 g of butter) has added a new concept that expands the scope of the relationships between lipid overload and obesity with cardiovascular diseases (Erridge et al. 2007). The clinical significance of this complex phenomenon has not been widely explored but could help to explain specific situations that have not yet found solutions.

At the global level of the relationships between lipids and cardiovascular disease risks, large epidemiological investigations seem to have closed the debate. Thus, the Nurses' Health Study carried out in 2005 by K. Oh, the United States, with 80,000 women, failed to establish a correlation between fat intake and these diseases, but there is no evidence that in some groups of individuals the situation may be different.

These examples acquired over nearly 70 years underline the apparent paradox between calories from fat and impaired health. This paradox begins to be partially explained by the role of genetic regulation in the individual response to the amount of ingested fat. In addition, the increasingly obvious control of individual energy homeostasis by the detection of fatty acids at the hypothalamus level must be considered (Migrenne et al. 2011). A possible disturbance at that level would be able to promote metabolic diseases as obesity or type 2 diabetes.

From the results of cardiac pathological events, it became gradually evident that it was not so much the amount of consumed fat but its quality that should be taken into consideration. So, what struck minds more than long academic discourses was the indisputable recognition that even with equivalent lipid-rich diets (approximately 40% of the total energy intake), Finnish and Greek people have very different incidences of cardiac diseases (heart attack). In the field of epidemiology, clinicians were unable to detect any significant relationship between fat intake and heart diseases by combining the results of a series of independent studies (study called meta-analysis). In addition, as shown in 2009 by C.M. Skeaff in New Zealand, a 5% increase of fat intake did not alter the incidence of these diseases. The same study has demonstrated that quality dominated quantity as the substitution of a portion of saturated fats for unsaturated fats, but not for carbohydrates, reduced the risk of cardiovascular diseases. The exploration of possible associations between dietary fats and heart attacks has been well resolved by the study of American nurses (about 83,000 women were followed for 20 years) (Halton et al. 2006). This survey has shown that low-carbohydrate diets, thus rich in fats and proteins, do not induce any additional risk of cardiovascular diseases. In contrast, fats of animal origin, therefore saturated, were more harmful than those unsaturated of vegetable origin. Many other studies have confirmed that the quality of ingested fat is indeed the main determinant in the onset of heart diseases, no additional benefit being obtained by reducing the energy participation of lipids to less than 35% of the intake.

In the field of cancer, the possible role of diet has been discussed for a long time but unfortunately not very accurately. Among all nutrients, the questioning of dietary lipids has become topical after the demonstration of a correlation between the incidence of various types of cancers and the fat intakes in 23 countries (Armstrong and Doll 1975). In this very large study, meat consumption was correlated to colon cancer and total fat intake was correlated with breast and ovarian cancers. Admittedly, breast and colorectal cancers have been the basis of many epidemiological studies. Despite the importance of inquiries and the wide variety of lipid consumptions in different populations, no definitive conclusion may at present be drawn for these two types of cancer. It is required to show great caution in this research area because the variation of a component of the diet is necessarily associated with changes in at least one of the other components, making it difficult to implicate one among several others. Thus, the fat content of a diet cannot be increased without decreasing the share of proteins or carbohydrates; this applies to animal as well as human experiments. This source of error has been often invoked when discussing the relationships between nutritional factors and cancers even in animal experiments where it seems easier to control each parameter.

For breast cancer, the largest meta-analysis carried out in 2003 (45 studies and nearly 600,000 people) has concluded that a 11% higher risk exists among heavy consumers of lipids compared to weaker consumers (Boyd et al. 2003). Many other studies have found no or only a weak correlation between fat intake and breast cancer prevalence. The latest report of the American Institute for Cancer Research (WCRF/AICR, 2007) has concluded that there is "very limited evidence that the amount of fat consumed is a cause of breast cancer." The importance of the hormonal status of women, variety of diets, and difficulty in acquiring accurate dietary data are certainly the source of many difficulties in evaluating the results.

For colorectal cancer, as for breast cancer, the relationship between lipid intake and cancer prevalence remains to be proved, the reported data remaining too heterogeneous. If the amount of ingested lipids has an impact on the risk of colorectal cancer, that relationship may be justified only by the contribution to the total energy intake. The WCRF/AICR report of 2007 "recognized that there was a reduced evidence, although suggestive, that high-fat diets increased the risk of colorectal cancer." The elevated energy of the Western diet, largely resulting from its high fat content, is probably the most important factor to be controlled to ensure efficiency in the various preventive recommendations proclaimed by the health authorities in many countries. An attempt to modify the risk of colorectal cancer by testing a 10% decrease of dietary lipids did not provide any positive results in a group of postmenopausal women. It is therefore too early to establish clear links between lipid intake and the development of colorectal cancer.

Numerous studies have reported that prostate cancer is often associated with overweight or obesity. This relationship has been apparently confirmed by the observation that a low-fat diet may have an inhibitory effect on the progression of this type of cancer (Lophatananon et al. 2010). Variability in epidemiological investigations and the variety of fatty acid species included in the diet did not allow the conclusion that ingested lipids may have an effect on this type of cancer. This is why the most recent works attempt to answer the question of whether prostate cancer depends more precisely on any specific fatty acid.

4.2 FATTY ACIDS AND HEALTH

The evolution of diet and physical activity in humans has led to the development of many diseases (metabolic syndrome, obesity, diabetes, atherosclerosis, and cancers) whose relationships with the quality of fatty acids are becoming increasingly evident. Among the fatty acids present in almost all foodstuffs, the emphasis is mainly on saturated, unsaturated (*n*-6 and *n*-3 series), and *trans* fatty acids (including conjugated fatty acids).

4.2.1 SATURATED FATTY ACIDS

In addition to being biosynthesized by humans, saturated fatty acids are involved in about one-third of the energy intake integrated into dietary lipids. Their abundance in many foodstuffs from animals (meat, cure meats, dairy products, and industrial preparations) and the hypothesis of their possible involvement in the genesis of cardiovascular diseases brought them quickly at the forefront of research. Gradually, their nutritional status became more complex; experiments in animals and epidemiological studies in humans have shown that besides the chemical diversity of these fatty acids there is a variety of physiological responses, previously unexpected. These responses depend in fact only on the carbon chain length. This structural parameter modulates the early stages of metabolism of these fatty acids in the intestine and also in the liver. The mechanisms modulating the differential effects of various fatty acids on the cardiovascular system are beginning to be better analyzed, but it is not the same for their involvement in the metabolic syndrome or in the process of carcinogenesis.

4.2.1.1 Long-Chain Saturated Fatty Acids

4.2.1.1.1 Cardiovascular Diseases

In the 1950s, a causal relationship between the amount of ingested saturated fatty acids and plasma cholesterol level was suspected. The cholesterol increase, which later proved to specifically characterize the fraction included in low-density lipoproteins (LDLs), has been rapidly linked to cardiovascular diseases. Thus, it was first shown in animals that the effects of the addition of saturated fatty acids in the pathology of atherosclerosis were equivalent to the addition of cholesterol. These observations are consistent with those tending to prove that plasma cholesterol is more dependent on the nature of dietary fatty acids than on the amount of ingested cholesterol. Thus, in 1965 the works done in the United States by A. Keys, who is often called "Mr. Cholesterol," clearly demonstrated that saturated fatty acids induced an increase in plasma cholesterol while polyunsaturated fatty acids induced a decrease. These findings should be considered as dietary recommendations for subjects concerned about lowering their plasma cholesterol. These recommendations were mainly taking into account the results obtained by substituting carbohydrates for saturated fatty acids in experimental diets or low-calorie diets. Although few changes in the indicators of cardiovascular diseases have been reported, beneficial effects have always been observed by substituting mono- or polyunsaturated fatty acids for saturated fatty acids. These observations led investigators to take into account the balance of dietary lipid groups before promulgating specific and predictive rules.

Following the great Seven Countries Study, which was performed on almost 13,000 people in seven different countries for 15 years, equations were established in 1980 by A. Keys to quantify the changes in cholesterolemia induced by changes in the composition of dietary fatty acids.

One of the equations established by A. Keys is as follows:

$$\Delta TC = 49.6 \times \Delta SFA - 23.3 \times \Delta PUFA + 0.6 \times \Delta C$$

where SFA = saturated fatty acids as a percentage of total calories, PUFA = polyunsaturated fatty acids as a percentage of total calories, TC = total plasma cholesterol (in micromoles per liter), and C = dietary cholesterol in milligrams per day.

From this equation, nutritionists have concluded that the ratio of polyunsaturated to saturated fatty acids is the crucial point to consider in forseeing cholesterolemia.

Later, this equation was stated more clearly after the analysis of numerous clinical trials carried out in 11 countries between 1970 and 1998. The relationship between total cholesterol and intake of various fatty acids became as follows:

$$\Delta TC = 32 \times \Delta SFA - 6 \times \Delta MUFA - 21 \times \Delta PUFA + 31 \times \Delta TUFA$$

where MUFA = monounsaturated fatty acids as a percentage of total calories and TFA = *trans* fatty acids as a percentage of total calories.

The authors have also determined the relationships governing triglyceride dietary fats, LDL-cholesterol, and high-density lipoprotein (HDL)-cholesterol (bound to HDLs). These equations are predictive and theoretical, but their value has been enhanced by the results of many intervention studies. Practically, it is therefore feasible to calculate that replacing about 20 g of butter (about 10 g of saturated fatty acids) in a dietary ration by the same amount of sunflower oil may lead to a reduction of about 5% of total cholesterol and LDL-cholesterol. These changes may then induce a 5% decrease in cardiovascular diseases in the tested population. Considering the overall impact of these diseases on mortality, this small experimental decrease in saturated fatty acid intake could prevent about 500,000 deaths annually in the world's population. This research clearly illustrated the benefits obtained for human health with a slight modification of plate content.

A meta-analysis of about 400 experimental studies have clearly shown that the isocaloric replacement of saturated fats (10% of the calorie intake) by complex carbohydrates induces a 0.52-mmol/L (or 200-mg/L) decrease in cholesterolemia and a 0.36-mmol/L (or 140-mg/L) decrease in LDL-cholesterol (Clarke et al. 1997). Disagreements remain about the interpretation of these results and the use of significant parameters (Hoenselaar 2012) despite the publication of nutritional recommendations in official reports edited in 2010 in the United States by the U.S. Department of Agriculture (USDA)/U.S. Department of Health and Human Services (http://www.cnpp.usda.gov/DGAs2010-DGACReport.htm) and in Europe by the European Food Safety Authority (EFSA) (http://www.efsa.europa.eu/en/efsajournal/doc/1461.pdf).

In the clinical field, epidemiological studies devoted to the relationships between dietary saturated fatty acids and cardiovascular diseases have sometimes shown a positive correlation, but often no reliable conclusions could be proposed. Thus, one of the largest studies, on nearly 60,000 Japanese followed for 14 years, has shown that the total cardiovascular diseases were inversely related to the amount of consumed saturated fats. Thus, the group of the heaviest consumers of saturated fats (average 20.3 g/day) had a risk 31% lower compared to the group of moderate consumers (average 9.2 g/day) (Yamagishi et al. 2010). In contrast, the large meta-analysis of 21 epidemiological studies carried out between 1981 and 2007, involving 347,747 people followed up during 23 years and totaling 11,000 cardiovascular events, has shown no significant correlation between these accidents and the amount of dietary saturated fats (Siri-Tarino et al. 2010). More recently (R. Chowdhury, United Kingdom, 2014), a meta-analysis of 32 observational studies (512,420 participants) did not allow clear support of cardiovascular guidelines encouraging a low consumption of saturated fats.

Therefore, it seems presently obvious to conclude that it remains difficult to compare the numerous studies devoted to this subject because the protocols and objectives are often different. Furthermore, it must be recalled that collecting accurate and reliable data in epidemiological surveys is very difficult and that the meta-analyses are not protected from criticism; several biases may exist at different levels, one of the largest being the preferential selection of the studies included in the final analysis.

A large country-size study carried out in Finland perhaps enabled to better assess the potential impact of a decrease in saturated fat intake on the incidence of cardiovascular diseases (Puska 2009). Indeed, in 1972, after establishing a national prevention program, first in the province of North Karelia and later throughout Finland, it was observed after 30 years that the fat intake decreased by 16%, the proportion of saturated fatty acids decreasing by 40%. During the same period, cholesterolemia decreased by 14%, and the most spectacular result was a drop of 14% in the mortality from cardiovascular diseases. If modification of the amount of ingested lipids cannot be the sole cause of these effects on public health, the intervention initiatives taken in Finland remain a major example of the importance of dietary lipids in the prevention of major causes of mortality. All the details of this experience at the level of one European country have been collected in a book that can be fully consulted on the Internet (http://www.thl.fi/thl-client /pdfs/731beafd-b544-42b2-b853-baa87db6a046).

The detailed analysis of the effects of various saturated fatty acids has shown that 12- to 18-carbon fatty acids, mainly from meat, are related to a higher level of cardiovascular risk than those with shorter chains (4–10 carbon atoms), the latter being mainly supplied by dairy products. Among the long-chain fatty acids, it is now established that stearic acid (18:0) is less harmful than palmitic acid (16:0) because after substitution it caused a significant decrease in total cholesterol without altering the triacylglycerol and HDL-cholesterol levels (Bonanone and Grundy 1988). It was even possible to specify that the replacement of 1% of the energy intake in the form of carbohydrates with 18:0 did not alter the plasma concentration of LDL and HDL but raised the total cholesterol to HDL-cholesterol ratio. The replacement by lauric acid (12:0) increased LDL- and HDL-cholesterol while lowering the total cholesterol to HDL-cholesterol ratio. Replacement by 14:0 or 16:0 also raised LDL- and HDL-cholesterol but did not affect the total cholesterol to HDL-cholesterol ratio.

Thus, contrary to popular opinion, the consumption of margarine and vegetal oils appears to be comparatively less harmful than the consumption of butter, lard, or fatty beef (lipids containing almost one quarter of palmitic acid) and largely less harmful than palm oil (containing almost half of palmitic acid). The situation of lauric acid (12:0) and myristic acid (14:0) with respect to cholesterolemia is poorly appreciated by nutritional experiments as foodstuffs also provide other fatty acids with longer chains. However, in contrast to lauric acid, myristic acid, supplied by butter, appeared to be as harmful as palmitic acid. In animals, the fatty acids 12:0, 14:0, and 16:0 have a hypercholesterolemic effect as opposed to the neutral position of the fatty acids 6:0, 8:0, 10:0, and 18:0. After the demonstration in animals of the atherogenicity of myristic acid, as palmitic acid, it might also be associated with an increased risk of heart attacks in humans (Micha and Mozaffarian 2010).

The recommendations deriving from several studies, although perfectible, on the relationships between saturated fats and cardiovascular diseases currently allow to advise a limitation of the amount of saturated fatty acids to 10% of the total energy intake. It is necessary to emphasize that reduction of the amount of dietary saturated fatty acids should be made, if possible, by a substitution with mono- or polyunsaturated fatty acids but not carbohydrates, the latter being deleterious for the cardiovascular system. This "toxic" relationship has been clearly demonstrated by a meta-analysis of 12 cohorts located in the United States and Europe involving nearly 344,700 people followed between 4 and 10 years (Jakobsen et al. 2009). During these studies, 7,404 cardiac events (including 2,155 deaths) were recorded. The main conclusion was that replacing only 5% of the energy intake as saturated fatty acids with the same amount of polyunsaturated fatty acids (mainly linoleic acid) decreased by 13% coronary accidents, the gain being 26% for heart attack deaths. In contrast, the same isocaloric replacement by carbohydrates was accompanied by a 7% increase in cardiac events, the total number of deaths remaining unchanged. One possible explanation for these phenomena lies in the increased level of saturated fatty acids in the blood caused by the ingestion of excess carbohydrates, preferential substrates of liver, and adipose tissue lipogenesis. The benefits of replacing some saturated by polyunsaturated fatty acids of vegetal origin have been supported by a recent meta-analysis of eight controlled clinical trials involving nearly 13,600 participants followed up for 8 years with the observation of 1,042 cardiovascular events (Mozaffarian et al. 2010). This study showed that an increase of 5%–15% of fatty acids of vegetal origin (mainly linoleic acid) in the dietary intake, replacing saturated fatty acids, decreased the risk of cardiovascular events by 19%. These results may be directly linked to the observation of a decrease in total plasma cholesterol of 0.76 mmol/L (0.29 g/L).

All these research efforts may leave nutritionists and consumers confused. The example of saturated fatty acids is the model on which all investigations on the possible roles of the main dietary lipids are designed. It will always be very difficult to assign a specific pathological effect to a lipid component or a lipid group even in attempting to correct the results from the interference of other nutrients. The studies of saturated fatty acids remain typical as clinical nutritionists, even experimentally, have difficulties to rule out an effect of one compound used to replace another or to eliminate a possible effect of a surplus of calories. To increase the complexity of the problem, it has been proved that if the replacement of an amount of saturated fatty acids with carbohydrates produced little effect the nature of these carbohydrates was not indifferent. Indeed, as shown in 2008 by G. Livesey, England, the replacement of lipids with fiber-rich carbohydrates (with a low glycemic index) is beneficial to health, whereas the use of refined carbohydrates (with a high glycemic index) may alter the cardiovascular system and insulin sensitivity. It is the same for replacement with monounsaturated fatty acids because the observed effect is highly dependent on other dietary components.

Therefore, it seems important to recommend a reduction in the intake of saturated fats, any compensation requiring to be carefully considered to reduce the risk of heart and vascular diseases.

4.2.1.1.2   Metabolic Diseases

It is now well established that dietary lipids with too much saturated fatty acids gradually lead to the emergence of a metabolic syndrome equivalent to the insulin resistance syndrome, a major characteristic of type 2 diabetes (formerly non-insulin-dependent diabetes mellitus). Numerous epidemiological studies have demonstrated a close relationship between the intake of saturated fatty acids and the resistance to insulin. Thus, B. Vessby showed in 2001 in Sweden that replacing butter and margarine with an oleic acid–rich oil may significantly reduce insulin resistance, except for high fat consumptions (>37% of the total energy intake).

The relationship between insulin sensitivity and saturated fats has been clearly demonstrated in patients with coronary artery disease, regardless of obesity. In contrast, the relationship with obesity has been demonstrated in subjects carrying the *FTO* gene, which is considered as one of the most important in common forms of obesity. Mechanisms involving palmitic acid in the regulation of glucose uptake by muscle cells or adipocytes are better known. As shown in England in 2009 by D. Gao, palmitic acid is involved at the level of phosphatidylinositol 3-kinase, an essential step in the enzymatic chain activated by insulin. The alteration of insulin stimulation is a property shared by palmitic acid, which is the most abundant saturated fatty acid in the diet, and also by the fatty acids 18:0, 20:0, and 24:0 unlike 12:0 and 14:0, which have no effect on that function.

Weight gain, insulin resistance, and dyslipidemia accompanying the metabolic syndrome disappear when saturated fatty acids are replaced by monounsaturated acids such as oleic acid. The confirmation of an antagonist effect of palmitic acid and oleic acid at the level of cellular synthesis of ceramides enlightens an important mechanism of insulin resistance in muscle cells (Hu et al. 2011). A confirmation of these effects can be found in the observation of a close relationship between saturated fatty acids and increased inflammatory activity in adipose tissue and muscles. Thus, it has been demonstrated that an excess of saturated fatty acids leads to a dysfunction of the energy balance naturally controlled by interleukins (IL-6, tumor necrosis factor [TNF]-α, and TGF-β) and adipokines (adiponectin, leptin, and resistin). An excess of saturated fatty acids also leads to adipocyte hypertrophy followed by cellular death (apoptosis) (Kennedy et al. 2009). Similar results have been previously obtained at the level of the vascular system of patients consuming too much saturated lipids, a situation known to promote cardiovascular diseases.

It should be noted that the close links between saturated fatty acids and type 2 diabetes are much more obvious when the observations are done from circulating saturated fatty acids (in plasma phospholipids or erythrocyte lipids) rather than from the estimation of dietary lipid amounts (Patel et al. 2010). One of the most remarkable points of that study was the observation of a highly predictive value of a high palmitic acid content for the incidence of type 2 diabetes. These findings are similar to those given earlier for the biochemical markers of cardiovascular pathologies.

In recent years, several nutrigenomics studies have shown that a diet rich in saturated fatty acids profoundly affects the expression of many genes (approximately 1500) in adipocytes. Among these, the transcription of genes involved in immune reactions is greatly enhanced. The inflammation that may result is further strengthened by a lack of affinity of the PPARγ receptors for the saturated fatty acids, a decreasing activity of these receptors being classically correlated with an increased inflammation process.

Nutritionists now take into account the latest observations and recommend a reduction of consumption of foods rich in saturated fatty acids, especially palmitic acid, to prevent insulin cellular resistance and therefore the metabolic syndrome. This objective may be obtained by increasing the simultaneous consumption of cereals, fruits and vegetables, fish, and lean meats where other types of fatty acids (mono- and polyunsaturated) are well represented.

4.2.1.1.3   Cancers

Saturated fatty acids were earlier suspected to be correlated with various cancers and gave rise to numerous experimental studies in animals and epidemiological investigations in humans. In terms of mechanism, the demonstration of a link between an increased synthesis of saturated fatty acids and cell proliferation is the source of numerous research involving the dietary intake of these fatty acids.

Several investigators have demonstrated a correlation between the consumption of saturated fatty acids and the incidence of breast or colon cancer, but many others have questioned these results due to biases or analytical difficulties. In fact, the main obstacle remains to attribute with certainty the results to saturated fatty acids in a complex diet in which the influence of other lipids and carbohydrates is difficult to assess. In addition, as in all similar studies, a significant correlation does not necessarily mean a causal relationship between the measured parameters.

4.2.1.1.3.1   Breast Cancer

Breast cancer is the leading cancer in women; more than one million cases occurred worldwide in 2002. It is the second most common newly diagnosed cancer among women in the United States, where about 300,000 cases were diagnosed in 2013 and approximately 40,000 women are expected to die from this cancer. It killed nearly 90,000 women in the European Union in 2008 and 11,500 women in France in 2011. In France, 1 in 9 women will be diagnosed with breast cancer during her lifetime and 1 in 27 will die.

The majority of our knowledge concerning the action of fatty acids on breast cancer originated from animal experiments. One of the admitted facts is the requirement of an n-6 fatty acid supply to initiate mammary carcinogenesis by dietary saturated fatty acids.

Numerous epidemiological studies have been undertaken to unravel the relationships between the dietary fatty acid composition and the incidence of breast cancer, the role of saturated fatty acids being isolated by appropriate statistical treatments regardless of the subjects' dietary habits. Thus, a large prospective study in Japan over 26,000 women followed for about 8 years failed to demonstrate a promoting effect of any dietary fatty acid on the incidence of breast cancer (Wakai et al. 2005). According to the authors, this observed lack of correlation may be related to a low consumption of saturated fatty acids in the investigated population (<13% of the total energy intake).

One of the largest meta-analyses, conducted from 45 studies in 20 countries, concerned 580,000 women and 25,000 cases of cancer (Boyd et al. 2003). This analysis showed that the incidence of breast cancer was related to the amount of the consumed saturated fatty acids, and also to the total amount of lipids. However, the risks were moderate, the group of the heaviest consumers of saturated lipids (mainly from meat) having a risk of cancer that was on average 19% higher than the group of weaker consumers. The authors emphasized that the risk of cancer associated with the consumption of fats was the highest in Asia, lower in Europe, and again lower in North America.

As for colorectal and prostate cancer, the notion of saturated fatty acids has lost any biochemical significance in that context because it mainly reflects dietary Western habits, classically rich in meat and low in fruits and vegetables.

In the present state of our knowledge, it seems that a reduction in fat intake as well as saturated fatty acid content can be recommended as a preventive measure against breast cancer. More detailed investigations will be necessary before women have the possibility to select proper nutrients for breast cancer prevention. The involvement of meat and dairy products (both rich in saturated fatty acids) should be defined more precisely to prevent this pathology from affecting about 90 women per 100,000 in France, as in several other countries.

4.2.1.1.3.2   Colorectal Cancer

Colorectal cancer is second in terms of occurrence in men, after lung cancer, and third in women, after lung and breast cancers. In the United States, where it is the second leading cause of cancer-related deaths, there were 143,000 estimated new cases of colorectal cancer and 50,800 deaths in 2013. This type of cancer killed 149,000 people in Europe in 2008 and 17,500 people in the French population in 2011.

Several laboratories have shown that in rodents saturated fatty acids accelerated the development of a colorectal cancer previously induced by carcinogens. This class of fatty acids would have a "promoter" effect. These experiments revealed that the source of saturated fatty acids is important because animal fats have a more powerful promoting effect than plant saturated fats (from coconut). Despite these observations, as for breast cancer, an excessive intake of that source of energy would be responsible for developing a colorectal cancer. These results suggest caution in the interpretation of the clinical surveys carried out in humans.

Interestingly, a large meta-analysis performed in 1997 in the United States by G.R. Howe, from 13 epidemiological studies completed in North and South America, Europe, Asia, and Australia and involving more than 5,200 cases of colorectal cancer, has shown no association between the risk of this type of cancer and the consumption of saturated fatty acids, after adjustment for total energy intake. In contrast, positive associations with energy intake were observed in 11 of the 13 studies, with findings confirmed by several subsequent studies. Furthermore, for yet unexplained reasons, the higher risk of colorectal cancer (+49% for men and +94% for women) observed in that meta-analysis for the highest energy intake compared to the lowest has not been regularly observed in other investigations.

A large Sino-American study, conducted over 61,000 people recruited in Singapore, has shown a positive correlation between saturated fat intake and the incidence of colorectal cancer (levels A or B of the Duke classification) but only in women (Butler et al. 2009). The authors attempted to explain these various differences in incidence by nutritional differences between sexes. Again, it seems clear that the statistics have some difficulty in analyzing the many parameters involved in the initiation of colorectal cancer. These results have been complemented by studies, including prospective ones, based on biomarkers analyzed in plasma, erythrocytes, and adipose tissue. This approach allowed the verification of a positive and significant association between cancer risk and content of saturated fatty acids (especially 16:0) in the erythrocyte membrane.

In summary, as reported by the AFSSA in its 2003 study on cancers and dietary fatty acids, it appears that dietary saturated fatty acids, like other types of fatty acids, are likely not directly associated with a risk of colorectal cancer, but most studies show an increased risk in relation to their contribution to the total energy intake. Thus, the prevention of this type of cancer may greatly benefit from food fortification in unsaturated fatty acids from fruits and vegetables, fish, and poultry.

4.2.1.1.3.3 *Prostate Cancer*

In terms of mortality, prostate cancer is the third most common cancer in men after lung cancer and colorectal cancer and is most common in men over 50 years. In the world, 14.1 million adults were diagnosed with prostate cancer in 2012 and there were 8.2 million deaths. In the United States, it was estimated that in 2013 there were about 240,000 new cases and 30,000 deaths from prostate cancer. In 2010, this cancer killed about 70,000 men in the European Union and 8,700 men in France.

Several studies have been carried out to investigate the relationships between the consumption of saturated fats and the risk of developing prostate cancer. These studies generally confirm this hypothesis and have suggested that a diet rich in saturated fats increases the risk of developing prostate cancer. As with other cancers discussed in this chapter, highlighting the risk of prostate cancer associated with saturated fats disappeared after correcting for the amount of energy provided by this type of fat. Some studies have indicated a close association with saturated fatty acids but only in the case of advanced forms of the disease. A follow-up of more than 500 Swedish patients suffering from prostate cancer allowed M.M. Epstein, the United States, to show in 2012 that mortality from the disease was closely related to the amount of fat ingested and also to the type of saturated fatty acids, particularly myristic acid (14:0) and short-chain fatty acids (4:0 to 10:0) (Epstein et al. 2012).

The frequently mentioned relationships between consumption of dairy products and prostate cancer have led to several studies and meta-analyses. The majority of these studies have indicated that the risk of cancer is higher, from 10% to 68%

depending on the country, in heavy consumers of dairy products compared to those who consume the least. The increased risk level has been placed to the credit of myristic and palmitic acids, abundant in milk and its derivatives, which are consumed in large amounts in some countries.

These results clearly ask the question of balance between the interest on dairy products for their calcium supply, which is essential for bone metabolism, and the increased risk of prostate cancer in men. As several authors have contradicted these relationships, more precise investigations are needed to clarify this situation and lead to precise recommendations to consumers.

4.2.1.2 Short-Chain and Medium-Chain Fatty Acids

For physiologists, short-chain fatty acids have 2 or 4 carbon atoms, medium-chain fatty acids contain 6–12 carbon atoms, and fatty acids with a long chain have 14–24 carbon atoms. Short-chain fatty acids are mainly characterized by butyric acid (4:0), present in milk fat and its derivatives (butter and cheese). Medium-chain fatty acids are supplied from milk, some vegetal oils (coconut and palm kernel), and chemically synthesized dietary supplements (Section 2.8.1).

Among the short-chain fatty acids, only butyrate (4:0) has received particular attention. Found in milk, it is also produced by bacterial fermentation in the human colon, as in other omnivorous animals and especially in ruminants. Many investigations have evidently been initiated in herbivores where it plays an important role in energy resources and thus in body growth, through the stimulation of digestive secretions and a modification of the intestinal flora. In humans, research was mainly oriented toward the anticancer properties of butyrate and secondarily to its growth factor properties. It was clearly demonstrated that butyrate is an inducer of apoptosis in colon cancer cells. Numerous studies have shown that the risk of carcinogenesis in the colon was associated with a decrease in bacterial fermentation and butyric acid production. This fatty acid is now well recognized as a primary source of energy for colonic epithelial cells and an actor in the protection of intestinal mucosa against inflammation and cell proliferation (Greer and O'Keefe 2011). This beneficial action emphasizes the importance of a fiber-rich diet generating butyric acid during its hydrolysis by the bacterial flora in the colon.

Besides advertising campaigns praising the nutritional benefits of fiber and probiotics, it is important that long-term studies involving risk markers of intestinal carcinogenesis are initiated in humans. These studies would undoubtedly clarify the dietary recommendations to limit the prevalence of colorectal cancer, which remains in the nonsmoking man the second leading cause of cancer mortality. The results already reported in 2009 by S.A. Vanhoutvin, Netherlands, in humans at the level of transcriptional regulation of several metabolic pathways by butyrate in colonic mucosa (fatty acid oxidation, electron transport chain, and oxidative stress) have suggested a research field promising new developments.

The metabolic pathways of triacylglycerols containing short-chain (short-chain triacylglycerol [SCT]) or medium-chain (medium-chain triacylglycerol [MCT]) fatty acids are different. In contrast to long-chain fatty acids, they are directly absorbed by intestinal mucosa and transported to the liver by the portal vein, without being incorporated into chylomicrons. This particular path is largely due to their solubility in the intestinal aqueous phase, which favors their hydrolysis, later their transport in the blood, and finally their oxidation in the mitochondria. It has been verified in *in vitro* experiments that MCT increases respiration in mitochondria, with a simultaneous reduction in reactive oxygen species' generation leading to a lowering of oxidative damage. This oxidation is also facilitated because these fatty acids do not require carnitine to be transported across mitochondrial membranes, unlike other fatty acids. It has been further shown that caprylic acid (8:0) inhibits the synthesis of hepatic apoprotein B, a protein required for the secretion of very-low-density lipoproteins (VLDLs) toward the adipose tissue. All these metabolic features explain clearly why SCTs and especially MCTs are less likely to be deposited in the adipose tissue than long-chain triacylglycerols (LCTs), thus contributing to their use in clinical nutrition often in the parenteral way.

In the 1950s, the specific metabolism of MCTs prompted investigators to explore their beneficial effects in the treatment of intestinal malabsorption and later in the prevention of overweight and obesity. After the initial research suggesting some efficacy in the prevention of obesity in humans, experiments in overweight individuals have clearly shown that the consumption of MCTs increased lipid oxidation and thermogenesis. These physiological effects resulted in a weight loss greater than that with the consumption of an equivalent amount of olive oil. A recent study in overweight individuals has shown that the inclusion of MCTs in a weight reduction diet may increase weight loss by reducing more strongly the fat mass (St-Onge and Bosarge 2008). MCTs are sometimes used mixed in a powder of dextrin or casein that may be added to foods or drinks. They can also be used as cooking oil, but preferably in an admixture with vegetal triacylglycerols. This mixture is obtained by the transesterification of MCT with rapeseed oil. The final product, marketed in Japan since 2002 under the name "Healthy Resseta," has a very high thermal stability and also nutritional properties close to those of the original MCT.

Naturally, MCTs are used in the treatment of many diseases characterized by a malabsorption syndrome (celiac disease, liver disease, digestive tract cancer, and intestinal resection) in adults, newborns, or even premature infants. In these treatments, as in cachexia, MCTs are generally introduced by parenteral infusion. It was further observed that food fortification of newborn food with MCTs leads to slow gastric emptying and is likely to improve the gastric digestion of these triacylglycerols.

On the other hand, it is remarkable that the consumption of foods enriched with MCT (mainly milk and its derivatives) was associated with a reduction in the telomere length, the result suggesting a deleterious effect of these fatty acids on cellular aging processes (Song et al. 2013).

MCTs from coconut oil have gained popularity in recent years by means of the athletes who use them as food supplements considering that they may increase energy levels, improve athletic performance, endurance, and even recovery. This practice, sometimes harmful at the digestive level, would only save the energy reserves of muscle without improving physical performances, and sometimes even decreases them. A negative effect has been confirmed in cyclists. However, a well-controlled study, published in 2009 by N. Nosaka, Japan, has shown that a 2-week ingestion of 6 g of MCT (74% of 8:0 and 26% of 10:0) was able to stop the increase in blood lactic acid while allowing a longer high-intensity exercise than with long-chain fatty acids. It therefore appears that further comparative studies are needed to optimize the experimental conditions (duration, frequency, quantity, and quality) before recommending a diet enriched with MCT oil to athletes. In waiting for these research, caution is required, especially, for prolonged high doses of these triacylglycerols of variable and often ill-defined composition depending on the supplier.

A favorable aspect of medium-chain fatty acids is their cholesterol-lowering effect, defined by comparison with long-chain saturated fatty acids. However, no recommendation has been proposed for a specific use of these oils to fight against hypercholesterolemia. The ingestion of large amounts of MCT has caused little side effects except a slight increase in cholesterolemia in individuals with already high cholesterol levels and an increase in plasma lipids. The clinical doses vary from 15 to 30 mL/day in children and 100 mL/day in adults. Lipid emulsions containing both MCTs and triacylglycerols of vegetal origin rich in n-6 fatty acids and of animal origin rich in n-3 fatty acids (SMOFlipid from Fresenius Kabi) are very commonly used in parenteral nutrition in patients in critical condition. The addition of long-chain unsaturated fatty acids provides adequate intake of essential fatty acids, and the variety of the fatty acid sources is recognized to be less harmful than a pure source of MCTs for liver function and immune and reticuloendothelial systems.

Several lipid emulsions enriched with MCTs are marketed in many countries for clinical applications. These products used in parenteral nutrition (venous infusion) are mixtures in variable proportions of soybean oil and MCT from coconut oil (Structolipid and Lipovenous from Fresenius Kabi, Lipofundin from Braun, and Critilip from Baxter), olive oil mixed with fish oil and MCT (SMOFlipid from Fresenius Kabi), or even fish oil and MCT (Lipoplus by Braun).

4.2.1.3 Branched-Chain Fatty Acids

Branched-chain fatty acids are saturated fatty acids whose carbon chain has one or more methyl groups (Section 2.2.1). Many molecular species in this lipid group are present in milk and ruminant tissues consumed by humans (beef, sheep, and goat). As explained in 2006 by B. Vlaeminck, Belgium, these compounds are synthesized by bacteria living in the rumen of these animals from amino acids such as leucine, isoleucine, or valine.

Given the average intake of 400 mg of these fatty acids in the normal diet (Ran-Ressler et al. 2011) and the importance of these compounds in bacteria, estimated by T. Kaneda, Canada, in 1991, it is natural to question whether they may have possible effects on several systems including the digestive organs. A more plausible hypothesis is the positive influence on the development of commensal bacteria from birth, these branched-chain fatty acids being highly concentrated in the walls of many bacteria and in particular many probiotics. This hypothesis was investigated in 2008 by R.R. Ran-Ressler, the United States, in preterm infants in relation to the development of necrotizing enterocolitis, a leading cause of morbidity in these infants.

In 2000, Z. Yang, the United States, determined that *iso* 15:0 (or 13-methyl tetradecanoic acid) was responsible for the anticancer effects of a fermented soy preparation traditionally used in China as a therapeutic supplement to prevent and treat various types of cancer (Yang et al. 2000). *In vitro* tests have shown that this branched fatty acid inhibited the growth of several cancer cell lines by inducing apoptosis but without toxic effects on healthy cells. These properties suggest that this lipid could lead to applications in human chemotherapy. The antitumor activity of *iso* and *anteiso* 16:0, determined in 2004 by S. Wongtangtintharn, Japan, on breast cancer cells has confirmed these initial findings. In contrast, the study of the action of polymethylated fatty acids (phytanic and pristanic acids), present in milk and beef meat, on the induction of a marker enzyme of prostate cancer (α-methylacyl-CoA racemase) has suggested a hypothesis of a significant role of these lipids in the genesis and progression of this type of cancer (Mobley et al. 2003). Further research should help to evaluate the importance of these branched-chain fatty acids of bovine origin in the etiology of prostate cancer.

4.2.2 *n*-9 FATTY ACIDS

The *n*-9 fatty acids are represented in the human diet primarily by one monoun-saturated fatty acid, oleic acid (18:1 *n*-9) (Section 3.2.1.3.3.2). As stated in 2008 by N. Moussavi, Canada, its impact on health still remains poorly known because it is likely too small to exceed the measurement errors in epidemiological studies and even after nutritional interventions. In addition, its presence mainly in olive oil (up to 80%) (Section 2.3.1.8), and also in meat, helps to increase the variability of results through other related factors and to blur the messages of studies that do not provide information on the origin of this fatty acid. Nevertheless, olive oil has been regarded as a source of benefits since ancient times and used traditionally in the treatment of many diseases (colic, rheumatism, and hypertension). Epidemiological studies can-not obviously settle the difference between the direct effects of oleic acid and the related effects of many antioxidants (tocopherols, polyphenols, and squalene) present in that natural product, which proves to be an extremely complex mixture.

4.2.2.1 Cardiovascular Disease

Among six major studies, three found no correlation between cardiovascular disease and intake of *n*-9 fatty acids, two found a positive correlation, and one study found an inverse relationship. These inconsistent results, also obtained in studies on the incidence of obesity, were probably the results of many differences in experimental protocols including energy intake, composition of the various lipid sources, and sex and age of subjects.

Since the study of A. Keys in 1980, monounsaturated fatty acids were convention-ally considered as having no influence on the blood markers related to cardiovascular risk. Their neutrality stems partly from their natural biosynthesis by endogenous desaturation of some saturated fatty acids, stearic acid being easily converted into oleic acid. When a part (5% of the total dietary intake) of the carbohydrate compo-nent was replaced isocalorically by monounsaturated fatty acids, no change could be observed in humans at the level of cholesterolemia as well as that of LDL-cholesterol. In contrast, it has been clearly established that a diet enriched with oleic acid, replac-ing saturated fatty acids, could significantly lower blood total cholesterol and LDL-cholesterol. These observations confirmed the validity of consuming oils like olive oil in place of saturated fats of vegetal origin (coconut and palm oil) or animal origin (butter and lard).

The beneficial effects of olive oil consumption have been clearly demonstrated by the French epidemiological Study of Three Cities (Bordeaux, Dijon, and Montpellier), which followed 9294 subjects (aged 65 years and more) for more than 5 years (Samieri et al. 2011). This study showed that the group of heavy consumers of olive oil had a stroke risk 41% lower than that observed in the group of individuals who did not consume that oil. Among the subjects who were analyzed for plasma oleic acid (1364 subjects), those with the highest levels had a risk 73% lower than that observed in subjects with the lowest levels.

A similar but larger epidemiological study carried out in Spain found that the risk of fatal heart attack may even be 44% lower. Thus, each 10 g/day of olive oil consumption was associated with a 13% reduction in death risk. Given the chemical complexity of olive oil, it is difficult to attribute these effects solely to its major component, oleic acid. However, since the 1990s, when the total lipid intake is moderate, this fatty acid has been considered primarily responsible for the beneficial effects of the Mediterranean diet on longevity and prevention of cardiovascular risk (Renaud et al. 1995). One of the main action mechanisms of this fatty acid would be to reduce the risk of thrombosis by inhibiting platelet aggregation or activation of fibrinolysis. As shown in 2011 by N. Papageorgiou, Greece, the consumption of 50 mL of olive oil was also able to induce a rapid and significant decrease in intercellular adhesion molecule (ICAM-1) and TNF-α, both molecules involved in atherogenesis.

In addition, the consumption of olive oil has long been associated with a reduction in blood pressure, although this effect, among others, has been attributed to some natural compounds (polyphenols) present in the oil. Some experiments in humans have shown that the consumption of olive oil may induce a modest decrease in blood pressure, even in individuals living in northern Europe. However, these effects were not regularly observed in all studies on these relationships.

The beneficial effect of the preferential consumption of oleic acid could also be related to its greater ability to be metabolized compared to saturated fatty acids. It could be oriented toward energy catabolism rather than its storage in fat reserves. It is also possible that oleic acid stimulates the sympathetic nervous system and thermogenesis.

Another monoene fatty acid but belonging to the n-7 series, palmitoleic acid (16:1 n-7), is present mainly in plants, the most common source being macadamia nut oil, which contains nearly 20% of this lipid. The oil has attracted the attention of physiologists for its effects on LDL-cholesterol, effects more similar to those induced by oleic acid than those induced by palmitic acid. More recent and well-controlled experiments in animal models and in humans (Griel et al. 2008) have shown that palmitoleic acid consumption (in the form of 40 g/day of macadamia nuts) may improve shortly the status of total cholesterol and LDL-cholesterol in subjects with slight hypercholesterolemia. Although it is still difficult to attribute these effects solely to palmitoleic acid, this research shows that macadamia nuts may participate in a diet contributing to limiting the development of cardiovascular disease.

4.2.2.2 Cancers

The link between cancer and oleic acid has been the subject of numerous studies and meta-analyses (Gerber 2009). Thus, in the countries of northern Europe, as in the United States and Canada, relationships between the amount of oleic acid intake and the risk of breast cancer were found to be the same as those for saturated fatty acids. In these countries, oleic acid originates mainly from meat. In contrast, when it came mainly from olive oil, as in the Mediterranean countries, its consumption was associated with a decreased risk. These various effects could result from the antioxidant compounds (squalene and phenolics) contained in olive oil, and also from the nature of specific diets in Mediterranean countries.

A large meta-analysis of 19 studies published since 1991, involving more than 37,000 people and associating cancer development and consumption of olive oil, shows the advantage of this oil on all types of cancer and mainly cancers of the breast and digestive system (Psaltopoulou et al. 2011). It should be noted that these effects did not depend on the geographic location (Mediterranean countries or elsewhere).

It is evident that intervention studies are needed to eliminate the potential role of secondary environmental parameters that remain difficult to remove by conventional statistical treatments. From a purely physiological point of view, it is currently impossible to assign the beneficial effects of olive oil on the cancer process exclusively to oleic acid. Too little epidemiological studies focusing on the relationships between oleic acid and prostate cancer have been performed to determine any conclusions. It is the same for benign prostatic hyperplasia. The lower prevalence of prostate cancer in Mediterranean countries may be correlated with the high consumption of olive oil, but other factors are not taken into account by epidemiologists and may also reduce the risk in this type of cancer, such as the consumption of tomatoes, wine, and fish.

The situation is also unclear for colorectal cancer. However, it appears that an increased consumption of olive oil is associated with a slight but significant reduction in the risk of cancer, at least for the colon but not for the rectum. As before, the specific role of oleic acid in a natural product that is as complex as olive oil remains to be determined.

4.2.2.2.1 Metabolic Diseases

The insulin resistance of our cells, causing diabetes and cardiovascular diseases, has been widely associated with a lipid overload and more specifically with a too much intake of saturated fatty acids, palmitic acid being the main actor of these disorders (Section 4.2.1.1). As previously stated, oleic acid proved to be able, in *in vitro* experiments, to offset the adverse effects of palmitic acid on inflammation and cell signaling pathways generated by insulin. In 2011, E.K. Vassiliou, the United States, showed that oleic acid, in both *in vitro* and *in vivo* models, may increase insulin secretion even in the presence of TNF-α, a cytokine that has been associated with the development of type 2 diabetes.

Rigorous clinical trials have shown that the replacement of a portion of saturated fatty acids by oleic acid improved insulin sensitivity and glycemic response among insulin-resistant subjects and also among healthy subjects (Gillingham et al. 2011). Thus, a Mediterranean diet, based on olive oil or peanut oil, may have a protective effect and may have beneficial effects on type 2 diabetes and obesity. Apart from this particular diet, it should be noted that regular consumption of peanut butter, common for North Americans, can also lower the risk of developing type 2 diabetes.

Palmitoleic acid (16:1 n-7), supplied by macadamia oil, is also biosynthesized in the adipose tissue. It is considered as the standard model of lipokines, a kind of lipid hormones modulating the systemic metabolism. This fatty acid is thus able to improve insulin sensitivity in skeletal muscle and liver, and the work reported in 2012 by T.A. Burns, the United States, has further shown that it inhibited lipogenesis and activated β-oxidation in adipocytes. Considering these properties, the commercial trend is to make it the "next omega." The development of research on the properties of this fatty acid, used primarily in cosmetics and isolated presently from macadamia oil or sea buckthorn oil, will quickly confirm or deny this metabolic hypothesis. Palmitoleic acid may in the future become a natural control agent against metabolic syndrome and obesity.

4.2.3 n-6 Fatty Acids

The main representative of this fatty acid series is linoleic acid (18:2 n-6), widely present in the diet not only in vegetal oils, commonly consumed (sunflower, corn, peanut, or soybean), but also in dairy products and meat (Section 3.2.1.3.4.1). The physiological effects of linoleic acid are still difficult to assess because sunflower oil containing about 65% of this compound is often used as the reference oil in many investigations. Encouraging results in animals have led to numerous clinical studies in two main fields, the cardiovascular diseases and various forms of cancers.

γ-Linolenic acid (18:3 n-6) is naturally produced from linoleic acid under the action of a key enzyme, Δ6 desaturase. The efficiency of this enzymatic step is limited by nutritional (saturated or *trans* fatty acid consumption), behavioral (age, smoking, and alcoholism), or pathological (infections, diabetes, atopic syndrome, multiple sclerosis, and arthritis) factors. γ-Linolenic acid–enriched oils (borage, evening primrose, and black currant) thus allow to skip that step leading to the synthesis of arachidonic acid, these oils being among the first ones to be used therapeutically. This fatty acid can also be metabolized directly into specific oxygenated derivatives such as prostaglandins of the first series, leukotrienes of the fourth series, and hydroxylated derivatives by the action of 15-lipoxygenase. These derivatives are known to have beneficial effects on blood vessels, platelet aggregation, and inflammation.

In many countries, medications containing γ-linolenic acid (e.g., Efamol®) are available for the treatment of dermatitis, atopic eczema, and diabetic neuropathies. Dietary supplementation with γ-linolenic acid has been proposed to improve the efficiency of the skin barrier in lowering epidermal hyperproliferation in subjects with dry skin and atopic dermatitis. A combination of borage oil and fish oil has also been used as a treatment for rheumatoid arthritis. All of these oils have also been proposed to fight against some breast pain (mastalgia), premenstrual syndrome, menopause symptoms, and Raynaud's disease.

As for n-3 fatty acids (Section 4.2.4), the discovery of variants for the gene encoding the desaturase system transforming linoleic acid (18:2 n -6) into γ-linolenic acid (18:3 n-6) and then into arachidonic acid (20:4 n-6) has changed our understanding of the relationship between nutrition and cell metabolism (Simopoulos 2010). Thus, in contrast to haplotype AA carriers, haplotype DD carriers can easily perform the conversion at the expense of n-3 fatty acids, especially if their dietary intake remains relatively too low. In this situation, the excessive accumulation of arachidonic acid in cell membranes may lead to an increased risk of cardiovascular diseases, cancers, and all diseases related to inflammatory reactions. The situation is even more acute for carriers of the haplotype AA who may be careful in limiting their consumption of n-6 fatty acids (sunflower and peanut oils) while consuming fish that provide them with n-3 long-chain fatty acids. The gene variability may contribute to the differences observed for certain diseases in different populations of the world. The extension of genetic tests in this area will certainly be an important part of future investigations to target groups of people most concerned with the prevention of certain diseases through nutrition. This approach opens the way to nutrigenetics, a new branch of nutritional genomics aiming to identify genetic variations in the effects of nutrient intake on the genome. This new discipline will help people in personalizing disease prevention advice based on genetic investigations.

4.2.3.1 Cardiovascular Disease

The vast majority of research in humans involving n-6 fatty acids is focused on their possible role in cardiovascular disease.

Linoleic acid is the major fatty acid in both food and human body. It has long been regarded as a cholesterol-lowering compound, a property later confirmed by many studies using vegetal oils. Investigations by A. Keys have been used to quantify the decrease in cholesterol induced by a given linoleic acid intake, the latter parameter being included in a formula in which it compensates the adverse effects of saturated fatty acids (Section 4.2.1.1.1). More detailed works have shown that linoleic acid is especially active on LDL-cholesterol.

Several intervention studies have strengthened the idea that a diet with high levels of n-6 fatty acids compared to saturated fatty acids was protective against cardiovascular events. Many studies have shown that increasing the proportion of linoleic acid in the diet is capable of inducing a significant reduction not only of cholesterolemia but also in the incidence of sudden death, heart attack, and stroke. Several epidemiological studies have confirmed these initial results by relating linoleic acid with cardioprotection. A recent meta-analysis of 11 cohorts involving nearly 345,000 people and about 7,400 heart attacks has clearly shown that the consumption of n-6 fatty acids is inversely correlated with the risk of cardiovascular events (Jakobsen et al. 2009). As an example of this relationship, it has been estimated that the substitution of saturated fatty acids up to 5% of the total energy intake (about 14 g/day) by an equivalent amount of linoleic acid decreased by 13% mortality from coronary heart disease. Despite these encouraging results, it must be noticed that intervention studies clearly demonstrating a benefit for an elevated linoleic acid intake have not been realized. After reanalyzing the data of the Sydney Diet Heart Study concerning a secondary prevention trial of coronary heart disease, C.E. Ramsden demonstrated that substituting dietary linoleic acid (9% of the food energy) in place of saturated fatty acids increased in men the rates of death from cardiovascular disease and coronary heart disease (Ramsden et al. 2013). The adverse effects of increasing the linoleic acid content are not necessarily generalizable to lower linoleic acid intakes, women, and subjects without established coronary heart disease. Unfortunately, the fatty acid content of blood or adipose tissue was not measured, preventing the validation of the whole data.

The direct beneficial effects of *n*-6 fatty acids on inflammation markers, which are typically increased during vascular injuries, have led to debates. Ingestion of very different amounts of linoleic acid has very little influence on the metabolism of arachidonic acid, the endogenous production of this fatty acid from linoleic acid being very strictly controlled. Contrary to what could be expected from these biochemical data, N. Papageorgiou, Greece, has observed in 2011 that a diet enriched with linoleic acid induces a decrease in the conventional inflammation markers in circulation and a significant decrease in intercellular adhesion molecules (ICAM-1), molecules typically involved in atherogenesis. This complex field but rich in future developments would deserved a major research effort to better define the optimal levels of the dietary *n*-6 fatty acids, mainly linoleic acid.

4.2.3.2 Cancers

Early animal models enabled the demonstration that lipids rich in *n*-6 fatty acids could influence the development of several types of cancer (breast, gastrointestinal tract, and prostate cancers). Thus, more than 40 years ago, it was described that corn oil, containing up to 62% of 18:2 *n*-6 (Section 2.3.1.10), induced a greater development of mammary tumors than coconut oil (rich in saturated fatty acids) (Section 2.3.1.9). This research has been developed over many years in animals, and all authors reached more or less the same conclusion, that is, linoleic acid has a facilitatory effect (or promoter effect) on the carcinogenesis of various organs (breast, colon, prostate, and stomach). Among the invoked mechanisms, the most active could be the intracellular conversion of linoleic acid into 13-hydroxyoctadecadienoic acid, a highly oxygenated derivative and mitogen that could increase phosphorylations, thymidine incorporation, and cell growth. Other approaches have been proposed such as an action on growth factors, eicosanoid biosynthesis or angiogenesis but their importance remains not clearly defined.

In humans, whether for breast, colon, or prostate cancer no general conclusion can be drawn from the numerous epidemiological studies performed in different countries. Rare comparative studies between groups of patients and healthy people, using serum or erythrocyte analyses, have suggested a protective effect of linoleic acid in the case of breast cancer but a deleterious effect in the case of prostate cancer. The disagreement between epidemiological and *in vitro* studies illustrates clearly the difficulty of identifying the possible but complex relationships between fatty acids and carcinogenesis. The involvement of reactions between the different kinds of fatty acids and the action of other dietary factors, such as vitamins or antioxidants, cannot presently be eliminated.

These conclusions have been confirmed by the large meta-analysis carried out in 2009 under the auspices of the World Health Organization (WHO) and the Food and Agriculture Organization (FAO). This work summarized the issue of such investigations by stating that "epidemiology itself has no power to provide firm conclusions about the relationships between cancer risk and the ingested amount of some type of fatty acids." Indeed, the importance of fatty acids other than linoleic acid and of vitamin E in foods has been well demonstrated in humans in 2001 by T. Hagenlaucher, Germany, who studied the formation rate of promutagenic DNA by leukocytes.

The study of the adipose tissue lipidome could also be a decisive approach emphasizing the simultaneous determination of several biomarkers rather than a particular fatty acid. This unique analytical and statistical approach has found an application in the research of groups of food with defined compositions in association with a high risk of breast cancer in the context of the European Prospective Investigation into Cancer and Nutrition (EPIC) study (Schulz et al. 2008). A major finding has been the demonstration of an increased risk of breast cancer primarily in connection with a diet low in fruits and bread but rich in fats from meat, fish, and margarine.

Besides linoleic acid, there is dihomo-γ-linolenic acid (20:3 *n*-6), a direct precursor of arachidonic acid (20:4 *n*-6) and also a source of oxygenated derivatives (mainly prostaglandin PGE1) known to be an inhibitor of tumor growth and to have anti-inflammatory effects. Although this fatty acid is active on cultured cells, there is no indication at present that it is of any help in the fight against cancer. The metabolic links between 20:3 *n*-6 and arachidonic acid suggest caution for any therapeutic supplementation with the former.

After all these works, suggesting a link between excessive intake of linoleic acid and cardiovascular disease or cancers, the health services of various countries have recommended that the intake of *n*-6 fatty acids should be between 5% and 10% of the total energy intake (about 14–28 g/day for a total of 2500 kcal/day), figures well above the average intake of the French population (about 4.2%, e.g., 12 g/day) (Section 3.2.1.3.1).

4.2.3.3   *n*-6 to *n*-3 Fatty Acid Ratio

In the second half of the twentieth century, following the advice of nutritionists and clinicians, the composition of lipids in the Western diet gradually changed. This trend was carried on by substituting vegetable oils, particularly rich in linoleic acid but low in linolenic acid, for saturated fats, favorite compounds in the past for the preparation of foods (butter, cream, lard, and fatback). In almost all Western diets, linolenic acid is the most represented *n*-3 fatty acid. Its daily intake is about 1.5 g, whereas the daily intake of linoleic acid is 10–20 times greater.

In this new nutritional status, physiologists have asked the question about the future of direct incorporation of *n*-3 polyunsaturated acids such as eicosapentaenoic acid (EPA) and docosahexaenoic acid (DHA) in the membranes or their biosynthesis from linolenic acid. Given our knowledge of the relationships between the metabolism of *n*-6 and *n*-3 fatty acids, it seems clear that a dietary excess of linoleic acid should be compensated by additional linolenic acid, possibly to promote tissue EPA accumulation. This hypothesis has been verified in animals and humans by maintaining the *n*-6 to *n*-3 fatty acid ratio constant in subjects ingesting three times more linolenic acid than a control group. A similar result can also be obtained by maintaining the proportion of linolenic acid constant, and by decreasing the proportion of linoleic acid in the diet. Curiously, the tissue DHA contents appeared insensitive to these dietary treatments, the bioconversion of linolenic acid seeming to be interrupted at the EPA level. A diet rich in linoleic acid is also able to inhibit the incorporation of EPA from fish oil in the cell membranes. All these data verify that the biochemical rules deduced from experiments in animals or even in cultured cells showing metabolic competition between essential fatty acid series may be applied also to humans. Indeed, all studies in animals have confirmed that the conversion of linoleic acid to arachidonic acid and linolenic acid to EPA was determined by the linoleic acid to linolenic acid ratio in the lipids of consumed foods. Thus, when this ratio exceeds 5 the tissues are enriched in arachidonic acid, and conversely when it is smaller than 5 this fatty acid tends to be replaced by EPA and sometimes by DHA. These conclusions have founded the recommendations edited by the ANSES in France (Section 3.2.1.3.2). These regulatory mechanisms have already been verified in 1992 by K.J. Clark, Australia, analyzing the blood of young children fed infant foods with n-6 to n-3 fatty acid ratios ranging from 3 to 19.

The excessive richness of our food in linoleic acid may therefore only contribute to increase cellular arachidonic acid levels and production of eicosanoids such as proinflammatory prostaglandins and leukotrienes, capable also of aggregating platelets. Furthermore, these derivatives are also required for communications between cells and for their defense against infections. Arachidonic acid is also a stimulator of adipocyte conversion and thus an inducer of obesity. It has also been verified that the content of n-6 fatty acids was higher in the blood and fat cells of young obese children than in control subjects. A large study carried out in the United States has even shown that the value of the n-6 to n-3 fatty acid ratio measured in the blood of pregnant women conditioned the adiposity of children 3 years after birth (Donahue et al. 2011). It is therefore essential to optimize the n-6 to n-3 fatty acid ratio in foods to promote the conversion of linolenic acid into EPA and DHA, and the formation of their derivatives (eicosanoids and docosanoids), which are antagonists of their counterparts bioconverted from arachidonic acid.

As emphasized by some authors, many studies have shown that humans need less essential fatty acids than what was previously assumed. Furthermore, the readjustment of the famous n-6 to n-3 fatty acid ratio to values close to those practiced by our ancestors (ratio of 4 to 5) should be done by reducing the daily intake of linoleic acid. Following these recommendations would perhaps avoid the wave of cardiovascular diseases, cancers, and inflammatory diseases characterizing the populations of developed countries associated with a Western diet type. To maintain adequate fat intake, such a readjustment may be complemented by a beneficial linolenic acid intake, plus a fortification in EPA and DHA, a method that has been validated in adult men and in children (Brenna et al. 2009).

The preferential use of vegetal oils such as rapeseed oil, walnut oil, and dietary supplements based on linseed oil, not to mention a proper intake of fish oil, is the answer to the need for changing the composition of our current fat intake.

4.2.4 *n*-3 Fatty Acids

At the end of the Second World War, our diet has evolved into an intensive production mode based on the culture across the globe of a small number of plant species providing large amounts of linoleic acid but very small amounts of linolenic acid. These crops have resulted in a huge development in the food industry, promoting the production and marketing of cheap foods rich in calories and poorly differentiated lipids. Intensive livestock oriented toward animals fed with agricultural products rich in *n*-6 fatty acids and a lower consumption of seafood have both contributed to the impoverishment of all our Western diet in *n*-3 fatty acids. Thus, the *n*-6 to *n*-3 fatty acid ratio of fat intake has increased over a century from a maximum of 5 to over 20 in Western countries (Section 2.2.1.3.2). Insidiously, the intake of essential fatty acids became progressively less and less riched in linolenic acid, this *n*-3 fatty acid deficiency leading to detrimental effects. It seems that the time has come to correct this situation by pointing out the pathologies that were developing in parallel to this nutritional change.

If in humans the most visible effects of an *n*-3 fatty acid deficiency were described in 1982 by R.T. Holman, the United States, the first major role assigned to a polyunsaturated fatty acid, EPA, has been in the prevention of heart disease in 1978 after the famous studies of Greenland Eskimos (Section 4.2.4.1). Since then, thousands of studies have been carried out in animals and humans on the impact of dietary *n*-3 fatty acids in the prevention of vascular diseases in the heart, brain, or whole body. These disorders are often related to the onset of inflammatory phenomena often caused by oxidative stress. Alteration of the vessel and capillary walls with the appearance of lesions triggering platelet aggregation and thrombosis has also been connected to this nutritional situation. Outside the circulatory system, these disorders are also at the origin of many diseases affecting the joints, immune system, adipose tissue, and mechanisms of carcinogenesis and even much higher brain functions.

The main *n*-3 fatty acid that humans find in their diet is linolenic acid (18:3 *n*-3). This compound contained in all green plants and in some seeds and oils, such as walnut, rapeseed, and soybean oils, is metabolized in the body into derivatives with longer and more unsaturated chains (EPA and DHA). This greater unsaturation results from the action of enzymes, the desaturases, encoded by genes (*FADS1* and *FADS2*) now known to be polymorphic (Simopoulos, 2010). As for *n*-6 fatty acids (Section 4.2.3), the recent discovery of two major variants of the genes encoding the desaturases, the haplotypes AA and DD, sheds a new light on the interpretation of clinical studies and the possibility of preventive or curative supplementation by *n*-3 fatty acids (Ameur et al. 2012). Thus, individuals with the variant DD will more effectively convert linolenic acid into EPA and DHA than those with the variant AA. This effect has been verified in the blood as the average circulating levels of DHA were approximately 24% higher in the former than in the latter. The value was even 43% for arachidonic acid levels. It can be inferred that people with the variant DD should be less prone to cardiovascular disease and other inflammatory diseases than the others. In contrast, an excess of *n*-6 fatty acids in their diet (high *n*-6 to *n*-3 fatty acid ratio) could be unfavorable in facilitating the biosynthesis of arachidonic acid, the precursor of many compounds highly active on inflammation. Meanwhile, vegetarian subjects with the variant AA should strive to enrich their diet with EPA or DHA (fish and shellfish) and plant products containing significant amounts of linolenic acid (walnut, canola, and soybean oils) to promote its conversion to more unsaturated n-3 fatty acids.

The knowledge of the variant type AA or DD of each individual in the future could therefore be a prerequisite for any clinical study and any attempt to correct an imbalance in essential fatty acids noticed in the diet or better in the blood. This knowledge will now be taken into account to ensure normal development, and also to prevent or control various chronic diseases related to food lipid imbalance. The impact of these genetic variants was demonstrated in lactating women consuming fish or a fish oil–supplemented diet (Molto-Puigmarti et al. 2010). As mentioned in 2008 by H.G. Joost, Germany, there is no doubt that these advances in human genomics will help to overcome the advances made so far. Genetic variations have already been taken into account to optimize therapeutic drug treatments, and now they open the field of personalized nutrition.

The determination of an inadequate intake of n-3 fatty acids is typically done by dietary survey, but the accuracy and validity are generally improved in the long term by biochemical analyses in blood or adipose tissue. Thus, W.S. Harris, the United States, proposed a biochemical marker called "omega-3 index" (Harris and Von Schacky 2004) as a faithful risk indicator for minor cardiac events or even cardiac arrest. The index is calculated as the sum of EPA + DHA as a percentage of total fatty acids in erythrocyte membranes. The author has verified that this index is closely correlated to the risk of various cardiovascular diseases, more effectively than other traditional parameters (C-reactive protein, LDL-cholesterol, and total cholesterol to HDL-cholesterol ratio). It has even been suggested that this index may be useful for identifying subjects who should receive a supplementation with n-3 fatty acids (Salisbury et al. 2011).

Despite several decades of research on the advantages that humans may draw for their health from a diet rich in n-3 fatty acids and the approval by many medical authorities, the overall benefits have not been the object of a general consensus. Opinions differ on the type of disease that is the most liable to a treatment often compared to a complementary medical practice. It is likely that a partial knowledge of the involved mechanisms, the too diverse experimental approaches, and the differences in physiological responses between animals and humans have led some investigators and scientific organizations to question permanently the benefits of a diet enriched in n-3 fatty acids.

Despite these reservations, and perhaps motivated by them, 548 studies of n-3 fatty acids were in progress and listed at the end of 2012 in the website www.clinicaltrials.gov. The explored pathologies belong to many areas, such as cardiology, inflammatory and metabolic diseases, cancer, and mental health disorders.

4.2.4.1 Cardiovascular Disease

The interest of the scientific community on n-3 fatty acids of marine origin began with a survey conducted in 1969 in Greenland by the Danish doctors H.O. Bang and J. Dyerberg when they noticed that Eskimos had a death rate due to heart disease much lower than an equivalent population living in Denmark (5.3% instead of 25%).

This investigation led the authors to correlate this feature to the high intake of fish and marine mammals by Eskimos (Bang et al. 1976).

Very rapidly, J. Dyerberg found that compared with control subjects of the same ethnic group but living in Denmark, Greenland Eskimos consumed the same amount of fat (40% of the total energy intake) but also had a higher cholesterolemia. It was also determined that the blood contained more saturated fatty acids and less poly-unsaturated fatty acids but exhibited 16 times higher EPA levels. These findings led him to the conclusion that "if the differences in diet are the main reasons for differences in mortality due to heart attack among Eskimos, these results argue in favor of a more qualitative than quantitative difference regarding dietary fatty acids." Subsequently, this author has hypothesized that the low incidence of heart disease in Greenland, Eskimos could be related to their high intake of *n*-3 fatty acids, leading mainly to EPA accumulation in their tissues.

Since these early works, a growing number of clinical and biochemical studies have provided more and more compelling evidence for the role of specific fatty acids of marine oils, EPA and DHA, in cardiovascular pathologies. The effects induced by these fatty acids are probably related to multiple mechanisms such as inflammation and platelet activation. The evidence from a wide variety of investigations relating an inadequate intake of *n*-3 fatty acids with risks of cardiovascular disease, including sudden death, has pushed many scientific organizations to put forward recommendations for the population. Thus, among many others, the American Heart Association, European Society for Cardiology, Scientific Advisory Committee on Nutrition (United Kingdom), and Australian Health and Medical Research Council all proclaimed 10 years ago to increase the dietary *n*-3 fatty acid intake.

In France, the AFSSA published an important report on *n*-3 fatty acids and cardiovascular disease (http://www.afssa.fr/Documents/NUT-Ra-omega3.pdf).

4.2.4.1.1 Coronary Heart Disease and Atherosclerosis

According to WHO data, ischemic heart disease is the leading cause of death worldwide (7.2 million deaths). In the United States, about 600,000 people die of heart disease every year and coronary heart disease is the most common type of heart disease, killing more than 385,000 people annually. In the European Union, cardiovascular disease causes over 1.8 million deaths every year. The incidence of this disease, a sudden manifestation of coronary attack, remains elevated in France with 120,000 cases per year. Its prognosis is serious since infarction is still responsible for 10%–12% of the total annual adult mortality, despite a significant decrease (about 30%) being found in the last 10 years. The economic burden associated with the management of cardiovascular disease is still considerable. Indeed, it is the cause of about 10% of hospital stays, the direct and indirect costs for the French society being around €28 billion per year. Risk factors are numerous, but all clinicians and nutritionists agree that diet plays a major role.

Several indications were provided by epidemiological studies carried out in Japan in 1980, but the first evidence of a beneficial effect of *n*-3 fatty acids was derived from the famous Zutphen study conducted in the Netherlands over 20 years (Kromhout et al. 1985). This study, which was widely distributed in all media, has shown that regular consumption of one to two fish meals per week was sufficient to prevent coronary heart disease, the death incidence being halved in subjects consuming an average of at least 30 g of fish each day compared to subjects consuming no fish. Comparable results were obtained by I.C. Torres, Portugal, in 2000 in Madeira Island, where the frequency of fatal cardiac events was 3.8 times lower among fishermen than among farmers. This lower mortality has been associated with a high fish consumption. The largest epidemiological study was conducted in 1990 by T. Hirayama throughout Japan (more than 3.8 million people were followed for 17 years). It has been estimated that in that country daily consumption of fish decreased the incidence of fatal myocardial infarction by 20% in men and 49% in women, compared to individuals who never consumed fish. A large meta-analysis done at the University of Harvard gave a summary of the results obtained with 13 cohorts, including more than 200,000 individuals followed for about 12 years (He et al. 2004). This work showed that for each additional consumption of fish equivalent to 20 g/day the risk of mortality from myocardial infarction decreased by 7%.

Most observational studies of EPA and DHA have assessed self-reported dietary intake rather than objective biomarkers, which may have led to measurement errors or bias. More difficult to realize but clinically more reliable, the measurement of circulating fatty acids likely reflects more exactly the dietary consumption and the biochemical processes at the organ level. To address these gaps, D. Mozaffarian has investigated the possible relationships between plasma fatty acids (EPA, DHA, and 22:5 *n*-3) as biomarkers and the risk for cardiovascular diseases. While following for about 11 years a cohort of 2692 adults without prevalent heart disease, the authors observed that higher plasma levels of phospholipid *n*-3 fatty acids were associated with lower cardiovascular deaths. More precisely, it has been calculated that individuals in the highest quintile of the *n*-3 fatty acid level lived an average of 2.22 more years after the age of 65 than did those in the lowest quintile (Mozafffarian et al. 2013). Using a validated semiquantitative food-frequency questionnaire, the authors could evaluate an average target dietary range of 250–400 mg of EPA + DHA per day.

To avoid the use of too many subjects and a too long and too expensive clinical follow-up, subsequent studies have been mainly focused on secondary prevention in patients who suffered a myocardial infarction. Of 14 published studies analyzed, 12 have highlighted the usefulness of a supplementation of *n*-3 fatty acids, from marine or vegetal origin, in the prevention of a recurrent coronary syndrome (Breslow 2006). All these results were confirmed by extensive studies with similar protocols in many countries across the world.

In contrast to these positive studies, other trials and a recent meta-analysis failed to demonstrate a reduction in cardiovascular events with EPA and/or DHA supplementation.

These negative results have been explained by differences in trial design, fatty acid dose, patient characteristics, duration of follow-up, and also previous pharmacological treatments (lipid-lowering drugs).

It is clear that more research is needed to answer the many questions raised by these studies that remain difficult to analyze, despite their large amplitude. No data are available, for example, on the incidence of coronary disease itself, as most studies have reported only the final phase of infarction and often its ultimate end, the death of the patient. Doubts will persist for a long time about the possible modes of action of n-3 fatty acids ingested with one fish serving in the form of capsules containing purified supplements, also under various chemical forms (triacylglycerols and fatty acids). It has been reported that with equal amounts of fatty acids different effects could be noted. Some authors have suggested that substances other than n-3 fatty acids (vitamins, essential amino acids, and trace elements) could be responsible for some of the observed effects, whereas heavy metals have been invoked to explain the absence of effect–dose relationships.

Very rapidly, the favorable impact of the ingestion of n-3 fatty acids on morbidity and mortality by heart attack was associated with less symptoms of atherosclerosis, the most common vascular disease in Western societies. This disease is mainly characterized by an accumulation made up of calcified lipid plaques (atheroma) in the arteries, causing injury (sclerosis) followed by thrombosis. These deposits may gradually obstruct the coronary arteries, thereby generating the vast majority of cases of myocardial infarction, characterized most often by the syndrome known as angina pectoris. Other locations may be injured by causing a variety of serious conditions such as stroke, aortic aneurysm, peripheral arterial occlusive disease, and kidney arterial hypertension. Although atherosclerosis is a multifactorial disorder, all specialists now accept that its initiation is based on two important risk factors, a high cholesterol level and an excess of blood triacylglycerols. It is precisely on both physiological parameters that the n-3 fatty acids are efficient, at least some of them.

Thus, in healthy subjects a daily intake of 2.5 g of EPA + DHA causes a significant decrease in blood triacylglycerol levels, without altering the values of total or lipoprotein cholesterol. In contrast, linolenic acid (18:3 n-3) produced no modification. Curiously, D.K. Banel, the United States, showed in 2009 that eating walnuts (30–108 g/day), containing about 9% of linolenic acid, proved to be efficient in significantly reducing the amount of total cholesterol and HDL-cholesterol. Although not verified by other investigators, this example once again raises the question of what are the n-3 fatty acids needed by humans to ensure maximum effect on physiological systems endangered by unbalanced diets. Unfortunately, the lowering effect of EPA and DHA on triglyceridemia remains poorly explained in humans, although animal experiments tend to prove that these fatty acids inhibit hepatic lipogenesis. Similarly, their effects on serum lipoproteins are, for reasons still unclear, variable from one experiment to another, likely explaining the difficulties encountered by clinicians using fish oils of different compositions or industrial preparations of fatty acids.

The anti-inflammatory effects of n-3 fatty acids contribute to the stabilization of atheromatous plaque, whose dramatic effect is to progressively slow blood flow or develop a tiny crack or a rupture on the inside surface of the blood vessel. This may trigger the formation of a blood clot (thrombosis) over the atheroma, which may completely block the blood flow. The thrombus may also be released, contributing to the formation of ischemia in other vessels of smaller sizes.

It has been clearly demonstrated that EPA and DHA reduce the endothelial production of inflammatory cytokines, such as interleukin (IL-1) and TNF-α, and inhibit the monocyte adhesion to vascular endothelium activated by these cytokines.

Numerous studies in humans have made it possible to verify the beneficial effects of fish oil on the blood markers of endothelial activity, showing an improvement in the health status of vessels even after coronary surgery. In combination with the inhibition of platelet aggregation by EPA and DHA, all these effects contribute to limit the development of atheromatous plaques, thus preventing their rupture, an event related to death by severe myocardial ischemia. Although not much explored clinically, fish consumption (more than 130 g/week) in subjects over 55 years has been associated with less coronary calcification (Heine-Bröring et al. 2010), a phenomenon also controlled by vitamin K_2 (Section 4.4.4.1). This effect was recently verified by comparing the incidence rate of the calcification in Japanese men and white American men (Sekikawa et al. 2014).

After numerous studies, experts from several countries have recommended as a preventive measure to ingest at least 450 mg/day of a EPA + DHA mixture, the quantity being increased to 1 g/day in people at risk or who have already suffered a heart attack (Section 3.2.1.3.2). In the United States, for the first time in 2010, the official health guide (*Dietary Guidelines for Americans*) published by the USDA recommended that the quantity and variety of consumed marine animals must be increased, replacing meat and poultry. The guide advised every citizen to consume between 230 and 340 g of seafood per week (http://www.cnpp.usda.gov /DGAs2010-PolicyDocument.htm).

As mentioned earlier, future studies on the prevention of cardiovascular diseases by providing *n*-3 fatty acids should take into account the progress in human genetics on the distribution of specific haplotypes of fatty acid desaturases (FADS A or D) in treated subjects (Martinelli et al. 2009). Thus, the holders of haplotype D may achieve a more efficient conversion of ingested linolenic acid into EPA and DHA than the holders of haplotype A, this representing a significant beneficial effect on the prevention of cardiovascular diseases related to inflammation. These genetic differences could also be the cause of variability of the results of surveys in large populations.

Despite some uncertainties, all the results have led national organizations and many national scientific and medical societies to publish recommendations for the prevention of cardiovascular disease. Thus, the American Heart Association has recommended that healthy people eat fish twice a week in association with linolenic acid–rich plants. Patients suffering from coronary artery disease should ingest 1 g/day of a mixture of EPA and DHA in the form of fatty fish or a nutritional supplement. The most recent recommendations issued in North America indicated a daily need of 250–500 mg EPA + DHA. Many countries have adopted these recommendations. So far, the proportions of each of these fatty acids are not taken into account. The participation of linolenic acid remains debated because its conversion into EPA and DHA is very low under normal conditions, except perhaps in young children and pregnant women, but it remains affected by the simultaneous intake of *n*-6 fatty acids.

4.2.4.1.2   Stroke (Cerebrovascular Accident)

According to the WHO, 15 million people suffer from stroke worldwide each year, the disease being everywhere a leading cause of long-term severe disability. Of the 15 million, 5 million die and another 5 million are permanently disabled. Mortality from stroke was the third leading cause of death in the United States, killing more than 140,000 people each year. Europe averages approximately 650,000 stroke deaths each year.

Approximately 130,000 strokes occurred in 2010 in France (Health Ministry), and a quarter of people died (mortality rate at 5 years is approximately 50% of patients). It is the second leading cause of death among women and the third among men. The other patients retain in most cases sequelae that constitute the leading cause of neurological disability in adults: in France, nearly 500,000 people (0.8% of the population) currently present sequelae of stroke (sources of the French Ministry of Health). These consequences represent a heavy burden on the health system. The importance of this disease has led the French government to establish a theme of prevention and care as a national action plan, CVA 2010-2014. Stroke is 80% of ischemic origin, that is to say, due to occlusion of a cerebral artery by a clot, locally generated or coming from another part of the upstream arterial tree. The interruption of blood flow will cause, as in the heart, a rapid destruction of the neighboring territory, interrupting any motor or sensory function whose control is located in the affected area. Treatment of stroke should be extremely fast and, as it proved to be most often inefficient, clinicians have naturally explored the nutritional way to ensure a possible prevention as in other areas of cardiovascular diseases. It became therefore important to determine whether, in addition to other risk factors (hypertension, smoking, and diabetes), it must be added a deficiency in dietary n-3 fatty acids or any other imbalance in the composition of dietary fatty acids.

Early studies in animals have confirmed the hypothesis that fish oil consumption could prevent the deleterious effects of experimental cerebral ischemia. Thus, O.A. Ozen discovered in Turkey in 2008 that rats fed a fish oil–enriched diet and subjected to an operation of ischemia and reperfusion of the prefrontal cortex had reduced oxidative stress and a lower number of apoptotic neurons in the injured area compared to control rats. These results have been repeatedly confirmed and supported by the demonstration of a beneficial effect of n-3 fatty acids on the functional recovery of the operated animals.

In humans, the first major ecological studies in various countries, especially Greenland and Japan, have highlighted the reduced risk of ischemic stroke among fish consumers. Numerous studies have been devoted to this subject, but the synthesis of results does not offer a clear picture allowing specific recommendations for potential prevention. The reason for this is likely the wide variety of experimental methods with respect to the assessment of n-3 fatty acid intake (inquiries and biomarker measurements), duration of the observations, and especially distinction between stroke types.

One of the first studies with a large survey was carried out in the United States on a cohort of 43,671 health professionals (men) who were followed for 12 years (He et al. 2002). From the dietary survey and the registration of brain injury (608 strokes, including 377 ischemic), the authors have concluded a 43% reduction in risk of ischemic stroke in people consuming one to three times fish monthly compared to those consuming less than once per month. In contrast, an increased fish consumption provided no further benefit. In addition, no effect on the risk of hemorrhagic stroke could be detected.

Many other studies have provided similar results, but more often less convincing. As for the exploration of possible relationships between stroke and n-3 fatty acids, it is tempting to explain the variability of results by invoking the intervention of other components in the flesh of the consumed fish or another nutrient. More complex experiments could certainly shed light on this problem that continues to sustain the discussion on the benefits, usefulness, and harms of a supplementation with fish oil in connection with vascular events.

What are the therapeutic or predictive capabilities of the precursor of all n-3 fatty acids, linolenic acid ingested with plant foods? Dr. T. Shimokawa, University of Nagya, Japan, is the first to show that a diet rich in linolenic acid increased life expectancy in a rat model with high risk of stroke, compared to a linoleic acid–rich diet (Shimokawa et al. 1988). In humans, the situation is less clear even after biomarker analysis giving evidence of the ingestion of specific fatty acids. Indeed, it has been observed that only linolenic acid in cholesterol esters and plasma phospholipids was less abundant in patients suffering from stroke than in controls, but another study failed to confirm these facts. These results obviously did not clarify the potential effects of a diet enriched in linolenic acid on stroke risk. Unfortunately, no experimental approach in humans is yet available in this area.

The analysis of existing studies suggests to clinicians to perform a strict selection of subjects to participate in epidemiological studies to eliminate as far as possible the maximum bias that may weaken the statistical comparisons. A few well-chosen subjects with a well-defined pathology have a greater statistical weight than a large number of subjects taken at random from a too heterogeneous population. The investigator must also necessarily take into account the difference in equipment of individuals in various forms of desaturases, in selecting participants for nutritional explorations involving the balance of n-6 and n-3 fatty acids. Many research efforts are needed to explore the impact of the association of major n-3 fatty acids and the n-6 to n-3 fatty acid ratio in foods and tissues on primary and secondary prevention of ischemic stroke. It is unfortunate that, similar to the works done in rats, investigations in humans on the possible effects of n-3 fatty acid supplementation using clinical procedures for recovering after a first stroke have not yet been undertaken.

In conclusion, these studies in animals and humans highlight the possibility of a nutritional approach of stroke prevention mainly based on the adoption of a diet containing vegetal oils rich in linolenic acid or purified supplements enriched in this lipid. Thanks to the acquired data, it is still relevant to recommend having two to four fish meals per week, an advice originally adapted to the fight against heart disease but remains valid for the prevention of ischemic stroke.

4.2.4.1.3 Cardiac Arrhythmia

Cardiac arrhythmia is the most common heart disorder. As a result of its various origins and manifestations, this disorder characterizes a family of heart diseases. Apart from benign manifestations such as extrasystoles, arrhythmia occurs when the heart is beating irregularly, sometimes when the pulse frequency is less than 60 (bradycardia) but more often when the frequency is greater than 100 per minute (tachycardia) without justification. The elderly and those who already have another important heart or lung disease might frequently have a rhythm disorder. One of the most serious forms of arrhythmia is cardiac fibrillation that corresponds strictly to different forms of cardiac arrhythmia, originating in the most common form in the atria (atrial fibrillation) or in the ventricles (ventricular fibrillation). The latter is a major cause of cardiac arrest and sudden death. From surveys carried out in Europe, it has been estimated that 2.5% of the subjects clinically explored presented an abnormal heart rhythm in the past year and that about 5% will have this anomaly at certain times of their life.

Very soon after the demonstration of the beneficial effects of n-3 fatty acids on heart function, investigators were also interested in their possible effects on cardiac arrhythmia and especially on ventricular fibrillation. Thus, evidence has been provided in rats that a diet enriched with fish oil reduces the incidence and severity of an arrhythmia previously experimentally induced by a cardiac ischemia. These encouraging results were then confirmed in the dog and also in various cellular or animal models. More recently, a suitable experimental model in dogs verified that supplementation with EPA + DHA reduced the susceptibility to atrial fibrillation and attenuated the formation of atrial fibrosis, the origin of electrophysiological disorders causing contraction disorders (Laurent et al. 2008).

How can biologists explain an effect of n-3 fatty acids on heart rhythm disorders? It is likely that this effect is generated through the increased biosynthesis of less arrhythmogenic eicosanoid derivatives and also by reducing the content of free fatty acids modulating the ion channels in cardiac cells.

The first epidemiological studies on the relationships between n-3 fatty acids and heart disease have already allowed the hypothesis of an antiarrhythmic effect of n-3 fatty acids as a result of a reduced risk of sudden death, despite the unchanged number of heart attacks. Unfortunately, many epidemiological research efforts on this subject did not allow the conclusion of an indisputable effect of n-3 fatty acids on the incidence of arrhythmia. The reason for this failure surely lies in the inaccurate determination of the forms and states of the disease, and also in the diversity of the used products and related pharmacological treatments that may affect the results. To address this complex problem, clinicians have attempted to use other physiological parameters related to the pathogenesis of sudden death or rhythm control by the balance between vagal and sympathetic activities. Thus, about 20 studies have been devoted to the relationship between n-3 fatty acids and arrhythmia considering heart rate variability. A meta-analysis of 15 controlled trials devoted to the short-term effects of fish oil supplementation on heart rate variability has revealed a favorable influence on the high frequency domain of heart rate variability (Xin et al. 2013). The observed enhancement of the vagal tone is likely at the base of the antiarrhythmic effect of fish oil.

One of the unresolved issues is the relative efficiency of EPA and DHA on the regulation of heart rate. Compared with EPA, DHA has often shown a superior efficiency in animals in the prevention of arrhythmia, regulation of blood pressure, and reduction of the atherosclerotic process. In humans, this trend was confirmed and emphasized by the demonstration of a strong association only between blood DHA levels and the risk of fibrillation. Similarly, a recent epidemiological study has clearly shown that individuals with the lowest levels of DHA in plasma phospholipids also have a higher risk of atrial fibrillation, with EPA showing no correlation (Wu et al. 2012).

Few intervention studies have been conducted in humans. The most interesting are those that explored the opportunity to acquire *in vivo* data from patients with implantable cardiac defibrillators. A meta-analysis of the three most controlled investigations in this area has unfortunately revealed heterogeneity in patient responses to supplementation with *n*-3 fatty acids (Jenkins et al. 2008). After 1 year of supplementation, one study showed a 26% reduction in the risk of ventricular fibrillation and another showed a worsening of incidents, the last recording no change. Apart from these works, few trials of *n*-3 fatty acids in secondary prevention have been undertaken, and they are far from an agreement on the effectiveness of that nutritional therapy. Compared to encouraging experiments in animals, it is certain that the used fatty acid doses are much lower in human trials, and the duration of treatment may also be insufficient to allow the full expression of antiarrhythmic effects.

In summary, despite inconclusive research on the benefits of increased fish consumption or supplementation with *n*-3 fatty acids in the battle against heart rhythm disorders, we cannot at present completely exclude an advantage of the nutritional approach. It remains important to follow the advice issued by competent medical authorities on the regular ingestion of *n*-3 fatty acids from marine animals or from dietary supplements preferably enriched with DHA.

4.2.4.2 Inflammatory and Immune Diseases

Long-chain *n*-3 and *n*-6 fatty acids, mainly from the diet, are incorporated in the membranes of almost every cell of the body. They are metabolized when necessary into many hydroxylated or oxygenated derivatives playing the role of lipid mediators involved in inflammatory reactions often coupled with immune responses. Inflammatory reactions may be tissue responses to injury or to bacterial or viral infections with several aspects such as activation of blood flow and capillary permeability to large molecules and cells, especially leukocytes. Immune responses are intended to eliminate toxic molecules or infectious agents. These reactions are generated in a particular system of cells originating in the bone marrow and migrating later to the immune system. All these reactions are triggered or inhibited by eicosanoids, originating mainly from arachidonic acid (20:4 *n*-6) and secondarily from dihomo-γ-linolenic acid (20:3 *n*-6). In contrast, these reactions may be inhibited by very similar derivatives from EPA and DHA. Since the discovery of these phenomena in the 1970s, intense research efforts have highlighted several biological mechanisms underlying these effects. Among the most important, it must be remembered that many derivatives of EPA and DHA (leukotrienes, resolvins, and neuroprotectin D1) are anti-inflammatory, whereas the prostaglandins derivating from arachidonic acid are proinflammatory. Furthermore, these derivatives may increase or decrease the expression or production of cytokines such as TNF-α, interferon-γ (INF-γ), and several interleukins. More and more data allow the consideration of changes in the organization of cellular membranes leading to the modulation of Toll-like receptors, plasma membrane receptors that are active in the recruitment of T lymphocytes.

Many investigations have been carried out in animals to test the effects of fish oil on inflammation. Differences have still been reported, probably caused by the use of different species, various diets, and measurement technologies. In humans, the results also diverged according to the protocols used, but high doses of EPA and DHA are generally capable of inhibiting the production of active substances such as interleukins and eicosanoids by mononuclear leukocytes. Several investigations have provided encouraging results on the clinical effects of n-3 fatty acids in slowing down the production of molecules involved in inflammatory response, even shortening the hospitalization period after surgery (Liang et al. 2008). The disagreements proclaimed by some authors probably originated in the diversity of *in vitro* tests used to assess the formation of various bioactive molecules, but they may also arise from the composition and amount of the added supplements.

Among the diseases resulting from inflammatory reactions, asthma, inflammatory bowel disease (Crohn's disease), and rheumatoid arthritis are the most frequently investigated.

Many studies have reported anti-inflammatory effects after treatment with a fish oil in asthmatic patients. Other studies were less enthusiastic and reported that subgroups sometimes appeared to be refractory to treatment for no apparent reason. A major report (Evidence Report/Technology Assessment No. 91) written in 2004 by an expert committee at the request of the Department of Health of the United States has established, after an analysis of 31 publications, an inability to definitively conclude on the value of a supplementation with n-3 fatty acids in the treatment of asthma (http://archive.ahrq.gov/clinic/epcarch.htm). Since the publication of this report, new results have shown that n-3 fatty acid intake was significantly inversely associated with the incidence of asthma in young adults (18–30 years old) studied during a 20-year follow-up (Li et al. 2013). DHA showed a greater inverse association than EPA.

Chronic inflammatory bowel diseases include Crohn's disease and ulcerative colitis, both characterized by an inflammation of the digestive tract wall (related to hyperactivity of the digestive immune system), being thus the source of ulcerations. In addition to many factors, the quality of food was often cited as contributing to the onset of the disease. Thus, its greater incidence in developed countries consuming industrial foods could be related to an imbalance of the n-6 to n-3 fatty acid ratio. From the earliest studies of possible relationships between these inflammatory diseases and n-3 fatty acids, it appeared that the latter were less concentrated in plasma and adipose cell phospholipids in patients than in control subjects. The relative excess of the abundant n-6 fatty acids likely intensified proinflammatory eicosanoids biosynthesis. Clinically, some beneficial effects have been obtained in patients with intestinal inflammation after supplementation with fish oil. Other studies are less conclusive, casting doubts on the effectiveness of this clinical approach in humans.

Rheumatoid arthritis is a chronic inflammatory disease with an autoimmune origin, but it is still poorly understood. Osteoarthritis is the most common form in the elderly in whom it is the most disabling joint disease. This disease is an arthritis complicated with inflammatory lesions of bone ends. It is mainly characterized by a loss of articular cartilage, causing pain and restriction of movement. It is not surprising that in the context of inflammatory diseases arthritis has been the object of a very large number of clinical explorations usually involving a supplementation with n-3 fatty acids from fish oil. They were the most compelling essays in the field of a "nutritional struggle" against inflammation as the majority of them have noticed beneficial effects. A recent meta-analysis was in agreement with these conclusions (Goldberg and Katz 2007). Indeed, the analysis of 17 reliable studies devoted to the therapeutic effects of n-3 fatty acids in rheumatoid arthritis has shown an indisputable analgesic effect contributing to relieve joint pain and shorten morning stiffness. Overall, all well-conducted clinical trials have shown a net profit improving the quality of life for patients and allowing a reduction of doses of anti-inflammatory products. These effects should be taken into account more seriously by medical doctors especially in the care of older people with musculoskeletal pain, these people being also known to be often n-3 fatty acid deficient. Although the doses used in these studies were very different, it seems that a daily intake of 3 to 4 g of EPA + DHA in the form of fish oil is necessary to obtain detectable effects after a few months of treatment. Encouraging results were also obtained more rapidly (2 weeks) with smaller amounts of krill oil, a product containing phospholipids enriched in EPA and DHA (Section 3.2.1.3.2). This therapeutic option, very widespread in the media, has not escaped the specialized companies that make extensive use of clinical results for the promotion of prescription products based on fish or krill oil.

The immune responses, usually coupled to inflammatory responses, are naturally designed to eliminate infectious agents or toxic molecules (viruses, bacteria, and fungi) that try to invade the body by inducing pathological conditions. These inflammatory responses are generated in a particular system of cells and tissues originating in the bone marrow and migrating later to a complex immune system, formed by spleen, thymus, and lymphoid tissues.

As fish consumption or EPA and DHA intake reduces the concentration of arachidonic acid in the cell membranes of the immune system, it also reduces the bioconversion of this n-6 fatty acid into derived eicosanoids. Furthermore, EPA competitively inhibits the biosynthesis of arachidonic acid derivatives at the cyclooxygenase step while being converted itself by the same enzyme into less active oxygenated derivatives. All these effects, observed in cell or animal models, naturally led to the hypothesis that n-3 fatty acids could enhance immune function. It has also been reported that supplementation of the diet of pregnant women with these fatty acids leads to an acceleration of the maturation of the immune system in newborns, likely reducing the sensitization phenomena to various allergens and the severity of potential atopic dermatitis. These effects may persist even in infancy by reducing the incidence of asthma, eczema, and pollen allergies (Kremmyda et al. 2011).

However, further studies are needed before making recommendations in clinical practice, especially for chronic treatments. The generally accepted safety of the products enriched in n-3 fatty acids of natural origin easily promotes advertisements among patients. It is certain that these natural substances, giving rise to an added tax of low value, will trigger with difficulty the development of objective and expensive research in this field of medicine.

Despite some conflicting results likely resulting from genetic differences, many clinical studies have led to the acceptance that n-3 fatty acids help in the protection or relief of all diseases involving inflammatory processes. Currently, arthritic phenomena seem to be more accountable to a long-term treatment by n-3 fatty acids such as EPA and DHA, mainly in the form of concentrated preparations.

4.2.4.3 Metabolic Diseases

The metabolic syndrome can be defined as a complex condition characterized by the presence of multiple risk factors for cardiovascular diseases and type 2 diabetes. This may include many symptoms or dysregulations such as hyperglycemia, insulin resistance, obesity, hypertension, hypertriglyceridemia, and low blood HDL-cholesterol. This concept is not new since it was set in 1988 by G. Reaven, the United States, under the name X syndrome. About 10 years later, the WHO named the metabolic syndrome and clinicians defined it by the presence in patients of at least three disorders of the following five: hypertriglyceridemia (≥ 1.17 mmol/L), low HDL-cholesterol (≤ 1 mmol/L in men and ≤ 1.3 mmol/L in women), hypertension (≥ 13 cm Hg), hyperglycemia (≥ 5.6 mmol/L), and high waist circumference (≥ 94 cm in men and ≥ 80 cm for women).

It has been demonstrated that the metabolic syndrome is directly related to the risk of cardiovascular disease frequently followed by mortality. In the United States, the prevalence of the syndrome is about one in three individuals. Because obesity is the major driver of metabolic syndrome development, it must be noted that about 30% of all American adults are presently overweight and about 32% are obese. In Europe, there is a great variability in the frequency of the syndrome between different populations. Thus, in nondiabetic subjects the frequency of the syndrome varied between 7% and 36% for men 40–55 years old and between 5% and 22% for women of the same age. In France, it has been estimated that 23.5% of men and 17.9% of women aged 35–64 years have the metabolic syndrome (http://www.agropolis.fr /pdf/sm/Dallongeville.pdf).

Through numerous epidemiological studies, consumption of fish or n-3 fatty acids from seafood has often been linked to a lower incidence of this syndrome. Some studies have reversed this idea by showing that the plurality of the analyzed diets could obscure the relationships between the metabolic syndrome and the presence of specific dietary fatty acids. It should be noted that the amount of ingested n-6 fatty acids did not appear to influence the inverse correlation between the intake of linolenic acid and the incidence of a metabolic syndrome (Mirmiran et al. 2012).

Nevertheless, it should be noted that recent developments in the enrichment in dietary n-6 fatty acids have often been related to a higher incidence of overweight and obesity. The influence of these new dietary practices may also be present during pregnancy and even from a young age.

The mechanisms underlying the action of n-3 fatty acids on various aspects of the metabolic syndrome are probably different from those contributing to the reduction of risk of cardiovascular disease. Thus, linolenic acid seems to rather modulate the lipoprotein synthesis, whereas EPA and DHA would act on triacylglycerol synthesis, thereby controlling adiposity (Poudyal et al. 2011). In contrast to experiments in animals, the results in humans for glucose homeostasis and insulin resistance are very disappointing. Thus, in obese subjects no effect of fish consumption or fish oil intake could be demonstrated on blood glucose or insulin levels. Similarly, insulin sensitivity could not be modified by this nutritional approach. It is possible that during these interventions over short periods the level of tissular n-3 fatty acids did not reach therapeutic levels sufficient to influence physiological and molecular systems. This explanation seems plausible as supplementation with 3 g of EPA + DHA daily for 6 months proved to be capable of improving the lipid profile and all parameters of insulin resistance after a lipid overload (Derosa et al. 2011). Other long-term experiments would be necessary to clarify the nature and duration of any preventive and curative treatment.

Given the extent of the spread of this disease in developed countries, many epidemiological studies have been undertaken in recent years. In Korea, the study of many people for 4 years has shown that those consuming fish daily (providing about 800 mg of n-3 fatty acids per day) had half the risk of metabolic syndrome than those consuming fish only once a week (40 mg of n-3 fatty acids per day). A high fish intake has been associated with low plasma triglycerides and high HDL-cholesterol. Curiously, in this work the conclusions have been drawn from the results obtained in men. Further investigations have detected a modest but positive effect of linolenic acid intake on the risk of developing diabetes. Most often, epidemiologists show only trends but no significant effects, even after collecting thousands of data. Thus, a recent analysis of the data collected in 4941 participants from the National Heart, Lung, and Blood Institute Family Heart Study did not support an association between dietary n-3 fatty acids and metabolic syndrome (Lai et al. 2013).

The diversity of lifestyles and diets is probably the cause of the difficulty in assessing specific physiological effects. These effects may be further influenced by the balance between the fatty acid series in foods and by a genetic polymorphism remaining to be discovered in this area as in many others. These influences could be detected when comparing studies that are similar but carried out in different countries. Thus, the association between fish consumption and a reduced risk of type 2 diabetes appeared to be more significant in the United States than in other parts of the world, whereas no correlation was found in Asia or Australia. It is the same for the ingestion of long-chain n-3 fatty acids.

A review of all published studies may leave the reader perplexed as to the wisest nutritional approach to prevent or fight against the metabolic syndrome, a recognized precursor stage of much more morbid cardiovascular disorders. It is remarkable that many intervention studies seem more convincing for the identification of a favorable effect of an *n*-3 fatty acid–rich diet on all the factors involved in the metabolic syndrome. The reason could be a better appreciation of the intakes during these experiments than during inquiries where the diet composition can be approximately assessed. We must not forget that a portion of steamed fish is not comparable to the same portion when fried.

Given the gravity of the increased incidence of metabolic syndrome frequently described as an epidemic, it seems appropriate to maintain a supply of *n*-3 fatty acids at values recommended by all international medical societies. The use of rapeseed and walnut oils must therefore be preferred while consuming fatty fish two to three times a week, prepared with the minimum of vegetal oil. In the absence of these natural sources of *n*-3 fatty acids, regular supplementation with purified EPA and DHA must be recommended at an early age and especially in the elderly, who are frequently deficient.

4.2.4.4 Cancers

Relationships between carcinogenesis and *n*-3 fatty acids have been the subject of much research in animals and have led to a consensus in favor of a positive effect of these compounds on the cancerization process both in cultured cells and in various organs.

Thus, rats supplemented with DHA exhibited a significant reduction in breast cancer and mice fed a fish oil–enriched diet showed a growth reduction of grafted tumor prostatic cells or a decrease of the incidence of colon cancer. These experimental results with animals have evidenced the existence of a close link between the production of prostaglandins and tumorigenesis. The anti-inflammatory effect of *n*-3 fatty acid derivatives has often been cited as one of the major aspects of their antitumor action. Although all the mechanisms involved in the action of these fatty acids in the prevention of cancers are not yet elucidated, it has been proposed that one of the main targets could be, as in cardiac cells, the ion channels on cell surface.

On a general level, many experimental models have enabled to show that supplementation with fish oil reinforced the toxic effect of chemotherapy on tumor cells (Baracos et al. 2004). Similarly, several animal studies have verified the usefulness of higher levels of tissue DHA in the potentiation of the effects of chemotherapy and radiotherapy as well as in slowing cachexia.

In addition, some clinical studies have also shown that the status of *n*-3 fatty acids was altered in patients with treated or untreated cancer, suggesting that supplementation with these fatty acids may improve patient health by strengthening the efficiency of traditional therapies and by reducing the side effects frequently observed. A recent clinical research by P. Bougnoux, France, verified that a pretreatment with DHA (1.8 g/day) improved the results of chemotherapy in women with breast cancer complicated with visceral metastases (Bougnoux et al. 2009). As emphasized by N. Hajjaji, France, in 2012, these few results should encourage clinicians to explore this hypothesis of a beneficial effect of a diet enriched with EPA and DHA before or during chemotherapy, an approach that does not alter the normal cells of the body.

In 1967, a study of epidemiological data collected in different countries led investigators to hypothesize that diet could be an environmental factor for the development or decrease of cancer risk. The relative importance of this factor has been rapidly estimated to 30%–40% of cancers in men and 60% in women. Studies in animals have specifically directed the interest of clinicians toward lipids and more accurately *n*-3 fatty acids, which could be a factor in preventing the cancer risk.

Extensive research undertaken over the past 40 years has shown that four cancers (breast, prostate, pancreatic, and colorectal cancers) may be influenced especially by *n*-3 fatty acids contained in dietary lipids.

4.2.4.4.1   Breast Cancer

The incidence of breast cancer in different countries around the world is discussed in Section 4.2.1.1.3.1.

A large epidemiological survey in 32 countries has shown a strong inverse correlation between the incidence of breast cancer and the consumption of fish. Among the first major studies in Europe, a large Norwegian survey found that women consuming poached fish at least five times a week had a 30% reduced risk of cancer compared to those consuming no more than twice a week. These results have been confirmed by a study carried out in Spain. One of the largest prospective studies carried out in 2003 by M. Gago-Dominguez, the United States, with more than 35,000 women came to the conclusion that there is an inverse relationship between consumption of *n*-3 fatty acids from marine animals and risk of breast cancer.

Analysis of fatty acids in the blood (serum and erythrocytes) led to a better assessment of fish consumption and enabled to verify the existence of a significant protective effect of *n*-3 fatty acids, the analysis of these markers in adipocytes showing that only linolenic acid is related to a lower risk of breast cancer. Without any nutritional investigation, the work done by Prof. P. Bougnoux's team has shown, in 241 patients with invasive breast cancer compared with 88 controls, that there is an inverse relationship between the cancer risk and the content of linolenic acid and DHA in breast adipose tissue (Maillard et al. 2002). Further works have emphasized the importance of the level of *n*-6 fatty acids on the protective effect of *n*-3 fatty acids. Again, the balance between the two essential fatty acid series must be seriously considered. The specific effect of linolenic acid is not very clear when analyzing the work of A.C. Thiebaut, France, in 2009, showing that the risk of breast cancer is inversely related to the amount of linolenic acid when it comes from plants and fruits but directly related if it comes from nuts and cooked foods. These results once again highlight the importance of the nature of the fatty acid food sources in the assessment of disease risks for the consumer. Unfortunately, for breast cancer, as in other cancers, few investigations have taken into account these data, as a result of ignoring this problem or, more often, availability of incomplete food composition tables to investigators working on this topic.

A long-term prognosis in patients with breast cancer remains difficult to establish but could still be found in some specific situations. Thus, a study carried out in 1993 by E. Lund, Norway, has shown that the wives of Norwegian sailors with breast cancer and frequently consuming fish have a mortality reduced by 30% compared to women consuming few fish. In contrast, a similar study carried out among American nurses followed for 18 years did not detect any effect of a diet rich in *n*-3 fatty acids on survival after initiation of cancer. Another large American survey, published in 2011 by R.E. Patterson, has shown that women diagnosed with and treated for breast cancer consuming an average of 365 mg/day of a EPA + DHA mixture had a mortality risk for all causes reduced by 40% after 7 years compared to women consuming an average of 18 mg/day. It is certain that the statistical results can be even sharper when the differences in consumption are higher. This nutritional and experimental effect is probably at the origin of the disagreements observed between the various investigations in this field.

It appears that the long-term interest of a diet rich in fish in the prevention of breast cancer is most evident in the Asian and Mediterranean countries than in the Nordic countries where fish is traditionally consumed. This evidence is often only a trend, statistically insignificant, despite the efforts of scientists to gather the highest number of subjects and avoid the most known interferences (age, age-associated diseases, endocrine status, diet, nutritional supplements, ethnicity, etc.). Other factors may unfortunately affect the measured parameters, often without the knowledge of the experimenter (diet, vitamins, antioxidants, pollutants, genotype, etc.).

Although supplementation with *n*-3 fatty acids did not provide expected results for the prevention of breast cancer, the results of epidemiological investigations led to the recommendation that women must maintain a diet balanced between meat and vegetables, with a fat intake containing polyunsaturated fatty acids with an *n*-6 to *n*-3 fatty acid ratio approaching 5 as recommended by international medical authorities.

4.2.4.4.2 Prostate Cancer

The prevalence of prostate cancer in several parts of the world and in France is discussed in Section 4.2.1.1.3.3. Its prevention in developed countries is facilitated by using a detection test with a specific antigen (prostate serum antigen [PSA]) in the blood that likely increases the figures of its prevalence but also allows early screening. This cancer is generally recognized as being hormone dependent, and its prevalence varies according to ethnicity.

Several studies in animals or in cultured cells have shown that the development of prostate tumors can be inhibited by the addition of EPA or DHA in food or in culture media.

Most published surveys taking into account the amount of ingested *n*-3 fatty acids from detailed marine fish consumption have shown an inverse correlation between the acid intake and the incidence of prostate cancer, although sometimes no effect has been reported.

The use of *n*-3 fatty acid analysis in whole blood as biomarkers allowed a more precise estimation of the fish consumption. Thus, it could be verified in nearly 15,000 men followed for 13 years that the higher the content of long-chain *n*-3 fatty acids the lower the risk of prostate cancer (Chavarro et al. 2007). In contrast, linolenic acid levels appeared to be not at all related to cancer risk. Nevertheless, some investigations have found a positive association between DHA levels in plasma phospholipids and the prevalence of high-grade prostate cancer (Bradsky et al. 2011). These findings suggest a greater complexity of the effects of n-3 fatty acids with respect to prostate cancer risk.

The literature review revealed a more pronounced effect of EPA and DHA on aggressive and well-developed tumors than on tumors in early growth stages. If these results are confirmed, more research should be focused on patients with tumors that are undergoing rapid development and have already spread.

One of the proposed mechanisms for the inhibitory effect of n-3 fatty acids on prostate carcinogenesis is their ability to inhibit the biosynthesis of eicosanoids derived from arachidonic acid (prostaglandins and leukotrienes) by means of a key enzyme, cyclooxygenase-2 (COX-2), known to be overexpressed in prostate tumors. The discovery of polymorphic forms of a single nucleotide of the *COX-2* gene (variant SNP) has helped in refining the relationships between n-3 fatty acids and prostate cancer risk. Thus, a detailed investigation of consumption in Sweden including 1500 prostate tumor–bearing subjects has shown a very significant inverse correlation between increasing intake of marine fish and prostate cancer risk, but this was only in individuals carrying one of the polymorphic forms of *COX-2* (Hedellin et al. 2006). This relationship was confirmed in 2009 by V. Fradet, the United States, when studying almost 1000 subjects, 466 having aggressive prostate cancer.

More accurate clinical work will naturally be necessary to clarify the intimate mechanisms underlying these effects and to define future treatment protocols. Six studies in progress on the effects of n-3 fatty acids and prostate cancer are listed on the international declarative website http://www.clinicaltrials.gov.

Recent nutritional research provides new hope in the fight against prostate carcinogenesis, a struggle already foreseen by many studies suggesting the usefulness of a preventive or curative treatment through a marine fish–rich diet. The interest of the new nutrigenic way resides in its perfect coexistence with traditional treatments which may get an increased efficiency and a decrease of the side effects of medications.

4.2.4.4.3 Colorectal Cancer

The prevalence of the colorectal cancer around the world and in France has been exposed in Section 4.2.1.1.3.2. France is one of the developed countries where the risk of colorectal cancer is high, but it is among the eight countries in the world with the best survival rate. The responsibility of lifestyle in the development of colorectal cancer is important. The protective effects of the consumption of fresh fruits and vegetables and physical activity are established as well as the detrimental effect of an excessive energy intake, especially one characterized by high intakes of fatty meats.

In 1975, a comparative study of the incidence of disease and the dietary habits in different countries implicated the lipids coming from excess meat consumption. Authors recommended the gradual replacement of much of the red meat (beef and sheep) in the diet by fish, and optionally by poultry. Numerous studies have been devoted to the possibility of reducing the incidence of colorectal cancer in increasing fish consumption. Although the majority of results have shown beneficial effects, their heterogeneity led the American Institute of Cancer to the conclusion that there was "only limited evidence that eating fish protects against colorectal cancer" in 2007 (WCFR/AICR 2007). The main source of difficulties of interpretation are, according to this report, the diversity of fish preparations consumed and a possible balance between fish and red meat on the participants' menu. Along with this report, a meta-analysis showed that fish consumption moderately reduced the risk of colorectal cancer, but the results seemed to be more evident in people eating fish every day (Geelen 2007). It has been calculated that each additional fish serving consumed per week added a 4% reduction in the risk of colorectal cancer.

Analysis of blood fatty acids as markers of fish consumption has permitted in Japanese subjects to better ensure that those with the highest blood levels of total n-3 fatty acids, linolenic acid, and DHA have the lowest risks of colorectal cancer (-76%, -61%, and -77%, respectively, for these fatty acids). This relationship was found again in the United States but only in subjects untreated with aspirin (Hall 2007). This last result emphasized the influence of the proinflammatory eicosanoids produced by COX-2 in a tumor whose synthesis has been blocked by aspirin. This inhibitor, commonly used in all pharmacopoeias and over the counter, is already known to inhibit the growth of colon cancer by gene expression mechanisms, as highlighted by S. Pathi, the United States, in 2012. This work will certainly help to explore the mode of action of n-3 fatty acids in their power to reduce the risk of carcinogenesis or tumor development in the colon. The example emphasizes the possible and often unexpected way that epidemiologists should avoid to increase the relevance of statistical results. Medication, sometimes required for another disease, may cause the inconclusive results reported by many studies.

To verify experimentally the effects of n-3 fatty acids, several supplementation trials with fish oil were undertaken early. For reasons that remain unexplained, even studies using analyses of fatty acids in tumor cells and proliferation rates have not delivered a coherent message that could guide clinicians to future therapies. However, it seems that the required amount of n-3 fatty acids ingested to slow the progression of colorectal cancer must be important, equivalent to at least one serving of fatty fish each day (Pot et al. 2009). A total of 11 studies in progress on the relationship between n-3 fatty acids and colorectal cancer are listed on the official website http://www.clinicaltrials.gov.

In conclusion, it seems increasingly necessary that the dietary rules already recommended for the prevention of colorectal cancer (less red meat and more fibers) should be completed with the recommendation to eat at least two to three servings of fatty fish per week. For exceptions to this rule, supplementation providing an equivalent amount of fish oil is strongly recommended.

4.2.4.5 *Nervous Diseases*

The idea that a nutrient can affect the mental functions of humans seemed surprising at first, even to the physiologist or the medical doctor, in the early twenty-first century. It is quite difficult to realize that intellectual capacities, such as learning and memory, may be dependent on the lipid diet. More than 20 years ago, J.M. Bourre summarized in his book all aspects of dietetics appearing with the progress of research increasingly inseparable from the intellect and the mind ("The dietetics of the brain, intelligence and pleasure") (Bourre 1990).

Knowledge of the fatty acid composition of the adult mammalian brain and its development from birth prompted early researchers to establish links between lipids and brain function. Indeed, since the early 1960s we know the relative abundance of DHA in brain membranes, this fatty acid constituting about one-third of the unsaturated fatty acids in the brain. It may be considered that the year 1967 is the first step of this story when B.L. Walker, Canada, showed that the DHA content of rat brain was dependent on the content of their dietary linolenic acid (18:3 n-3). These encouraging data for future investigations naturally led investigators to consider the n-3 fatty acid content of the brain in the development of its higher functions. Numerous studies have explored the influence of n-3 fatty acids, and especially the optimal value of the n-6 to n-3 fatty acid ratio on the learning capacity of animals. In humans, the investigations aimed primarily at possible relationships between intake of n-3 fatty acids and behavior, cognitive performances, and their alteration during aging or in psychiatric disorders. It is highly interesting to note that the recent discovery of a relative elongation of telomeres in leukocytes collected from subjects supplemented with EPA + DHA opened up the possibility of a direct influence of n-3 fatty acids on the aging process (Kiecolt-Glaser et al. 2012).

It should be emphasized that psychiatric diseases are among the most common diseases in the world and are at the forefront of the major causes of disability (WHO, 2001, http://who.int/whr/2001/chapter2/en/index3.html). Published studies have reported that the 12-month prevalence of any mental illness excluding substance use disorders is 24.8% among adults (42.6 among adolescents) and that nearly 50% will develop at least one mental illness during their lifetime (U.S National Comorbidity Survey 2014, http://fas.org/sgp/crs/misc/R43047.pdf). The economic burden of mental illness in the United States is substantial; it was about $300 billion in 2002. In Europe, one in four people is affected, with an involved cost estimated at nearly €800 billion in 2010. In France, one in five people suffers, representing an annual cost of more than €100 billion. We now understand the interest of researchers on treatments that would be based on small changes in eating habits.

As mentioned in 2006 by L.C. Reis, the United States, it is remarkable that modern data on the importance of certain lipids mainly from fish are strengthening the old religious and medical practices of several millennia, practices attributing to fish the symbols of happiness and peace in almost all cultures.

4.2.4.5.1 *Development of the Nervous System*

n-3 Fatty acids, especially DHA, have proved very early to be molecules required for the growth and development of the nervous system in relation to their abundance in the brain cell membranes. For over 20 years, studies in mammals have shown the involvement of *n*-3 long-chain fatty acids in neuronal function and in neuro-physiological and behavioral manifestations as well. It is now well proved that a deficiency in *n*-3 fatty acids during key phases of development significantly alters, to varying degrees, cognitive functions in all experienced animals. In humans, the accumulation of DHA in the brain is mainly in the last trimester of pregnancy but continues during the first 2 years of life, and hence the importance of a regular intake of this fatty acid in pregnant women and young children. Finally, in adults, DHA will approach 30% in some brain lipids and nearly 40% of retinal lipids.

4.2.4.5.1.1 *Paleoanthropological Studies*

The need for an intake of *n*-3 polyunsaturated fatty acids, especially DHA, to ensure normal brain development has been known for over 40 years through the work of M.A. Crawford at Imperial College in London. His thesis has prompted paleoan-thropologists to investigate what could be the dietary conditions required to facilitate the encephalization process that has characterized the development of the cerebral cortex of hominids especially in the last million years. The most widely accepted hypothesis regarding their evolution proposed that *Homo erectus*, originally hunter and gatherer in the African savannah, migrated from the Rift Valley in East Africa toward lacustrian or coastal areas where they could evolve into a bipedal position progressively upright. During this geographical evolution, the diet of our ancestors likely became increasingly diversified toward animal preys rich in *n*-3 polyunsatu-rated fatty acids (EPA and DHA), molecules absent in plant resources.

It is interesting to note that bipedalism appeared in *Australopithecus* probably 5 million years ago and settled permanently at the time of *Homo habilis*, 3 million years later. This evolution, which was accompanied by the discovery of tools, was contemporary with the significant increase in brain size of our ancestors, a phenom-enon known as encephalization. Bipedalism, probably favored by attending aquatic areas, could mainly facilitate the capture of fish, mollusks, or crustaceans rich in EPA and DHA, in addition to land animals. Thus, *Australopithecus*, with a brain not exceeding 500 cm^3, remained in the African savannah while the genus *Homo* gradually shifted during its migration to the coasts of Africa and Eurasia in areas rich in aquatic animals. At the same time, it evolved into *Homo sapiens*, character-ized by a brain three times larger than its bipedal ancestors. As M.A. Crawford pointed out, it is curious to observe that the cultural development of early *H. sapi-ens* was exclusively confined to lakes or areas near marine environments. Although controversial, the "aquatic primate theory," initiated in 1948 by the German anthro-pologist Max Westenhöfer, has recently been developed by integrating other anatom-ical and physiological characteristics (Morgan 1997). The fish bones and shellfish detected in 2010 by D. Braun, South Africa, in prehistoric sites near the great lakes of East Africa, where the early *Homo* was discovered, provided serious credit to the Crawford hypothesis. The varied aquatic food (fish and shellfish), even from freshwater, mixed with other terrestrial preys could supply throughout the evolution of hominids the necessary amounts of DHA, and also of arachidonic acid. Current knowledge may only stress the importance of these compounds in the mechanisms of encephalization that will condition the future cultural development of our species. As highlighted by the experts of this problem, "the specific brain nutrition had and still has the power to affect the evolution of hominid brain" (Broadhurst et al. 1998). The evolution of the modern human could thus be dependent on maintaining or, as its ancestors, increasing the consumption of foods rich in *n*-3 fatty acids.

The hypothesis of a close relationship between the pre-hominid brain size and the consumption of aquatic animals rich in *n*-3 fatty acids is also compatible with the discovery of a genetic modification (haplotype D) promoting the conversion of linolenic acid into EPA and DHA, a modification that could appear more than 400,000 years ago at the time of the great migration of hominids out of Africa (Ameur et al. 2012).

4.2.4.5.1.2 Maternity and Brain Development: Infant Foods
Shortly after the discovery of the role of DHA in retina function, the positive influence of linolenic acid–enriched diet on learning ability was demonstrated for the first time in young rats by M.S. Lamptey, Canada, in 1976. These initial findings were later confirmed with more sophisticated tests in young animals fed a diet enriched with linolenic acid (perilla oil) or in their mothers (7 weeks before birth), or in rats fed a food containing soybean oil for two generations. Numerous studies of the relationships between cognitive performance and *n*-3 fatty acids, mainly DHA, have been carried out in animals. Multiple experiments have also shown that a deficiency in *n*-3 fatty acids induced disturbances of brain function, especially when the deficiency was occurring at critical development periods such as neurogenesis, synaptogenesis, and myelination. The conditions to go back to a normal brain function could be related to the age and the duration of postdeficiency feeding (Denis et al. 2011).

In humans, the first studies in the early 1980s by R.T. Holman, the United States, based only on unique clinical cases have shown that linolenic acid is essential for nerve function, as in animals. It is now well known that during the last trimester of pregnancy, and also during the first 2 years of life, DHA is the most important fatty acid for the normal development of motor, sensory, and cognitive abilities of the individual. Unfortunately, the relative importance of EPA compared to DHA in these phenomena is not yet clear. It would appear, since the studies carried out in 1996 in infants by S.E. Carlson, the United States, that an excessive intake of EPA (as fish oil) tended to decrease the content of phospholipid arachidonic acid, this fatty acid being essential for cell signaling. In addition, premature infants have a high risk of DHA deficiency because of a too low intake *in utero*. Because linolenic acid supplementation was inefficient, maternal DHA supplementation seemed to be the only way to ensure its supply to the child through breast-feeding (Francois et al. 2003). Therefore, it is now recognized that all at-term and preterm newborns require dietary DHA; evidence has been provided that it ensures a normal visual function. Although cognitive performance is difficult to assess in very young children, a review of major published studies suggests a positive correlation between DHA and cognitive function, obviously without the guarantee of a precise causal correlation between the two variables. One surprising approach was that of a large survey in 28 nations of the possible correlation between breast milk levels of DHA and children's cognitive performances estimated with the Programme for International Student Assessment math score (Lassek et al. 2013). The study has shown a positive and highly significant contribution of DHA to math scores. Together, dietary fish (positively) and total fat amount (negatively) explained 61% of the variance in maternal milk DHA.

Although not definitive, these investigations are in favor of breast-feeding, breast milk supplying DHA amounts consistent with the known infant requirements. Though often questioned, this natural practice is described as being the most beneficial for the cognitive development of the child, an advantage perceptible even in adulthood. Admittedly, as with all psychological approaches, some doubts remained and a detailed analysis of the results led to the question of their scientific value. Certainly, these research efforts have helped to encourage parents and the manufacturers of children's food to opt for DHA supplementation, an approach already adopted in many countries, including the United States in 2002. In 2008, a group of international experts concluded that the benefits of adding DHA (and arachidonic acid) to foods were evident for children (Koletzko et al. 2008). This practice has been confirmed by many scientific and medical societies in various countries.

Further research is needed to better assess the benefits of an n-3 fatty acid supplementation in the pregnant woman or in the newborn, in waiting for a clarification of the required doses and the most favorable periods for that supplementation. The incidence of n-3 fatty acid supplementation in the mother needs to be explored in the context of improvement of the cognitive performances of the future child. An epidemiological investigation in this area was carried out in Denmark in a large number of children from mothers consuming close to 500 g of fish per day during pregnancy. It has been found that an increased consumption of fish was associated with higher motor and mental development at 6 and 10 months old. A similar relationship between duration of breast-feeding and development was also highlighted. These positive aspects of mother–child interactions have recently been supported by the observation of an inverse correlation between the dietary n-6 to n-3 fatty acid ratio and the child's neurodevelopment (Bernard et al. 2013). Furthermore, the results suggested that maternal diet influenced the child's brain during the fetal life but not during or by breast-feeding. As emphasized by the authors, the data argue in favor of promotion of breast-feeding and limitation of intakes of n-6 fatty acids during pregnancy.

A Norwegian study on premature infants supplemented for 9 weeks with 32 mg of DHA and 31 mg of arachidonic acid per 100 mL of milk has clearly shown that a supplementation after 6 months is associated with better mental and motor development as well as better recognition memory (Henriksen et al. 2008). Motor development seems little changed after a regular intake of DHA from birth as the ability to sit down without the help of a support happened only 1 week before control subjects. A recent review of the literature over the past decade (over 100 works) summarized the demonstrated benefits obtained with a maternal diet enriched with DHA (Morse 2012). This maternal preparation proved to be recommended within permitted doses for its effects on the lengthening of gestation in cases of high-risk pregnancy, increase in weight and size of the newborn, increase in visual acuity, and improvement of various neurophysiological parameters.

The different kinds of testings and analytical procedures (psychological and psychomotor tests) probably explain the uncertainty that still exists among health professionals on a preventive treatment in altering the diet. Many investigations are needed to assess the long-term effects, such as improved school performances and lowering of learning disabilities. Despite the still imprecise results actually obtained, official recommendations on the intake of *n*-3 fatty acids in pregnant women and young children should be actively spread in the population. Recognition of the diversity of DHA-enriched products for pregnant women available in drugstores shows the general awareness of the population about a possible and easy improvement of the intellectual performance of future children. A better understanding of the metabolism of these fatty acids in development would be desirable to better assess the requirements of different age groups without forgetting the prematures, more and more numerous in all developed countries (about 13% of births in the United States, from 5.5% to 11.4% in Europe, and 6% in France).

> The European Union has approved a claim relating DHA to eye and brain development in fetuses and breast-fed infants. In France, the AFSSA agreed in March 2010 that DHA is an essential fatty acid for its role in the structure and function of brain and eyes and set a minimum intake of 0.32% of the total fatty acid intake. Thus, despite varying results from different studies, experts recommend that artificial milks must contain 0.2%–0.3% of fatty acids as DHA, a level similar to that observed in breast milk.

4.2.4.5.2 Mental Health: Cognitive and Sensory Performances

The influence of *n*-3 fatty acids on cognitive functions has often been discussed in animals. Generally, a deficiency of these compounds induced a decrease in learning abilities while DHA enhanced them. In humans, the situation is more complex due to a poor control of food intake and the complexity of neurobiological mechanisms. At the nutritional level, clinicians have few accurate results based on consumption surveys and sometimes they have more reliable results based on *n*-3 fatty acid contents in plasma and erythrocytes.

4.2.4.5.2.1 Cognitive Performance in Children and Adults

The role of *n*-3 fatty acids demonstrated during the brain development in very young children may be also operating in adults, considering the well known neuronal plasticity. This latter property is based on both the rapid turnover of synaptic membranes and the neurogenesis, a mechanism now recognized by all neurobiologists. In rats, the recent demonstration of a close link between the intake of *n*-3 fatty acids (linolenic acid and DHA) and the improvement of neuronal and behavioral plasticity underlined some of the mechanisms involved (Bhatia et al. 2011).

Low blood *n*-3 fatty acid concentrations have been reported several times in children with various behavioral or learning difficulties. It was recently verified in a population of 493 healthy schoolchildren aged 7–9 years that lower blood DHA concentrations were significantly associated with poorer reading ability and working memory performance measured using the British Ability Scales (II) (Montgomery et al. 2013). These findings suggest that dietary supplementation with DHA and/or EPA could be beneficial not only for children suffering from heavy learning difficulties but also for the general school population.

It is well known that long-chain n-3 fatty acids rapidly accumulate in the human brain in the perinatal period, but this accretion continues during late infancy and early childhood. Thus, clinicians have hypothesized that DHA and/or EPA supplementation could have an effect on brain development during these periods. The main difficulty is that the common psychological test (Bayley Scales of Infant Development) provides only global indicators of cognitive functioning but does not reveal subtle differences in very young children. A recent investigation of fish oil supplementation in infants 9 months of age used a nonstandardized laboratory measurement (free-play paradigm) to appreciate specific cognitive processes through the context of toy play, known to change markedly across brain development (Harbild et al. 2013). This research has identified in supplemented infants a positive change in the attention scores in free-play tests (increased number of looks) associated with an increase in erythrocyte EPA content and a decrease in blood pressure.

An intervention study with young indigenous Australians aged 3–13 years supplemented with fish oil enabled the observation of significant improvements in Draw-A-Person, testing nonverbal cognitive development, but without effects in reading or spelling (Parletta et al. 2013).

Further works are needed using several methodologies to appreciate more clearly the importance of possible changes induced by various n-3 fatty acid supplements in infants. Moreover, several clinicians suggest that the influence of these supplementations on cognitive development should be examined in the long term and not only in infants, as some cognitive abilities emerge later in early childhood.

Some studies have even explored the benefits derived from the supplementation of healthy adults with EPA and DHA. Thus, in 2005 G. Fontani, Italy, found that a daily intake for about a month of 4 g of fish oil improved the mood profile with increased vigor and cognitive performances such as attention and responsiveness. In 2012, A. Nilsson, Sweden, demonstrated that an intake of 3 g/day of EPA + DHA for 5 weeks may significantly improve the performance of working memory in patients of average age 63 years having no memory complaints; however, the treatment did not change the spatial perception.

A beneficial effect of DHA on cortical functions has been determined on the basis of neurophysiological measurements. Similarly, the daily intake of 1 g of fish oil for 3 months in young subjects caused better blood flow in the cerebral cortex during cognitive tests, but no apparent improvement in response scores occurred (Jackson et al. 2012). The investigations of R. Narandran, the United States, have confirmed these results and even offered the prospect of a possible improvement of memorial performance after the ingestion of EPA + DHA (2 g/day) by young adults who already had an excellent working memory (Narandran et al. 2012).

More recently using functional magnetic resonance imaging (MRI), it has been found in young adults that a 30-day supplementation with EPA was more effective than DHA in enhancing neurocognitive functioning (Bauer et al. 2014). Furthermore, these authors have also done a meta-analysis showing that, in applying the principle of neural efficiency as established by R. Haier ("smart brains work less hard"), an increase in EPA intake was more advantageous than DHA in reducing "brain effort" relative to cognitive performance.

Recent research has confirmed that supplementation with EPA and/or DHA may modulate brain function, at least on a physiological basis. Further works on a longer term will be required to accurately examine with specific and constantly evolving techniques the physiological effects and especially the cognitive effects of each n-3 fatty acid.

4.2.4.5.2.2   Age-Related Decline

Numerous studies have shown a reduction in the concentration of polyunsaturated fatty acids in the brain with age. The cause of this reduction was attributed to the lower efficiency of both their transport from the blood and their metabolism in the brain itself. To complicate the picture, the progressive slowing of some purely nutritional steps such as nutrient intake or intestinal absorption with age may be added. These changes may lead to a significant drop in the level of brain DHA. It is most significant that, at least in animals, the most important change in lipid composition was observed in the hippocampus, an area known to play a central role in memory and spatial orientation.

The benefits induced by fish consumption, or EPA and DHA ingestion, in the fight against cognitive decline associated with age remain difficult to estimate. Some works concerned with mild cognitive impairment (MCI) have described in detail the benefits that healthy adults could derive from a frequent consumption of marine fish. MCI, which affects almost exclusively the memory, is a predictor in 10%–15% of cases of a progression toward dementia in the year following diagnosis.

Some epidemiological and psychological studies in many subjects have shown that fish consumption, and therefore n-3 long-chain fatty acids, was associated with a reduced risk of cognitive function loss (memory, comprehension, and psychomotor speed), but some other studies did not confirm completely this association. An original approach was recently made by a Swedish team who positively associated the nutritional intake of EPA and DHA with global cognitive performances and especially with the volume of brain gray matter measured by MRI (Titova et al. 2012). The objective of these measurements by brain imaging demonstrated at least one indisputable anatomical effect, even if the exploration results are subject to understandable reservations.

The analysis of many studies done in this field allowed the conclusion that there is a more or less important benefit of n-3 fatty acids on cognitive activity during aging. The variability of results observed in all observational studies may result from unreliable information provided by the questionnaires of food consumption. As pointed out by G.L. Bowman, the United States, in 2011, probably the best approach to correct this inaccuracy is to use the analysis of n-3 fatty acids in plasma or erythrocytes. Thus, a French study with 246 people aged 63–74 years followed for 4 years has shown that a higher proportion of DHA (but not EPA) in erythrocyte membranes was significantly associated with a lower risk of cognitive decline, assessed by tests of orientation, attention, calculation, memory, language, and visual construction (Heude et al. 2003). A U.S. study showed similar results, the relationship being stronger when the cognitive decline was appreciated by measurements affecting the verbal field. Many other studies have been devoted to this problem, but the variability and diversity of the used psychological tests again required prudence in the conclusions and possible recommendations. A point under discussion remains the influence of dietary n-6 fatty acids in the magnitude and diversity of results. It seems that, even in the absence of a direct causal relationship, the blood n-3 fatty acid composition is a reliable biochemical indicator of cognitive deterioration associated with aging in the absence of dementia and cardiovascular disease (Tan et al. 2012). Despite these results, the significance of the blood DHA status remains unclear, is it a reflection of food intake or metabolism? This question is fundamental for studies targeting aged people. Future research in this area should take into account this alternative and perhaps explore the alteration of the catabolism of n-3 fatty acids during aging and how to correct any deviations induced by nutritional changes.

Besides these observations, what are the results of the experiences of EPA and DHA intake on cognitive decline? It must be recognized that these interventions are far from clarifying the problem, probably because of the wide variety of protocols and the difficulties in assessing neuropsychiatric symptoms. For example, some authors were able to detect improvements in the learning function while others were unable to show any change. A recent meta-analysis of the Cochrane center underlined the fact that presently a supplementation with n-3 fatty acids provided little benefit on cognitive function in mentally healthy elderly, but no study has examined the effects on the incidence of dementia (Sydenham et al. 2012). The report concluded with the need of long-term works to identify more important changes during aging. It must be noticed that in the analyzed works the supplementations were performed with fish oils and also with various types of purified n-3 fatty acids.

The discovery of a close association between cognitive decline and the risk of Alzheimer's disease and the presence of the allele of the apolipoprotein epsilon 4 (*APOE* isoform) recently modified the approach to these complex relationships. As mentioned by A.M. Minihane, England, in 2000, this allele could indeed interact with n-3 fatty acids and thus promote the genesis of cardiovascular disease and secondary brain function disorders. The validity of this explanation is based on the observation of a closer association between the n-3 fatty acid concentration in erythrocyte membranes and the preservation of cognitive performance in elderly carriers of the *APOE* ε4 allele (Whalley et al. 2008). All the mechanisms involved in this gene–food interaction and cognitive performance are still poorly known, the involvement of nonlipid compounds remaining possible. Future research should take into account not only the nature and quantity of ingested fatty acids but also the genetic status of *APOE* in the examined subjects. If these approaches are confirmed, any attempt to prevent the loss of cognitive performance related to age must be preceded by a genetic study to increase efficiency.

A plan of federative research, COGINUT, was launched in 2007 in France by the INSERM to study "the impact of the nutritional status in polyunsaturated fatty acids and antioxidants on brain aging (dementia, cognitive decline, mood disorders) in the elderly. This project will help to delay the pathological brain aging by nutritional specific recommendations, adapted food products or nutritional supplements." The project was nested in the Study of Three Cities, which included more than 9000 volunteers aged 65 years and over living in Bordeaux, Dijon, and Montpellier and followed up every 2 years for 7 years. It was demonstrated that older persons who have at least one fish meal per week and consumed fruits and vegetables on a daily basis had a 30% decreased risk of severe cognitive deterioration in the following 4 years. With the regular consumption of olive oil, these foods are basic components of the Mediterranean diet, which is also associated with better cognitive performances.

What could be the biological links between *n*-3 fatty acid intake and memory consolidation? Outside the field of genes, the positive impact of *n*-3 fatty acids on the lowering of brain disturbances associated with aging could be at the level of protection of glutamate receptors and also at reducing inflammation linked to memory loss (Kelly et al. 2011). In this last field, it has been suggested that the neuroprotective action (as for the cardioprotective effect) of *n*-3 fatty acids lies in the inhibition of the synthesis of TNF-α and interleukins and in the increase in acetylcholine synthesis. The demonstration of an altered brain glucose metabolism in animals deficient in *n*-3 fatty acids has raised the hypothesis of a beneficial role of DHA in the fight against brain aging and even Alzheimer's disease by increasing neuronal glucose metabolism (Cunnane et al. 2011).

The vast majority of works published in the last 5 years are in favor of maintaining and improving even modest cognitive performances in older people by consuming fish or dietary supplements rich in EPA and/or DHA. Given the low cost and safety of these fatty acids, current data allowed to incite the population to have a diet rich in *n*-3 fatty acids or at least DHA at a level of at least 500 mg of DHA per day, as recommended by health authorities (Section 3.2.1.3.2).

4.2.4.5.2.3 Dementia, Alzheimer's Disease

Some consequences of brain aging are among the most devastating. This aging is thus often characterized by dementia, most commonly caused by Alzheimer's disease. It affects more than 25 million people worldwide, 5.2 million Americans, and about 7 million Europeans (EU 27). In France, about 860,000 people currently suffer from this disease, with an incidence of 225,000 new cases per year. The evolution for the coming years is quite dark because considering the current trend it may be expected to have in France more than two million people in 2040 (e.g., 3% of people with dementia). Dementia has significant social and economic implications in terms of direct medical costs, direct social costs, and costs of informal care. The total global societal costs of dementia have been estimated by the WHO to be $604 billion in the world. In the United States, Alzheimer's costed the nation $203 billion in 2013. This number is expected to rise to $1.2 trillion by 2050 (http://www.alzdiscovery.org). In France, the global cost has been estimated as $26 billion (€18.8 billion), for example, 0.6% of GDP.

Alzheimer's disease is the most common dementia in elderly patients (approximately 70% of the neurodegenerative diseases), the statistical risk of developing the disease doubling for every 5 years after age 65. This chronic disease is characterized by a progressive deterioration of cognitive functions including memory, judgment, decision making, language, orientation, and so on. Neuropathologists add other symptoms such as neuronal plasticity alteration (selective loss of neurons and synapses), intracellular deposit of insoluble fibrillar proteins, and formation of extracellular senile plaques. These plaques, called amyloid plaques, are mainly formed of isoforms of β-amyloid peptide (Aβ), the most abundant being Aβ40 and Aβ42, the latter characterizing advanced forms of the disease. In most cases, and as for cognitive decline (see earlier), possession of the epsilon 4 allele of the *APOE* gene (15% of the population) is a genetic susceptibility factor that increases by about four times the risk of developing this disease.

In the absence of pharmacological treatments, it is possible, as with cognitive decline, to consider that the lipid intake may modulate the expression of this genetic factor in Alzheimer's disease. With the development of animal models and transgenic cell cultures, it has become increasingly clear that supplementation with EPA and DHA appears to be able to reduce the deposition of β-amyloid, a harmful peptide for the nervous system. Numerous studies have also shown that in addition to its anti-inflammatory role DHA could enhance neuronal survival. These positive developments have caused clinicians to naturally decide to explore the relationships between *n*-3 fatty acid dietary intake and incidence of Alzheimer's disease. Given this ongoing epidemic progression, it seems urgent to define simple and inexpensive processes to be applied rapidly and safely, as early as possible and preferably in genetically predisposed subjects.

Autopsy examinations have shown that the concentration of DHA in brain tissue was reduced in patients with this disease. It is very noticeable that these patients frequently display a reduction of brain volume and hippocampal atrophy, these changes being themselves associated with a decrease in the erythrocyte omega-3 index. These relationships are particularly corroborated by recent studies using MRI (Pottala et al. 2014).

Some epidemiological studies have been carried out to establish a correlation between DHA and prevention of symptoms. It must be recognized that before 2005 the number and quality of works did not encourage clinicians to make relevant conclusions so that no clear recommendation could be made by the American Agency for Healthcare Research and Quality on the effects of *n*-3 fatty acids on cognitive functions and their pathological degradation (http://www.ncbi.nlm.nih.gov/books /NBK37650/, Evidence Reports/Technology Assessments, No. 114, 2005).

Since this critical review, nine large epidemiological and convincing studies in a dozen countries have established that *n*-3 fatty acid intake, verified and assessed by their blood content, decreased the risk of cognitive decline, dementia, or Alzheimer's disease (Cunnane et al. 2009). Adding other works published elsewhere, 19 reliable studies on this topic have been carried out between 1997 and 2008. Although 2 of them have shown no effect, 17 others have reported an inverse relationship between the risk of Alzheimer's disease and the *n*-3 fatty acid dietary intake. Several other epidemiological studies published recently came to the same conclusions. It is interesting to note that in the elderly the highest blood concentration of *n*-3 fatty acids was associated with lower plasma concentrations of peptide Aβ42, an undisputed marker of a lower risk of cognitive decline including Alzheimer's disease (Gu et al. 2012).

In France, two studies have contributed to the understanding of this problem. The PAQUID study has shown an inverse association between the frequency of fish consumption and the risk of developing dementia or cognitive decline in 7 years, results confirmed by the Study of Three Cities (Bordeaux, Dijon, and Montpellier). This latest survey made clear that weekly fish consumption was associated with a marked reduction in the incidence of dementia, but only in individuals not carrying the epsilon 4 allele of the *APOE* gene.

Unfortunately, it must be recognized that the supplementation experiments with EPA and DHA have brought up few convincing results, the favorable interpretations raising several questions on the experimental protocols and on the result analyses (Cunnane et al. 2009). The studies suggested that the effects of *n*-3 fatty acids were closely linked to the state of development of the disease, these fatty acids being more efficient in the early stages. The accurate assessment of early symptoms seems to be necessary to ensure the success of this nutritional therapy, as for the other pharmacological or behavioral treatments known to date. This is the position taken by one of the greatest experts on the subject G. Cole, the United States, who recognized that DHA had the ability to slow down several deleterious molecular mechanisms in the nervous system at the beginning of the disease (prodromal stage) (Cole and Frautschy 2010). This clinician openly militates for early nutritional intervention especially in patients at risk, including the administration of DHA and antioxidants. To improve future research in this complex field, P.A. Dacks of the Alzheimer's Drug Discovery Foundation, New York, provided in 2013 a practical overview of the use of *n*-3 fatty acids for dementia prevention. He suggested that the supplementation experiments have not targeted people deficient in EPA and DHA, a population at high risk of dementia or cognitive decline. He emphasized also that positive results could be limited to *APOE4* noncarriers and that the supplementation must be applied prophylactically at or before early stages of decline. Indeed, the beneficial role of DHA has been clearly demonstrated in the inhibition mechanisms of amyloid plaque deposition and in slowing cognitive decline only in Alzheimer's patients who did not carry *APOE4* (Quinn et al. 2010). All new prospective research must now take into account this *APOE* polymorphism.

Under the authority of the French Alzheimer Plan 2008–2012, a large prevention study of memory disorders among those aged 70 years and older is under way, led by the University Hospital of Toulouse (the MAPT study, Multidomain Alzheimer Preventive Trial). Its objective is to evaluate the protective role of additional *n*-3 fatty acids and several recommendations concerning mainly memory (http://www .chu-toulouse.fr/IMG/pdf/cp_etude_mapt_28_fev_2011.pdf). Another multidisciplinary research program has been initiated at the national level, the COGINUT program led by the INRA; its purpose is to analyze the relationships between diet and brain aging, emphasizing the role of *n*-3 fatty acids and antioxidants. E.J. Johnson could indeed show in 2012 that carotenoids (lutein and zeaxanthin) may play a role in the maintenance of cognitive abilities in the elderly. Future research will certainly contribute to the determination of new preventive and therapeutic strategies for various forms of age-related dementia, detailing the biological mechanisms and treatment modalities at each disease step. Globally, 10 projects are under way or starting out in various research centers targeting neuropsychiatric effects of supplementation with *n*-3 fatty acids in patients with dementia of varying severity (http://clinicaltrials.gov/ct2/search).

In the absence of investigations on primary prevention, many basic and clinical studies would be needed before considering the prescription of DHA or EPA to prevent or especially treat Alzheimer's disease. These advances will definitely benefit from the related progress of research on the molecular diagnosis of the disease, allowing an early identification of subjects at risk. They may strongly encourage governments to support research in the long term in well-defined conditions to rapidly consider a solution of prevention or improvement of dementia, including Alzheimer's disease.

Vascular dementia remains a poorly defined syndrome, although numerous research were undergone in this field during the last 80 years. It has been described most often as a cognitive decline associated with intracerebral vascular lesions caused by stroke, embolism, or bleeding vessels in the brain. It is the second leading type of common dementia after Alzheimer's disease, 15%–30% of cases of dementia, but it should not be forgotten that this condition is certainly underestimated in clinical diagnostics and that the risk factors are common to both forms of dementia.

Unfortunately, few works are devoted to the effects of *n*-3 acids on the vascular damages at the origin of these dementias. Most epidemiological studies still show an inverse relationship between fish consumption and incidence of the disease. A large epidemiological study has concluded that increasing fish consumption (from one serving per month to four servings per week) significantly lowered by 35% the risk of vascular dementia, but only when the consumed fish was not fried (Huang et al. 2005). The large Italian experimental GISSI study has detected no effect after a daily supplementation of 850 mg of *n*-3 fatty acids.

The small number of studies and the too small number of dementia cases reported in each one do not allow with certainty definition of any therapeutic potential. Further works on a larger scale will be needed to explore the complex field of vascular dementia, a disease that is always diagnosed with difficulty. Currently, 10 studies have begun or are under way on the relationship between *n*-3 fatty acids and Alzheimer's disease (http://www.clinicaltrials.gov).

In summary, after convincing experiments on animals or cultured cells, epidemiological studies lead us to recognize a strong link between the EPA and DHA intake and the prevention of Alzheimer's disease. Clinical trials are very limited, probably due to their long duration and their difficulties in inducing costs poorly compensated by a subsequent sale of natural products minimally processed. Suggestive results of these tests should still encourage people to follow the recommendations of health authorities regarding the consumption of foodstuffs rich in *n*-3 fatty acids (Section 3.2.1.3.2).

4.2.4.5.2.4 *Visual Functions*

It has long been known that DHA is the most abundant fatty acid in the mammalian retina, as it makes up 30% of all the fatty acids in retina phospholipids. As for the whole nervous system, DHA accumulates preferentially in membranes but especially in those of retina photoreceptor cells (up to 60% of the total fatty acids) where it plays a key role. As for the nervous system, the retina is sensitive to only a lengthy *n*-3 fatty acid deficiency, complicating the realization of nutritional experiments.

Mechanisms underlying the involvement of DHA in the transduction of visual signals in the retina seem to be related to the biophysical properties of membranes rich in this fatty acid. In 1973, it was reported that the electrical responses of rat retinal photoreceptors were affected by DHA depletion. Later, it was shown that severe DHA deficiency causes retinal functional deficits and a reduction of visual acuity.

Despite the difficulty of clinical research in humans, works done in 2001 by S.M. Innis, Canada, have shown a significant association between DHA content of breast milk and visual acuity in children aged from 2 months to 1 year. In contrast, no improvement of visual function could be detected in breast-fed infants receiving DHA for 4 months. This lack of effect was likely caused by a too short treatment or by the prior DHA status of the mothers. Furthermore, in these experiments the fatty acid amounts ingested by the infants and the tissue modifications remained unknown. An experimental study, carried out in 2008 by L.G. Smithers, Australia, has shown that the DHA enrichment of breast milk or infant formula (1% of total fatty acids instead of 0.3%) induced after 4 months better visual acuity in premature infants. This study has also allowed the author to conclude that premature infants had a high DHA requirement, which is insufficiently provided by conventional milk.

Age-related macular degeneration (AMD or ARMD) is a disease of the retina that is the leading cause of visual impairment in people over 50 years in Western societies. One out of four people after 75 years and two people from 80 years would be affected by AMD.

The prevalence of AMD appeared to be similar in Caucasian populations from the United States, Australia, and European countries, despite major geographical and lifestyle differences.

AMD affects more than two million individuals in the United States. In France, the number of people affected has been estimated at over one million, with 3000 new cases of blindness being reported each year. Two forms of AMD have been described, the atrophic form (or dry form) and the exudative (or wet or neovascular) form, which have the same effects on vision but evolve differently.

Many studies have suggested that *n*-3 fatty acids may be beneficial in the fight against retinal neovascularization, a disorder common to diabetic retinopathy and especially to AMD. In 2005, 16 works were selected by the experts of the American Agency for Healthcare Research and Quality to establish an inventory of the effects of *n*-3 fatty acids on eye health. Their analysis did not lead to any recommendation for either primary or secondary prevention (Evidence Reports/

Technology Assessments, No. 117, AHRQ Publication No. 05-E008-2, 2005). Since that report, a dozen works resulting from rigorous protocols, although different, allowed to conclude that EPA and DHA are obviously involved in the prevention of AMD and that a too low intake of these fatty acids promotes the occurrence of this disease (Lecerf 2010). In addition, it appeared that a high linoleic acid intake was associated with a significant reduction in the protective effect of n-3 fatty acids. We thus find again the importance of the n-6 to n-3 fatty acid ratio in dietary lipid intake. One of the most relevant studies is the one included in the large American Age Related Eye Disease Study (AREDS) (Report No. 20). It has clearly shown a protective effect of the long-chain n-3 fatty acids in the prevalence and progression of AMD, their supplementation being able to decrease the risk of neovascularization up to 40%. These results are supported by a more recent French study (Nutritional AMD Treatment 2 Study Group) evaluating the efficacy of a mixture of DHA (840 mg/day) and EPA (270 mg/day) for 3 years in preventing the exudative form of AMD (Souied et al. 2013). The authors have appreciated a 68% lower risk in participants having the highest content of n-3 fatty acids in red blood cell membranes. Similar observations have been reported in 2013 by B.M. Merle after an analysis of results collected during 5 years in a French population in Bordeaux (ALIENOR study).

These promising results have also been confirmed in 2011 by a large epidemiological survey conducted by W.G. Christen, the United States, including nearly 40,000 women's health professionals aged over 45 years. This study has shown that a regular consumption of EPA and DHA or fish was associated with a lower risk (35%–45% lower) of developing the AMD disease. In contrast, no relationship could be detected with saturated, monounsaturated, and *trans* fatty acids.

Following many other studies of various sizes that confirm these effects, it is reasonable to suggest consumption of n-3 fatty acids as an additional way of preventing AMD. These remarkable advances caused U.S. authorities (National Institutes of Health [NIHs], Bethesda, Maryland) to undertake in 2008 a new clinical trial, AREDS2 (http://www.areds2.org/), to extend their first program. The results should probably not be known until 2014.

As part of the French study ALIENOR ("cohort study in general aged population"), launched in 2006, it has already been shown that a high intake of n-3 fatty acids is associated with a significant decrease in AMD risk, the significance being stronger for early and neovascular forms (Merle et al. 2011).

Studies of the mechanisms underlying this protection have shown that EPA and DHA, as well as their metabolites (eicosanoids), exercise their anti-angiogenic properties at the level of intracellular signaling, through the activation of transcription factors, or through inflammation mediators. A natural DHA derivative, neuroprotectine D1, may also protect retinal cells fighting directly against oxidative stress and apoptosis. Undoubtedly, this compound will be used for clinical studies when its mechanism of action becomes more fully known.

Ten clinical studies concerning the relationships between AMD and n-3 fatty acids have been officially declared (http://www.clinicaltrials.gov).

All investigations consistently indicate that consumption of fish or preferably supplements rich in n-3 fatty acids, besides some carotenoids such as lutein and zeaxanthin (Section 4.4.1.1), may contribute to the prevention of the worst retinal diseases such as AMD. The clinical trials under way might surely deliver in the near future precise data on the most effective intakes of n-3 fatty acids of marine origin, possibly combined with carotenoids, to protect the retina against severe degeneration.

4.2.4.5.2.5   Auditory Functions

Little work has been devoted to a possible effect of n-3 fatty acids on hearing loss (presbycusis) and its genesis, a health problem often related to age as about a third of people over 65 years have trouble with audition.

On the basis of an inquiry on dietary intake and auditory capacity, a study monitored nearly 3000 people over 50 years for 5 years and was able to detect a clear correlation between the consumption of fish, thus n-3 fatty acids, and a reduced risk of presbycusis (Gopinath et al. 2010). This advantage was significant in subjects consuming fish at least twice a week, having a 42% reduced risk of hearing loss, compared to subjects consuming less than one fish serving a week. Further research is needed to confirm these initial results and determine the doses and nature of the n-3 fatty acids that may effectively prevent the hearing loss related to age.

4.2.4.5.3   Mental Health: Mood Disorders

The term "mood disorders" is officially used in the International Classification of Diseases (ICD-10); it brings together a set of mental disorders, the most important being the depressive disorders classified in depression (or major depressive disorder) and bipolar disorder (or manic depression). European epidemiological studies have shown that nearly 10% of the population suffered from depressive disorders, and that nearly a quarter of the population will present a mental disorder during their lifetime. In 2011, a report of the European Brain Council revealed that mood disorders induced low direct costs (about 23% of a total of €113 billion), the other part (77%) covering mainly sick benefits and sometimes suicide attempts. This society burden is steadily increasing as a result of the aging of a population frequently taking several medicines.

Numerous epidemiological and clinical studies have supported the hypothesis that the abundance of n-6 fatty acids in Western diets, correlated with a lowering of n-3 fatty acids, was at the origin of various mental disorders. Indeed, a deficiency of long-chain n-3 fatty acids has been frequently reported in patients with various types of psychiatric disorders. Among these, hyperactivity, aggression, suicide, simple and bipolar depression, schizophrenia, and dementia have been noticed. The Global Summit on Omega-3 held in Bruges, Belgium, in March 2011 has emphasized that these diseases resulted from a deficiency of EPA and DHA and were among the most important challenges for the future of humanity (http://www.omega3summit.org/). The experts have claimed that the costs associated with these disorders could quickly bankrupt the national health systems, given that mental health problems are already more severe than those related to the current obesity epidemic.

A great number of clinical studies have been performed but with reduced success. Among all these mental disorders, only those that are the subject of intense research on their possible links with n-3 fatty acids will be considered.

Simple depression (dysthymia), commonly known as depression, is characterized by the chronic presence of at least one symptom of the following three: depression, anhedonia (loss of pleasure), and decreased energy. When two of these symptoms are present, for at least 2 weeks, the disorder is called major depression. Two other forms of depression are also studied, postpartum depression and seasonal depression. A particular form of mood disorder is the bipolar disorder. It is characterized by alternating periods of depression followed by manic episodes, the latter showing periods of euphoria and overexcitation.

For a review of studies on the effects of n-3 fatty acids on mental health and carried out before 2005, a major report published by the Department of Health of the United States (Evidence Reports/Technology Assessments, No. 116, Agency for Healthcare Research and Quality, July 2005) may be consulted (http://www.ncbi .nlm.nih.gov/books/NBK37689/).

4.2.4.5.3.1 *Depression*

Depression is characterized by a state of deep sadness for no apparent reason and lasting longer than 2 weeks. This condition results in a sadness that may often be accompanied by minor disturbances such as loss of sleep and appetite and impaired memory and concentration. Sometimes, these disorders may lead to suicidal thoughts. Depression is two times more frequent in women than in men. The WHO has amounted to about 120 million the number of people suffering from depression, which is considered the leading cause of disability worldwide. Estimates indicated that about 15 million American adults, or about 6.7% of the population aged 18 years and older (about 9% of the population), met the criteria for current depression. A recently reported study revealed an overall prevalence of depressive disorders in Europe of about 8.6%. In France, about three million people are suffering from depression.

In 1996, a reverse relationship between the severity of clinical symptoms of depression and the blood concentration of EPA was demonstrated for the first time by P.B. Adams in Australia, and this was supported by many other studies. Analyses of fatty acids in the blood and adipose tissue have linked high concentrations of EPA and DHA with a low incidence of depression symptoms. The study of a population in Bordeaux aged on average 74.6 years has revealed that among plasma fatty acids only low levels of EPA characterized the subjects with depressive symptoms (Féart et al. 2008). Research based on erythrocyte fatty acids has reached similar conclusions. Despite this consensus on n-3 fatty acids, particularly EPA, the cause of this relationship is still poorly known: is it of internal origin such as a deviation of a cellular fatty acid metabolism or is it of nutritional origin?

The nutritional hypothesis seems to be preferred if one considers the many epidemiological studies from around the world that have shown an inverse relationship between fish consumption and prevalence of depression. In Europe, two large studies carried out in Finland among several thousand people have supported these conclusions but with sharper results in women than in men. When people from different countries were compared, it is significant that the prevalence of depression decreased with increasing consumption of fish. Thus, the Asian countries (Japan, Korea, and Taiwan) consuming more fish had the lowest rate of major depression in the world. This relationship was probably the basis for the absence of seasonal depression in the Icelandic population living above the 64th parallel but consuming large amounts of fish. Similarly, it has been shown that postpartum depression was less common in women frequently consuming fish or having high DHA levels in breast milk. It must be borne in mind that such studies cannot provide conclusive and definitive answers because, in addition to the imprecision of consumer surveys, many genetic, economic, and sociocultural factors are interfering with the physiological data. The fish-cooking method also seems to be important on the intensity of depression because it has been observed that the consumption of fried fish increased the risk of developing more severe symptoms (Hoffmire et al. 2012). The negative influence of the method of cooking, especially baking and grilling, on the levels of EPA and DHA has already been highlighted in 2008 by H. Chung.

The administration of EPA and/or DHA in the form of dietary supplements to patients with depression was attempted shortly after the demonstration by P.B. Adams in 1996 of a relationship between the blood levels of EPA and the severity of the disease. In 1999, A. Stoll, the United States, first demonstrated a positive effect of fish oil in reducing depressive illness. This work was followed by that of B. Nemets, Israel, in 2002 confirming that EPA administration decreased significantly depressive disorders in more than half of patients with major depression after 4 weeks. In 2003, these encouraging results were popularized in the book by D. Servan-Schreiber (*Guérir*, Robert Laffont). The book spread in France the hypothesis of a benefit of fish oil intake in the restoration of psychic balance, in assuming a stabilization of brain membranes. Prudently, the author wrote "it will take several years before a sufficient number of studies of this type is achieved. In fact, the omega-3 fatty acids are natural products, thus impossible to be patented. They have no interest for the big pharmaceutical companies that fund most scientific studies on depression." These conclusions are somewhat pessimistic, but obviously they may be applied to all areas of medicine where these lipids occur; on the other hand, the academic nature of these research efforts ensures their impartiality for future users.

Although still controversial, after many studies completed before 2004 the results have been prudently analyzed by the American Agency for Healthcare Research and Quality (Evidence Reports/Technology Assessments, No. 116). This expert panel acknowledged that some serious studies suggested that *n*-3 fatty acids are beneficial in treatments, sometimes of short duration, but recommended further research in developing better experimental protocols. Since 2002, a review of the literature on this subject has shown the uncertainty still lingering on this type of treatment, despite the large research effort under way in that direction in comparison with other possible therapies for depression (Bloch and Hannestad 2012). As an example of positive results, a recent investigation relating *n*-3 fatty acid intake with depressive symptoms in 1746 American adults has shown that mainly in women higher intakes of *n*-3 fatty acids (absolute or relative to the *n*-3 to *n*-6 fatty acid ratio) were associated with a lower risk of elevated depressive symptoms, specifically in the domains of somatic complaints and positive effects (Beydoun et al. 2013).

The importance of the quality and quantity of the fatty acids used as supplements was examined by several meta-analyses. One of them, summarizing 15 essays published in several countries, has shown a specific and beneficial effect of fish oil EPA on depression, but only if it represents over 60% of an EPA + DHA mixture and with an intake higher than 0.2 g/day and up to 2.2 g/day (Sublette et al. 2011). In contrast, a recent investigation with fish oil supplementation (2 g of DHA and 0.6 g of EPA/day) has found an inverse correlation between erythrocyte DHA content and scores of depression after a 16-week period (Meyer et al. 2013).

In the case of perinatal depression (or postpartum depression), the majority of research works has indicated a more or less marked tendency of a link between disorder frequency and low levels of dietary *n*-3 fatty acids. A recent study published in 2012 by C.M. Rocha, including 100 Brazilian women, has indicated a 24% higher prevalence of these disorders when the *n*-6 to *n*-3 fatty acid ratio in the diet exceeded a value equal to 9. This topic deserves controlled trials with purified *n*-3 fatty acids taken alone or mixed and especially with an estimate of their status through blood tests.

In all cases, more accurate conclusions about the mechanisms of mental health regulation could only be drawn from investigations concerning larger samples and suitable biochemical analyses. The selection criteria should also be a rigorous choice by psychiatric specialists to clarify the people who could best benefit from supplementation with *n*-3 fatty acids. This area of research is interesting many teams as currently 8 large trials reported in the international clinical trials registry of WHO are recruiting subjects in several countries (http://apps.who.int/trialsearch/) and 40 are listed on the official website (http://www.clinicaltrials.gov).

Much research in animals has allowed a better understanding of the mechanisms underlying behavioral disorders induced by brain DHA deficiency. It has been found in rats that DHA deficiency induced dysfunction characterized by a deficiency of dopamine in cortical areas and an increase in dopamine reserves in the nucleus accumbens. Reduced dopamine reserves observed in the cortex could thus be related to a cognitive impairment, whereas an excess of this neurotransmitter in the nucleus accumbens could be the cause of hyperactivity in deficient animals. Significant changes were also observed in the serotonergic and cholinergic systems. Furthermore, several studies have suggested that these changes at the level of synaptic transmission could result from a direct action of fatty acids on the expression of genes involved in neurotransmission processes, membrane plasticity, and neurogenesis. Several other mechanisms may also be involved, such as the modulation of the receptor B linked to tyrosine kinase/BDNF or the increased expression of the neurotrophic factors BDNF and synaptophysin.

All the previous basic research may only justify the consumption of foods rich in *n*-3 fatty acids for the prevention of mood disorders in general. In patients with depressive disorders, EPA supplementation may be recommended, either alone or in addition to more conventional therapies to improve their efficiency. Future research should better define the type of treatment and the necessary doses of these natural products that can be combined with traditional treatments to moderate the overload of synthetic substances in patients.

4.2.4.5.3.2   Bipolar Disorder

Bipolar disorder (or manic-depressive disorder) is a psychiatric diagnosis that defines a category of mood disorders characterized by repetitive fluctuations between periods of marked excitement (mania) and periods of sadness (depression). More severe cases may lead to dangerous behavior, even to suicide. Thus, this psychiatric disorder is associated with a suicide rate that can go up to 60 times that measured in the normal population. Two main types of bipolar disorders have been described; they are distinguished by a characteristic manic episode (mania, type I) or an attenuated manic episode (hypomania, type II). Many patients with bipolar disorder have psychotic symptoms often confused with schizophrenia, which may lead to a dispersion of results in clinical trials. Worldwide, the disease affects more than 1% of adults and is among the 10 most costly and debilitating diseases. The WHO has found that prevalence and incidence of bipolar disorder are very similar across the world. The prevalence per 100,000 individuals ranged from about 421 in South Asia to 490 in Africa and Europe. In the United States, the prevalence of this disease is increasing in adults, rising from 905 patients in 1995 to 1,679 patients in 2003 per 100,000 inhabitants. This increase is even higher among young people under 20 years. In France, more than 600,000 people of all ages and both sexes are affected. Although symptoms may lighten with time, generally the disease persists for very long periods. Studies have estimated that the annual cost of hospitalizations in France represents about €1.3 billion. If this condition is partly genetic, it is recognized that environmental factors may be involved. Among these, the quality of dietary fats has been repeatedly mentioned.

As for depression, a comparative study in several countries on the prevalence of bipolar disorder according to the consumption of fish was published in 2002 by S. Noaghiul, the United States. This work has clearly shown that the prevalence of the disease, particularly the classified type II, decreased (from 6.6 per 100,000 for Germany to 0.2 per 100,000 for Iceland) according to an exponential law when fish consumption increased from 5.4 to 100 kg/year. It should be noted that the consumption threshold below which the prevalence is rapidly growing is about 23 kg/ year. This amount corresponds to about 440 g/week, or two servings of fish, as has already been recommended for the prevention of cardiovascular diseases by medical organizations (Section 4.2.4.1). The authors have estimated that this fish consumption corresponded to a minimum intake of approximately 1.25 g/day of EPA + DHA when it came from salmon, or about 0.11 g/day when it came from cod, the amount recommended by clinicians being about 0.50 g/day.

Like other mood disorders, it is significant that the analysis of autopsy samples of prefrontal cortex in patients with bipolar disorder have shown lower DHA concentrations than in controls. Similar results were obtained by analyzing the erythrocyte membranes in adult or adolescent patients.

One of the first nutritional investigations in patients with bipolar disorder types I and II was carried out in 1999 by A.L. Stoll, the United States. A significantly greater beneficial effect has been obtained with a prescription of 6.2 g of EPA and 3.4 g of DHA daily for 4 months. Subsequently, four out of five studies led to the same conclusions in using only a supplementation of EPA. A recent meta-analysis based on five studies, comparing treated groups with a placebo group of at least five subjects, concluded that a beneficial effect of the consumption of fish oil on bipolar symptoms, unlike mania episodes, exists (Sarris et al. 2012). All these studies have shown that intakes of purified EPA or marine oil fractions are the most active, linolenic acid having no effect. The specificity of EPA recalls similar results obtained in depressed subjects.

The mechanisms involved in the effects of *n*-3 fatty acids on bipolar disorder are probably similar to those explored to explain the effects on depression, plus a possible action at the level of neuronal functions linked to endocannabinoid-dependent signals.

The overall results obtained in the area of bipolar disorder in 2006 appeared so convincing that the American Psychiatric Association decided to recommend to patients with psychiatric disorders, especially unipolar and bipolar disorders, to take at least 1 g/day of a mixture of EPA + DHA or to eat at least three fish servings per week, these products having only negligible biological risk.

4.2.4.5.3.3   *Suicide*

Suicide is a serious public health problem in a number of countries. In the United States, suicide is the tenth leading cause of death, 36,909 in 2009, or 12 suicides per 100,000 people per year. The range of suicide rates is very large, between 3 and 30 in Western societies. In France, suicide is responsible for about 11,000 deaths per year, and suicide attempts are probably more than 15 times higher (Department of Health statistics). It is established that the importance of suicide in premature mortality has more than doubled in 30 years. Psychiatric studies have shown that 25%–50% of patients with bipolar disorder had at least one suicide attempt and 15%–20% died. The second condition is the depression risk associated with alcohol and drug abuse. It should also be noted that 10%–13% of patients with schizophrenia die by suicide.

In 1990, shortly before comparable studies on depression were performed, an extensive survey on the lifestyle of the Japanese showed that a daily consumption of one fish serving was associated with very low suicide rates. Later, this relationship was found in a Finnish population. Subjects with the least suicidal ideations consumed at least two fish servings per week. The Japanese, known for their high suicide rate (about 22 per 100,000 people) and also for their high consumption of fish, have been the subject of a large epidemiological study. Thus, a relatively high number of deaths by suicide has been observed among women eating very few fish. It is clear that in such a large study the origin of *n*-3 fatty acids and their actually consumed amount cannot be readily determined. In China, the study of fatty acid composition of erythrocytes in patients hospitalized for attempted suicide compared to control subjects hospitalized after accident revealed a significant association between a low EPA concentration and a high risk of suicide. A similar relationship with DHA has been established in the U.S. military died by suicide, compared to other mentally healthy militaries (Lewis et al. 2011). In this study, the authors found that soldiers in regions of the front (between 2002 and 2008) with the lowest DHA blood levels had a risk of suicide increased by 62% compared to those with the highest levels. The interest generated by these results prompted the medical service of the U.S. Army to launch a major research program on this subject. It is remarkable that in Belgium, S.R. De Vriese noted in 2004 that annual changes in the rate of violent suicides followed changes in the blood levels of long-chain polyunsaturated fatty acids (EPA, DHA, and arachidonic acid) as well as the level of serotonergic markers.

Consistent with such observations, it is noticeable that as for major depressive disorder lower DHA concentrations were detected in the cortex of depressed suicide victims compared to controls without cardiovascular disease (McNamara et al. 2013).

If these results are confirmed by larger studies, it could have involvements in the neurobiology of suicide and in the means that psychiatrists will implement to reduce the risk of suicide or at least the possibility of renewal of a previous attempt. To date, no study on the prevention of suicide or treatment by supplementation of n-3 fatty acids has explored their possible effects on the frequency of suicidal ideation.

4.2.4.5.3.4   Anxiety Disorders
Anxiety disorders are part of a group of mental disorders characterized by exaggerated fear responses leading to a state of distress. This condition is most often divided into generalized anxiety disorders; social phobia; and other phobias such as agoraphobia, panic disorder, and obsessive compulsive disorder. If we consider the general population aged 18–65 years, approximately 10% of individuals will show an anxiety disorder during a year and 27% will have an anxiety disorder at some point in their lives. Although experts separate anxiety disorders from mood disorders, there is a relationship between these two diseases as people suffering from severe depression are, in the following year, eight times more likely to have anxiety disorders than the normal population.

These observations help to explain why some therapeutic approaches for one of these pathologies may also be beneficial to the other. Thus, clinicians have explored the various relationships between the presence of phobia and the n-3 fatty acid status in the same patients. Their research have revealed that the erythrocyte levels of EPA and DHA were lower in subjects with social phobia than in normal subjects, the reduction being correlated with disease severity. It is obvious that these observations cannot yet provide clarification on the origin of this deficit: is it the result of an inadequate intake or a specific deficiency of the erythrocyte membranes? From this observation, few investigations have been undertaken and with a very small number of patients. Tests on a large scale should be conducted to offer patients with anxiety disorders a new therapy different from that based on the administration of various antidepressants, without eliminating the possibility of an association between the two types of treatment.

4.2.4.5.3.5   Schizophrenia
Schizophrenia is a severe psychosis occurring in early adulthood, more or less chronic, characterized by signs of dissociation, emotional dissonance, and incoherent delusional activity. This disorder usually causes a loss of contact with the outside world and sometimes an autistic withdrawal. The prevelance for schizophrenia is approximately 1.1% in the population over the age of 18 (U.S. National Institute of Mental Health), for example, 51 million people worldwide are suffering from

schizophrenia. The disease is very costly for families and society. The overall U.S. cost of schizophrenia in 2002 was estimated to be 62.7 billion corresponding to 2.2 million people. In France, approximately 400,000 patients are affected, who represent about 20% of full time psychiatric hospitalizations and 1% of the total health expenditure (INSERM data).

The first mention of a possible link between schizophrenia and a disorder of fatty acid metabolism was made about 30 years ago by D.F. Horrobin in England. Subsequently, several studies have shown that n-3 fatty acids were less abundant in various tissues collected from patients with an initial form of the disease. Since then, the source of this condition, consequence or cause of the disease, has remained unknown. The observation made in 2010 by M.M. Sethom, Tunisia, of the restoration of normal DHA levels in erythrocyte membranes in schizophrenic patients treated with neuroleptic drugs remained totally unexplained. Several attempts to treat schizophrenia by ingestion of n-3 fatty acids have led to a reduction in the severity of clinical symptoms. The same conclusions were drawn from a large Swedish epidemiological study. It was noted that among women with psychotic symptoms eating three to four fish servings per week halves the risk of developing the disease compared to women eating no fish.

The hypothesis of a close link between n-3 fatty acids and disease has been strengthened by the test of preventing psychotic disorders in young subjects through the ingestion of 0.7 g of EPA and 0.48 g of DHA daily for 12 weeks (Amminger et al. 2010). At the end of the experiment, the authors found a reduction in symptoms (positive and negative) in the treated subjects. The most amazing observation was that the beneficial effects were maintained 1 year after the cessation of treatment, contrary to the observations made after treatment with antipsychotic drugs. In patients with a first psychotic episode, observations of the brain (hippocampus) by MRI have verified that EPA induced changes that could confirm its neuroprotective effect. However, it appeared that EPA supplementation produced no effect in patients with established schizophrenia. The recent distinction of at least two endophenotypes of schizophrenia will certainly improve future therapeutic trials in terms of targeting the most suitable subjects to receive an effective supplementation of n-3 fatty acids.

This type of dietary treatment of schizophrenia and other psychoses with n-3 fatty acids is an additional weapon in disease prevention, especially for the treatment of young people with a net risk of psychotic disorders. Several studies have insisted on a greater efficiency of the treatment at first signs of the disease (prodromal phase). More recent data have suggested that supplementation with 1–3 g of EPA could reduce the doses of antipsychotic drugs in critically ill patients.

What might be the mechanisms underlying the effects of n-3 fatty acids on psychotic disorders? The therapeutic effects of n-3 fatty acids were attributed, as in many other cases, to changes in membrane fluidity and also to interactions with dopaminergic and serotonergic systems, typically associated with the pathophysiology of schizophrenia.

The frequently unconvincing nature of the results obtained so far has often been accounted for by subjects previously treated with one or more neuroleptics interfering with the intake of n-3 fatty acids. In addition, patients with schizophrenia often have a poor diet, sometimes combined with alcohol, tobacco, or illicit drugs, that can influence the results observed by clinicians. Other better controlled research will be needed to improve our understanding of the therapeutic action of n-3 fatty acids, which is already seen as a promising approach in the treatment of psychotic diseases such as schizophrenia.

Currently, 3 large trials in this area, announced in the international clinical trials registry of the WHO, are recruiting subjects in Slovenia, Germany, and the United States (http://apps.who.int/trialsearch/) and 14 trials are listed in the website http://www.clinicaltrials.gov.

It may already be considered that the aforementioned results may only encourage clinicians to appreciate in subjects at risk of schizophrenia their frequency of EPA and DHA consumption. In case of doubt, it may be convenient to determine their blood fatty acid status. If necessary, it is recommended to supplement the diet of these deficient subjects to reach the intake of n-3 fatty acids recommended by medical authorities.

4.2.4.5.3.6 *Hostility*

Hostility is a form of rejection of others or oneself that is expressed by anger, violence, or even aggression. In psychological terms, hostility is often defined as an absolute refusal to accept the obvious. It has been shown that this attitude is, as are depression and anxiety, associated with a high sympathetic tone and especially with the risk of cardiovascular disease. Hostility remains a complex phenomenon and several aspects (cynicism, anger, and aggression) should be taken into account by various psychological tests, which are sometimes adapted to the countries where the research is undertaken.

It is generally accepted that angry or hostile behavior doubles the risk of developing heart disease within 10 years. Conversely, a treatment leading to a reduction of hostility also reduces the risk of suffering from cardiac ischemia. Given the close relationship that has already been established for many years between the consumption of n-3 fatty acids and cardiovascular disease (Section 4.2.4.1), it seems obvious to investigate whether the beneficial effects of these compounds may also improve the hostility state, even without any cardiac involvement. Early research in this area has revealed very low DHA levels and very high n-6 fatty acid levels in the plasma of violent men known for their antisocial personality. These observations were completed among men guilty of domestic violence by a demonstration in 2004 by the American specialist J.R. Hibbeln of an inverse relationship between plasma DHA and corticotropin-releasing factor levels, hormone known to be associated with violent and defensive behaviors.

Several epidemiological studies have attempted to demonstrate an effect of dietary n-3 fatty acid intake on violent behavior. Reinforced by the results obtained in the 2000s, clinicians have initiated several studies to explore possible associations between anger, violence, or homicide and the level of consumption of n-3 fatty acids or their blood content. An analysis of the literature from multiple geographical locations has shown, despite a variety of protocols, great unanimity in the conclusions:

fish consumption is consistently associated with a reduction in violence and even murder; it makes men quieter and perhaps even happier. What are the data available in this area?

Since 1992, the possibility of a beneficial effect on aggressive behavior and depression has been reported in a nutritional trial with a fish-rich diet that should lower cholesterol. In 1996, it was shown by T. Hamazaki (Japan) that ingestion of a dietary supplement (1.5 g of DHA + 0.2 g of EPA per day) by students for 3 months prevented the increasing aggression measured by psychological tests under conditions of mental stress during exams. These results were confirmed in a similar experiment and supplemented by the demonstration of a decrease in norepinephrine blood levels only in subjects ingesting DHA. These tests, although carried out on a small number of subjects (50 students between 19 and 30 years), encouraged this type of treatment in people very sensitive to psychological stress. One should of course be cautious in drawing general and definitive conclusions as it has been shown that the level of education could change the results, a higher level of education inducing more significant results after DHA ingestion. Very similar tests conducted with placebo groups were also conducted in prisons in the United Kingdom and the Netherlands (Zaalberg et al. 2010). In both cases, supplementation of the usual diet with fish oil for 1–3 months had significantly reduced antisocial behavior, leading to a decrease in incidents between incarcerated subjects and prison staff. From these studies, it is unfortunately difficult to attribute the benefits of treatment only to n-3 fatty acids as preparations also contained a lot of vitamins and trace elements. Separate trials should be planned to test the individual effect of EPA or DHA, regardless of other substances. Despite their cost, the potential benefits of these tests may provide not only economic progress in the long term but also an improvement of social relations in prisons or in locations under surveillance.

Parallel to the first experimental demonstration of an antihostility effect of n-3 fatty acids, several epidemiological observations have been made worldwide, the vast majority confirming the previous results. The calming effect of n-3 fatty acids has been confirmed by the work of J.R. Hibbeln, which was carried out in several Western countries. This author found that a greater consumption of marine animals was correlated with a lower mortality rate from homicide, after an analysis of data obtained in 36 countries. In addition, he has verified in five countries that mortality rates were proportional to the ingested linoleic acid amounts. These results further emphasized the importance of the n-6 to n-3 fatty acid ratio of dietary lipids in the orientation of behavioral responses.

As for cardiovascular disease or cancers, the data acquired so far have allowed the hypothesis that many diseases, organic or psychiatric, originate from the excessive fraction of vegetable oils rich in linoleic acid in the Occidental diet. It is therefore desirable that the relationships between dietary lipids and violence are promptly identified, to provide a simple prevention of violent behavior that affects our societies and severely strains state budgets. As mentioned by J.R. Hibbeln in 2007, must we go back to the millenial traditions of the Christian church and the wise Chinese who, early on, have associated fish to calm and peaceful behavior?

4.2.4.5.3.7 Autism

Autism is a complex mental disorder mainly based on a genetic component and secondarily on an environmental influence. Studies conducted in several continents (Asia, Europe, and North America) have reported a prevalence rate of approximately 1%. In France, this disorder affects approximately 30,000 children. It grows mainly in children aged 5–8 years and is characterized by inappropriate social interaction, communication problems, and a very limited interest in the environment. The causes are still unknown, but in recent years some research has led to the involvement of changes in the composition of membrane lipids in nervous tissue of individuals with this disease.

As with other psychiatric disorders, the concentration of n-3 fatty acids, especially DHA, has often been found to be lower in children with autism than in control children or even children with mental retardation. In plasma or erythrocyte phospholipids, only the arachidonic acid to EPA ratio was altered. This imbalance facing an excess of n-6 fatty acids in children with autism could be restored by providing fish oil. Similar changes were also observed in free fatty acids in the plasma, mainly for DHA, but also for arachidonic acid and several other fatty acids. In all cases, the part played by diet alongside other metabolic intrinsic factors has not been assessed. Some authors have hypothesized that the deficit often found in n-3 fatty acids may be related to metabolic slowdown in the mother when the pregnancy occurred too late or in the presence of multiple pregnancies. Prematurity was also cited as a possible cause of autism, given the late fetal uptake of polyunsaturated fatty acids.

Several attempts have been made to find curative effects of a supplementation with n-3 fatty acids for autism. The tests are few and often performed on small numbers of children of very different ages, which may be the cause of the dispersion of results. Thus, improvements in stereotyped behaviors and hyperactivity were observed in 2007 by G.P. Amminger, Austria, in some boys who were supplemented daily with 0.7 g of DHA and 0.8 g of EPA for 6 weeks. Some investigators have also observed with similar treatments significant improvements in behaviors related to the disease for some of the explored children.

It seems that the benefits obtained with a supplementation of n-3 fatty acids are not any more observed after a certain age, as supplementation in autistic subjects beyond 18 years is usually followed by any detectable effect.

If the aforementioned works provided some hope for an autism treatment by nutritional supplementation, it must be recognized that a review of the literature prior to 2008 did not allow any definitive conclusions on this subject (Bent et al. 2009). We may therefore expect better controlled results obtained over longer periods and with more subjects to reach definitive recommendations about the role of n-3 fatty acids, and even n-6 fatty acids, in autism treatment. Currently, eight large trials in this area, reported in the international clinical trials registry of the WHO, are recruiting in Australia, the United States, Canada, and Europe (http://apps.who.int/trialsearch/) and eight studies are reported on the website http://www.clinicaltrials.gov.

4.2.4.5.3.8 Hyperactivity

Experts of mental disorders in children and adolescents consider hyperactivity, named "attention deficit hyperactivity disorder" (ADHD) (or hyperkinesis), as a behavioral disorder characterized primarily by a difficulty to focus attention and a lack of consistency in activities that require cognitive involvement. ADHD is often associated with other disorders, and without treatment it may lead to many psychological complications. ADHD in children persists in 70% of cases in adulthood. This is the most common disorder in child psychopathology, as indicated by a prevalence of 3%–5% of children in Western countries. Thus, one to two children per class are concerned, about 400,000 children of 4–19 years and up to 7% of adults in the French population. The etiology of ADHD is not known, although biological and environmental factors are undoubtedly at the origin of this disease.

It appeared very early that hyperactive children are more often thirsty than healthy children and frequently suffered from polydipsia, eczema, asthma, and other allergies. The relationship of these symptoms with those observed after an essential fatty acid deficiency has been confirmed by several investigators who have also found that they were accompanied by lower blood levels of arachidonic acid and DHA. An attempt to supplement these patients with evening primrose oil rich in γ-linolenic acid, a precursor of arachidonic acid, did not change the symptoms. Without prejudging the mechanisms involved, some investigations have shown that the severity of symptoms in some young subjects and adults was accompanied by a greater deficiency in essential fatty acids.

On the basis of these works, several clinical trials were initiated with the ingestion of n-3 fatty acids in various forms and the evaluation of symptoms using suitable and accepted psychological protocols. While indicating a trend, the published results are very variable, probably because of too much diversity of the administered products, treatment times, and tests used and perhaps badly defined forms of the disease (Gow and Hibbeln 2014). Thus, the biochemical composition of products enriched with EPA and/or DHA appeared to be important as the majority of clinical trials reporting results interpreted as positive have been obtained only after EPA administration. In addition, N. Vaisman, Israel, discovered in 2008 that a phosphatidylserine rich in EPA and DHA was twice as effective in improving scores of visual attention than a fish oil of similar n-3 fatty acid composition. An intervention study has demonstrated that administration of DHA, unlike EPA, was able to improve after a 4-month treatment the symptoms of a subgroup of children 7–12 years old with ADHD difficulties in reading and spelling knowledge (Milte et al. 2012). The majority of reviews on this topic emphasized clearly that children with ADHD-related symptoms may benefit from supplementation with n-3 fatty acids, especially those exhibiting difficulties with attention and learning. New findings were able to bring a different approach to the debate in demonstrating that lower plasma n-3 and n-6 fatty acid levels in ADHD subjects were associated with abnormal emotion processing (Gow et al. 2013). The authors have tested by electroencephalography the neuronal responses to facial stimuli depicting four emotions (fear, sad, happiness, and anger). Unfortunately, there are to date no published data from the structural neuroimaging of omega-3 intervention trials in children or adults with ADHD.

The relative importance of the various n-3 fatty acids remains unclear, as well as the possible synergistic role of n-6 fatty acids. The variability of the results provided by these numerous research efforts could be partly explained by malnutrition and functional deficits experienced by many patients with ADHD. Although there is no causal link between these nutritional disturbances and the disease, all experts agree on their aggravating role. The essential fatty acid deficiency may be a factor modulating the symptoms of the disease in patients, without taking part in their determinism. There is therefore enough data to suggest the necessity for larger clinical intervention trials with n-3 fatty acids combined with neuronal activity measurement and new neuroimaging techniques.

Despite the pessimistic conclusions drawn from a recent analysis of 13 publications, selected from 366 contributions on the subject by the Cochrane Institute (Gillies et al. 2012), recent investigations have suggested that in the near future an effort of more intense research with larger numbers of subjects should lead to the proposal of an effective nutritional therapy. It is significant that 17 studies in this area are declared on the official website of NIH clinical trials (http://clinicaltrials.gov).

4.2.5   *trans* AND CONJUGATED FATTY ACIDS

Besides the most unsaturated fatty acids having *cis* double bonds, a very small proportion of them has at least one *trans* double bond. Historically, these fatty acids have a natural origin, but currently some have an artificial origin through industrial or domestic food preparation. Among the monoene fatty acids, vaccenic acid is the most abundant natural product of biohydrogenation in ruminants, whereas its isomer elaidic acid is derived from the industrial processing of dietary lipids (Section 3.2.2.1).

Among the *trans* fatty acids derived from linoleic acid, the most abundant have two conjugated double bonds (conjugated linoleic acid [CLA]), one with a *cis* conformation and the other having a *trans* conformation. The position of the double bonds is variable depending on the origin of these fatty acids, either natural for rumenic acid or artificial for some of its homologues, formed by heating or catalytic processing of oils (Section 3.2.2.2).

Although the maximum levels in fats do not exceed 5% for *trans* monoenes and 0.7% for conjugated dienes (CLA), their original properties justify a special consideration of their possible involvement in human health.

4.2.5.1   *trans* Fatty Acids

Only the *trans* fatty acids (monoenes or nonconjugated dienes) (Section 2.2.3) are examined here; the *trans* fatty acids with two conjugated double bonds (CLA) (Section 2.2.4) are examined separately due to their special physiological properties.

The problem of *trans* fatty acids in food appeared around 1984 and has constantly grown with time. Thousands of articles have been published in all fields of biochemistry and physiology, and numerous epidemiological and intervention studies have been devoted to the potential hazards arising from their use. *trans* Fatty acids mainly originate from dairy products and, some of them, especially from industrial fat processing. Among the analogues of oleic acid having a *trans* conformation, elaidic acid (or *trans* 9-18:1) prevails in foods containing industrially processed fats and its isomer, vaccenic acid (or *trans* 11-18:1), is mainly present in the fat-containing products (meat and milk) from ruminants.

It appeared first that when the diet is low in essential fatty acids the addition of *trans,trans* unconjugated linoleic acid (*trans* 9, *trans* 12-18:2) exacerbated the deficiency symptoms, probably due to a competition at the level of desaturases leading to the biosynthesis of arachidonic acid and also EPA and DHA. This effect, observed so far in animals, could affect hemostasis through the metabolism of prostaglandins. It has even been reported that a high consumption of *trans* fatty acids can induce irritable and aggressive behavior as a result of a secondary deficiency in *n*-3 fatty acids. The low level of *trans* fatty acids observed in current human consumption, in most developed countries, has fortunately minimized the practical significance of these results.

The decrease in membrane fluidity induced by *trans* fatty acids (elaidic and vaccenic acids) has been well established as a result of *in vitro* experiments with high supplementation, but the practical effect of the ingestion of modest amounts of *trans* fatty acids on cell function remains to be demonstrated. A high intake of *trans* fatty acids in humans has been associated with elevated oxidative stress, which could be linked to the development of diseases such as atherosclerosis, diabetes, or cancer.

The long-term toxicity of a consumption of about 5 to 6 g/day of *trans* fatty acids is real because in the United States the mortality due to this factor was estimated to be about 3% of the total deaths.

4.2.5.1.1 Cardiovascular Disease

It is widely accepted that the largest survey carried out in seven Western countries (the Seven Countries Study) has shown a positive correlation between *trans* fatty acids and mortality from coronary heart disease. Obviously, these dietary *trans* fatty acids were of mixed origin, originating mainly from dairy fats processed by the industry and used in the manufacture of prepared foods. In several clinical studies, no correlation could be found between the intake of *trans* fatty acids (between 0.5% and 11% of the total energy intake) and plasma LDL-cholesterol.

However, epidemiological studies have shown that *trans* fatty acids are significantly more harmful than saturated fatty acids. Several investigations have shown that the risks of cardiovascular disease were on the account of the consumption of hydrogenated fats, so artificial, more than of dairy products. The risks induced by the different types of *trans* fatty acids are difficult to assess in epidemiological studies. However, a distinction according to the origin of *trans* fatty acids has been attempted in a recent study that has shown a lack of effects when these fatty acids are from ruminant origin but a significant presence of cardiovascular risks when they are of industrial origin (Mozaffarian et al. 2006). This differential effect is probably justified by the discovery in 2012 by M. Minville-Walz, France, of an enhanced lipogenesis expression by elaidic acid in smooth muscle cells from human aorta. The resulting lipid accumulation may be a cause of atherogenesis known to be related to disorders of lipid metabolism. Among the many tests done to verify the impact of the ingestion of the natural *trans* fatty acids on cardiovascular disease risks, one of the newest observation was done in healthy subjects by several teams at the University of Clermont-Ferrand, France (Malpuech-Brugère et al. 2010). Authors have shown evidently that after 1 month with a diet enriched up to four times with *trans* fatty acids of milk origin (mainly vaccenic acid) the changes in blood markers (cholesterol,

HDL, and LDL) reflected a decreased risk of cardiovascular disease, even at the maximum dose tested (8% of the fat intake).

In the same way, experts forming the Task Force Trans Fats Free Americas from the WHO estimated in April 2007 that a significant reduction in heart disease could be achieved by eliminating *trans* fatty acids from industrial sources (elaidic acid and derivatives of linoleic acid). They have found that in South America up to 62,000–225,000 deaths could be avoided by reducing the intake of these fatty acids to 9 g/ day. The toxicity of *trans* fatty acids was estimated accurately with well-controlled clinical trials. These tests have indeed shown that for diets rich in *trans* fatty acids obtained by hydrogenation, the replacement of 2% of them by the same amount of saturated fatty acids lowered by about 20% the risk of cardiovascular accidents and even by 32% for a replacement by polyunsaturated fatty acids.

Taking into account the latest studies, it may be assumed that no cardiovascular risk is induced when the ingestion of *trans* fatty acids (from milk or industrial products) is less than 2% of the total energy intake. This limit has also been recommended by the AFSSA in 2005 and corresponded to the level observed in the average French diet. In contrast, a consumption of *trans* fatty acids from industrial sources of around 10 g/day significantly increased the risk of cardiovascular disease.

AFSSA conclusions (April 2005): epidemiological studies have shown that excessive consumption of *trans* fatty acids (over 2% of the total energy intake, or 5 g/day) was associated with an increased cardiovascular risk. These adverse effects are mainly due to an increase of "bad" cholesterol and a decrease of "good" cholesterol. However, an increase in cardiovascular risk has not been demonstrated with the consumption of naturally occurring *trans* fatty acids, especially at the levels currently observed in France.

Some studies have suggested that *trans* fatty acids of industrial origin could be specifically responsible for an increased risk of cardiovascular disease. This cause–effect relationship is still being evaluated. However, the increased risk observed in these studies could, at least, be related to an excess intake of artificial *trans* fatty acids that would be added to the natural ones (Http://www.afssa.fr/Documents/NUT-Ra-AGtrans.pdf).

4.2.5.1.2 Inflammatory Diseases

Epidemiological studies have shown that *trans* fatty acids can be the cause of an increase in inflammatory markers. Thus, according to the large American survey Nurses' Health Study, the consumption of *trans* fatty acids may be related to a greater amount of TNF-α receptors, C-reactive protein, and interleukin IL-6. Meanwhile, nutritional experiments have confirmed a close association between inflammatory markers and intake of *trans* fatty acids from industrial sources. An investigation of the effects of different *trans* fatty acids on cultured human endothelial cells has shown that elaidic acid and linolelaidic acid, both present in hydrogenated fats, induced cellular inflammation with generation of free radicals and reduction of NO production, a vasodilation mediator (Iwata et al. 2011). In contrast, the authors have clearly shown that vaccenic acid, present in dairy products and ruminant meats, had no effect on these markers. These results have shown well the differential effects of various *trans* fatty acids on inflammation and undoubtedly explained the variability of the findings of previous investigations. It is certain that clinical verifications of these effects will be difficult to realize but are absolutely necessary to determine with precision the consumption standards and limits for hydrogenated fats.

4.2.5.1.3   Metabolic Diseases

The large American study Nurses' Health Study has found that the ingestion of *trans* fatty acids was positively correlated to the development of type 2 diabetes. Contrary to these results, nutritional experiments in healthy or obese individuals with type 2 diabetes have failed to demonstrate a significant effect of *trans* isomers of oleic acid. An extensive review of the works devoted to this issue until 2010 has shown that there is no convincing evidence of an adverse effect of *trans* fatty acids on insulin resistance and diabetes risk, at least with the amounts commonly consumed in western Europe. Furthermore, in 2010 D. Mozaffarian, the United States, demonstrated that plasma *trans* palmitoleic acid levels (*trans* 7-16:1) were associated with low insulin resistance and reduced dyslipidemia. Although a causal relationship between these parameters cannot be now specified, the presence in the blood of this natural fatty acid originating specifically from ruminants could indicate a reduced risk of metabolic syndrome. *trans* Palmitoleic acid may also be at the origin of the beneficial effects of dairy intake on metabolism.

The sometimes conflicting results obtained in animals and humans have highlighted the complexity of the mechanisms involved and the difficulty of decision making by official health authorities. Numerous experimental studies will be needed to shed light on the possible associations between insulin and dietary *trans* fatty acids.

4.2.5.1.4   Cancers

Some surveys were conducted to identify possible relationships between consumption of *trans* fatty acids and several types of cancer localized in various organs, especially prostate, breast, and colon.

Prostate cancer: an international study (the EURAMIC Study) has revealed a correlation between *trans* fatty acids from adipose tissue and incidence of prostate cancer, but other studies have found no association. These disagreements are arising probably from a lack of control over the nature of the *trans* fatty acids, their relative importance with respect to other fatty acids, and the fat energy intake.

Breast cancer: for prostate cancer, a correlation between *trans* fatty acids from adipose tissue and incidence of breast cancer has been demonstrated in some studies in postmenopausal women. In contrast, several epidemiological studies have noted only low or even negligible effects of the consumption of *trans* fatty acids on the risk of breast cancer. So, it seems urgent to clarify with further experiments all possible relationships between *trans* fatty acid storage in breast adipose tissue and breast cancer incidence.

Colorectal cancer: a review of the literature regarding the association between consumption of *trans* fatty acids and colorectal cancer has shown mixed results, some works describing a positive correlation only in women and others a positive correlation in both women and men, but some studies have found no effect. It seems obvious that these confusing results may depend on various dietary habits difficult to assess in the examined populations, but they also depend on the absence of knowledge concerning other dietary lipid components.

Considering the results of epidemiological and clinical studies, many expert panels have recommended limiting the consumption of *trans* fatty acids. Several scientific societies and national organizations have adopted an ingestion limit of 1% of the energy intake (WHO/FAO report, International Society for the Study of Fatty Acids and Lipids, Netherlands Health Council, and American Heart Association), whereas others have advised to remain below 2% (U.K. Ministry of Agriculture, EURODIET 2000) (Section 3.2.2.1). It therefore seems wise to take the general advice of the AFSSA given in 2005 (see Section 4.2.5.1.1).

Insofar as nearly 96% of food products containing fats and sold in France have less than 1% of *trans* fatty acids, all alarmism about it becomes irrelevant at the national level.

4.2.5.2 Conjugated Fatty Acids

Since the publication in 1987 by Y. Ha, the United States, of an antitumor effect of CLA in mice, a multitude of studies have made their contribution to constantly developing knowledge in different areas of health. The conjugated derivatives of linoleic acid were the most explored, but some works were also dedicated to the conjugated derivatives of linolenic acid. These will not be examined in this chapter as they are present in very low levels in the diet and give rise only to pharmacological trials on cultured cells.

CLAs have a *trans* double bond adjacent (conjugated) to at least one *cis* double bond (Section 2.2.4). Thus, during the biohydrogenation process of linoleic acid from plants in the ruminant stomach there is a *cis–trans* isomerization of the double bond at C-12 accompanied by a migration at C-11, the double bond at C-9 remaining unchanged. Thus, the most abundant isomer, *cis* 9,*trans* 11-18:2, named rumenic acid is formed. This natural CLA is found in milk and dairy products and to a lesser degree in ruminant meat. Rumenic acid may also be generated, besides other isomers, by heating vegetal oils (refining and frying).

The AFSSA report in 2005 has indicated that in France the average CLA intake is 0.2 g/day in men and 0.17 g/day in women, less than 0.1% of the total energy intake (Section 3.2.2.2).

As for *trans* fatty acids from industrial sources, several studies have been devoted to the specific roles of CLA in many areas of health such as atherosclerosis, inflammation, insulin sensitivity, cancer, and immune reactions. The diversity of the observed biological effects may be connected to the multiplicity of the tested isomers despite the presence of rumenic acid in natural products.

4.2.5.2.1 Cardiovascular Disease

Several studies with various animal models have shown that CLA acted in a very variable manner on lipoprotein metabolism and development of arterial lesions. The often conflicting results may be put to the account of the variety of experimental conditions. It should be noted that, in the rabbit as in the hamster, CLA induced a lower formation of atherosclerotic lesions or even a regression of preformed lesions. Little research has been undertaken in humans. It was nevertheless shown in 2000 by H. Blankson that among overweight subjects a mixture of CLAs caused a weight drop and decreased cholesterol (total, LDL, and HDL). These favorable effects could be mainly on account of rumenic acid. A review of clinical studies published in 2001 by N. Combe, France, has concluded a deleterious effect of *trans* 10,*cis* 12-18:2 on atherogenesis markers, whereas the other isomer, rumenic acid, is rather beneficial.

4.2.5.2.2   Inflammatory Diseases

Experimentally, the effects of CLA on inflammation appeared to vary according to the condition of patients and the various trials. Thus, in healthy subjects they may cause a rise in C-reactive protein while remaining without effect on TNF-α and its receptors, whereas in diabetic subjects they have no effect on the blood level of C-reactive protein and IL-6. Despite results showing that in mice CLAs induced gene expression leading to an increase in inflammation indicators, the mechanism of action remains poorly defined and more specific tests are needed to clarify this issue. More recent research on human adipocytes have demonstrated that *trans* 10,*cis* 12-18:2 induced inflammatory signals (cytokines) capable of activating neighboring preadipocytes. This reaction may be related to the decrease of lipid depots in adipose tissue.

4.2.5.2.3   Immunomodulatory Effects

Numerous studies in animals or isolated cells have shown a modulating effect of CLA on immune function. The few research efforts in humans are not as convincing. A study of responses to vaccination against influenza has found no effect, whereas another study on responses to vaccination against hepatitis B has shown an increase in the protection level. Broader experimentations and a better understanding of the molecular mechanisms involved would certainly allow therapeutic developments for some CLA isomers in the field of vaccination.

4.2.5.2.4   Metabolic Diseases

In animals and humans, the analysis of many experimental results has shown that CLA had no effect or could have adverse effects on insulin sensitivity and glucose metabolism. Clinical trials with two purified isomers (rumenic acid and *trans* 10, *cis* 12-18:2) carried out on obese patients have shown that both caused an increase in insulin resistance. Tests on human cells in culture have shown that CLA may activate insulin secretion by pancreatic cells, an action that could cause some side effects of these fatty acids but could also be a future treatment of type 2 diabetes (Schmidt et al. 2011).

4.2.5.2.5   Effects on Body Composition

A total of 10 years after the discovery of CLAs' anticancer effect, the highlighting of their effect on body composition in mice has boosted the interest of physiologists on these fatty acids. The first results in this area have been obtained on mice fed a diet supplemented with CLA and were characterized by the observation of a loss of about 60% of fat reserves. This effect is the result of a reduced deposit and also an increase in lipolysis. It was then demonstrated that only the *trans* 10,*cis* 12-18:2 isomer was responsible for the slimming effect, probably due to its ability to inhibit the adipocyte lipoprotein lipase and to increase the outflow of glycerol.

Experimental tests in humans have not led to the same conclusions, CLA supplementation, usually in the form of a mixture of rumenic acid and *trans* 10, *cis* 12-18:2, having caused a small reduction in body fat in half of the studies. However, it seemed that large doses of CLA (above 3 g/day) were required to demonstrate a moderate and sustained effect on fat loss.

These results could be accounted to the differential oxidation of CLA, rumenic acid being more rapidly oxidized in the human body than its homologue *trans* 10, *cis* 12-18:2, unlike rodents. These physiological data, explaining the weight loss effects of some commercial preparations, clearly showed the importance of considering the nature of the molecular species involved and the composition of the mixtures used in clinical trials. The complexity of the underlying mechanisms explains the currently existing confusion in this research area (Kennedy et al. 2010).

It is certain that with the advancement of knowledge in this area new related products belonging to this group of lipids will be experimentally tested and advanced practice in the treatment of overweight or obesity will soon become available to clinicians.

In terms of bone physiology, it must be mentioned that during the weight loss induced by CLA no net effect was found on bone mass. The few clinical trials in this area are inconclusive; highly controlled trials in subjects affected by this issue will be necessary to elucidate the possible influence of CLA on skeletal health.

Unlike the Food and Drug Administration (FDA) of the United States, these inconclusive results led the AFSSA to consider the supplementation of diets with CLAs as scientifically unjustifiable regardless of the proportion of the different isomers. The use of these CLAs in dietary supplements proposed for weight control therefore requires clear information to be given to consumers before other works are completed in this field.

4.2.5.2.6 Cancers

The antineoplastic effect was the first physiological property of CLAs demonstrated in 1987 in inhibiting the development of skin tumors induced in mice. Since then, numerous studies have clearly demonstrated that a CLA mixture or even pure rumenic acid exerts cytotoxic and antiproliferative effects on several types of tumor cells in culture or on various induced cancers in animals. The situation appears more complex when one examines the modes of action of the two isomers rumenic acid and *trans*-10,*cis* 12-18:2 in the models explored. Furthermore, if these two compounds effectively have inhibited the development of induced tumors, an ineffectiveness or some activating effects have been also observed on the growth of spontaneous tumors. On the other hand, variable activities according to the isomers have been found on cancer cell lines in culture. The difficulty of using these results may stem from the diversity of modes of action of the isomers; *trans* 10,*cis* 12-18:2 induced apoptosis and inhibited the lipoxygenase pathway, whereas rumenic acid had no effect on apoptosis while also inhibiting the cyclooxygenase pathway. Other differential influences on the generation of reactive oxygen species (Pierre et al. 2013), metabolism of fatty acids, and expression of certain genes also occured.

Unfortunately, epidemiological studies in humans have not brought out results as convincing as those obtained in animals. Whereas two studies have provided different results for breast cancer, a prospective study has found a decrease in the incidence of colorectal cancer in relation to rumenic acid consumption. Thus, in the absence of intervention studies on the action of CLAs on tumorigenesis in humans there is no evidence of any protective effect of these fatty acids against any type of cancer.

4.2.6 INFLUENCE OF THE STRUCTURE OF TRIACYLGLYCEROLS

Very early, as soon as biochemists have been able to analyze the molecular structure of triacylglycerols, nutritionists have shown in animals that the absorption of fatty acids was dependent on their distribution on the glycerol molecule (Section 3.2.1.4).

Indeed, the position of a fatty acid in one ingested molecule of triacylglycerol conditions its absorption either as free fatty acid or in the form of 2-monoacylglycerol. Subsequently, these pathways will determine the composition of chylomicrons in the blood after their biosynthesis in intestinal mucosa. Thus, it has been determined in animals that lauric, myristic, and palmitic acids in the *sn*-2 position induced increased intestinal absorption. This explains the better absorption of lipids of breast milk in infants or preterm babies in comparison with other sources of lipids. This hypothesis has been explored in preterm babies, in whom the ingestion of structured triacylglycerols with palmitic acid in position *sn*-1,3 had a negative influence on the absorption of saturated fatty acids. In adults, it has been verified that the enrichment of the *sn*-2 position of dietary lipids in palmitic acid contributed to the reduction of postprandial lipemia, a parameter known to be directly related to atherosclerosis (Sanders et al. 2011). In the same nutritional conditions, a decrease in insulinemia and plasma coagulation factor FVIIa was observed. The situation is very similar with DHA, located predominantly in the *sn*-2 position of the triacylglycerol molecules in fish oil and more rapidly incorporated into the lipoproteins that EPA acylated mainly in positions *sn*-1 and *sn*-3.

If the influence of the structure of triacylglycerols on the bioavailability of fatty acids could be explored physiologically in humans, the influence of that structure on cardiovascular diseases was analyzed experimentally only in animals. Thus, many experiments using defined molecular species or mixtures of interesterified triacylglycerols compared to naturally occurring mixtures have highlighted the atherogenic character of palmitic acid when it was in position *sn*-2, in contrast to the situation in humans (Section 3.2.1.4). These differential effects demonstrate the difficulty of extrapolating in the field of nutrition the physiological effects observed in animals to humans.

Recent advances in nutrition research have revealed that fatty acid bioavailability was dependent on the position of fatty acids on the glycerol backbone and also on the supramolecular arrangements of lipid molecules in foodstuffs. Among the complex parameters influencing lipid digestion, the most important are the type or size of the fat droplets and their interfacial composition either in their native forms or after the various manufacture processes (Michalski et al. 2013).

4.3 STEROLS AND HEALTH

Humans obtain from their food two types of sterols, one represented by cholesterol, a steroid alcohol characteristic of animals, and phytosterols, represented by a large number of compounds related to cholesterol but characteristic of the vegetal kingdom. If a human ingests about the same amount of each of these two sterol groups (between 200 and 500 mg/day), their metabolism in the body is very different. It was only recently known that the ingested phytosterols inhibited, in a manner still poorly understood, the intestinal absorption of cholesterol, but the metabolism of the latter was amply explored for nearly a century. Its importance in cell biology and physiology of the body has attracted the attention of clinicians in many fields and is now indisputable. In contrast, the role played by foodstuffs in that economy now appears increasingly reduced, despite the numerous investigations, claims, and recommendations that were made in the near past.

4.3.1 Cholesterol

If cholesterol has been involved from the early twentieth century in cardiovascular disease, other areas of medicine such as cancer and neurological diseases have been explored for their potential links with cholesterol metabolism and transport disorders.

4.3.1.1 Cardiovascular Diseases

In 1913, the involvement of fats, and more particularly cholesterol, in the generation of atherosclerosis was demonstrated experimentally in animals. This theme owed its success to the use of rabbit as the experimental animal, a species more sensitive to dietary cholesterol than rats or dogs and the vegetarian diet of the rabbit likely explaining the results.

The observation of cholesterol deposits in atherosclerotic lesions in humans quickly led to the recommendation of a reduction of blood cholesterol levels. Nearly half a century later, this discovery allowed the demonstration of a parallelism between the amount of ingested cholesterol and cholesterolemia. These results pushed A. Keys to try to anticipate changes in cholesterolemia according to the diet (Section 4.2.1.1.1). The numerous investigations done in the 1950s after the discovery of familial hypercholesterolemia have permanently credited the idea of strong links between high blood cholesterol and development of atherosclerosis. Despite these assurances, several investigators have discussed the responsibility of other nutrients (proteins and carbohydrates) included in the diet. In contrast, few investigators were concerned about whether cholesterol was the primary agent of vascular alteration or could have been secondarily deposited in vessels that were previously highly injured.

Since then, causal relationships between cholesterol and cardiovascular disease have been the subject of many controversies, but are now well accepted thanks to numerous epidemiological and clinical explorations. Thus, in 1980 the Seven

Countries Study showed a positive correlation between cholesterol and coronary heart disease independent of other risk factors. These findings were later verified in populations that migrated to countries with various incidences of this type of pathology. The large Framingham Heart Study, which began in 1948 and is continuing today with the third generation of patients, gives precisions on the occurrence of heart attack and stroke risks in relation to increases in plasma cholesterol.

While remaining aware of the complexity and variety of the human diet, many epidemiological studies have attempted to validate the lipid hypothesis of atherosclerosis (Steinberg 2002). Several of these studies tend to show that a decrease in dietary cholesterol may reduce the incidence of cardiovascular diseases. Among these surveys, there was the Honolulu Heart Program on 8000 Japanese who were followed for 10 years, which established that the ingestion of cholesterol is clearly associated with a risk of cardiovascular disease. It was the same for the Western Electric Study performed in Chicago, Illinois. Many investigators have highlighted the existence of large differences in composition between the analyzed diets and the variety of experimental protocols, these biases minimizing the scientific value of these findings. In addition, it remains to be established that blood cholesterol is causally associated with the disease. Could that parameter be only an indicator?

In 1954, dissident claims were published against this "all cholesterol" hypothesis in the field of cardiovascular disease. Thus, the famous physiologist Irvine H. Page (1901–1991), the United States, discoverer of angiotensin and serotonin, emphasized that the amount of ingested fats is the main factor influencing cholesterolemia much more than the amount of cholesterol. Among the arguments used to validate this hypothesis, I.H. Page underlined that unlike animals, the addition of cholesterol in the human diet (as eggs) did not affect the concentration of circulating cholesterol. In addition, he concluded that no blood parameter could be a diagnostic of atherosclerosis. He recalled that "normal" cholesterolemia values are valid only for statistical studies of large populations.

As the investigators developed their analytical and statistical tools, many studies incited doubt on the impact of dietary cholesterol on the blood cholesterol level and the relationship between the latter and atherosclerosis. This long history, the cholesterol controversy, deserves analysis to understand one of the most debated clinical problems of the last half century.

Historically, several epidemiological studies have reported a positive relationship between dietary cholesterol, cholesterolemia, and the incidence of cardiovascular disease. For example, after many others, the large and famous Seven Countries Study showed a close correlation between cholesterol intake and mortality after 5 years of follow-up. Subsequent analysis of the results has shown that these findings should be counterbalanced by the amount of dietary saturated fats. Taking into account this parameter, any relationship between dietary cholesterol and heart disease disappeared. This kind of interference from other nutrients had already been detected in the Study of the Twenty Countries in 1988 by D.M. Hegsted. This question might seem finally settled after a careful analysis of 13 epidemiological studies, which showed that the difference in cholesterol intake between patients with cardiovascular disease and healthy individuals was only about 16 mg/day (Ravnskov 1995). The author pointed out the lack of credit that must be given to the 16 mg/day compared to the 1000 mg daily synthesized by the body. Many epidemiologists in various countries have addressed this problem, but owing to the diversity of lifestyles (smoking, alcohol, physical activity, hypertension, etc.) they could not establish coherent links between cholesterol intake and risks of infarction or fatal cardiovascular event.

In contrast, experimental approaches based on the absorption of varying amounts of eggs, foodstuff particularly rich in cholesterol, have apparently provided valuable information for understanding this important issue. These research first had the merit to highlight a natural inequality of the population face to different dietary cholesterol levels. This heterogeneity was characterized by the existence of individuals who may be considered "hyporesponders" (70% of the population) with a stable cholesterolemia despite the ingestion of large amounts of cholesterol and "hyperresponders" (30% of the population) with a cholesterolemia slightly increased by the consumption of eggs. It is clear from these studies that the consumption of eggs, up to seven per week, did not significantly increase the risk of coronary heart disease in nondiabetic individuals. A recent Spanish study on a large cohort (the SUN project) has confirmed these findings. Regarding the sensitivity of diabetics to cholesterol, the Health ABC Study carried out in the United States by D.K. Houston showed in 2011 that subjects with type 2 diabetes had a relative risk of cardiovascular disease increased nearly four times after the ingestion of cholesterol and even five times after egg consumption. This point was confirmed by other studies and remained troubling and yet unexplained. Despite our current ignorance, the evidence of the mechanisms should encourage nutritionists to recommend low intakes of cholesterol, and therefore eggs, to diabetics without neglecting traditional therapeutic methods to control any hypercholesterolemia. The large issue of the relationships between dietary and blood cholesterol may be considered as solved by reading the only meta-analysis done to date in humans. Indeed, the analysis of nearly 400 nutritional experiments by R. Clarke, England, in 1997 has shown that for a daily lowering of 200 mg of cholesterol in meals cholesterolemia decreased only by 0.13 mmol/L (50 mg/L), knowing that it must not exceed 5.16 mmol/L (2,000 mg/L). Therefore, it clearly appears that lowering blood cholesterol in this case is unrelated to the imposed nutritional endeavor.

As pointed out in 2010 by J.D. Spence, the Center of Atherosclerosis Research, London, Ontario, cholesterol has many deleterious effects on arteries, including induction of LDL oxidation and increased postprandial lipemia. The advice to limit egg consumption may be applied especially to patients at risk of cardiovascular disease and diabetes. For this author, the advice to remove any egg consumption after cardiac events has the same scope as that of not smoking during a lung cancer. Dietary cholesterol remains a serious matter that should be naturally taken into account by the population but especially after consulting health professionals. The recent demonstration by the same author of a close link between egg consumption and importance of carotid atherosclerotic plaques detected by ultrasound underscores the potential danger of too much cholesterol intake, especially in patients at risk of cardiovascular disease (age, smoking, poor physical activity, hypertension, and hypercholesterolemia) (Spence et al. 2012).

The great disparity between studies and the individual variability are certainly causing trouble to highlight an indisputable correlation between heart disease and cholesterol intake. The differences are undoubtedly the result of differences in protocols and also in statistical processing. It seems difficult to distinguish the effects of dietary cholesterol independently of other known or suspected dietary factors that are associated with risks of cardiovascular disease. As J.M. Lecerf, France, pointed out, studies based on egg consumption cannot avoid the interference of other factors such as low levels of saturated fatty acids and the presence of other nutrients considered to be able to minimize the cardiovascular risk (minerals, vitamin B, vitamin D, and *n*-3 fatty acids) (Lecerf and de Lorgeril 2011).

The origin of the individual differences in sensitivity remains obscure but unfortunately could result from the variability in the intestinal absorption of cholesterol or the ability to control its endogenous metabolism. The influence of dietary fatty acids on cholesterol metabolism should also be taken into account in the studied diets. To complicate matters, few studies have considered the relative importance of the cholesterol transported by LDL and HDL ("good and bad" cholesterol), but in 1951 these parameters were already considered to be the most significant.

It is clear that the scale of these studies and their impact are the consequence of the predominance of cardiovascular disease in the general population mortality. The WHO has estimated that about a third of all deaths worldwide are caused by cardiovascular disease. If nothing is done, the extrapolation to 2015 has shown that more than 20 million people will be victims of these causes (http://www.who.int/topics/cardiovascular_diseases/fr/index.html). In France, cardiovascular disease kills about 150,000 people per year, which over the years is in fact the leading cause of death in the country.

> Two major recommendations adopted by Western medicine to prevent heart disease are to maintain a nutritional cholesterol intake below 300 mg/day and cholesterolemia lower than about 5 mmol/L (2 g/L). It should be noted that these values are set more as a precautionary measure than by taking into account indisputable clinical works. From the aforementioned surveys and clinical works, it follows that these objectives may be achieved partly by reducing cholesterol intake but mainly by observing a set of dietary measures such as reducing saturated fat intake and increasing the consumption of unsaturated oils. These measures must be naturally accompanied by the removal of other well-known risk factors (smoking, alcohol, physical inactivity, hypertension, etc.).

4.3.1.2   Nervous Diseases

Apart from cardiovascular disease, the incidence of dementia, Alzheimer's disease, or depression has recently been associated with the circulating or cellular cholesterol status. For the first time in 1994, a possible link between cholesterol and Alzheimer's disease was discussed in nutritional studies carried on rabbits. These animals fed a cholesterol-enriched diet had significant development of the disease marker β-amyloid peptide in the hippocampus. From this observation, many studies in animals using a nutritional or a pharmacological approach have confirmed and clarified the relationship between cholesterol and cerebral accumulation of this peptide.

In humans, the various announced hypotheses are only the result of epidemiological studies attempting to link the incidence of the disease to various risk factors, including cholesterol or cholesterol-lowering therapy. A close correlation was demonstrated in 2003 by M.A. Pappolla, the United States, between cholesterol and the presence of amyloid plaques found at autopsy in patients less than 55 years old, but this relationship tended to disappear in older patients. This trend suggested the intervention of additional factors involved in the deposition of amyloid. Among many epidemiological studies, some have failed to establish a relationship between elevated cholesterol levels and the risk of Alzheimer's disease. Other investigations made by K. Rockwood, Canada, in 2003 have shown that subjects treated with statins, potent inhibitors of cholesterol biosynthesis, had half the risk of developing cognitive problems than control subjects. These disagreements have not yet been explained; they could be due to incomplete measurements of all types of cholesterol (total, LDL, and HDL) or to various stages of the disease. In addition, an effect of statins on other undiscovered mechanisms remains possible.

Biochemical studies have shown that the metabolism of amyloid precursor proteins was sensitive to cholesterol. This phenomenon has been confirmed in 2003 by K. Rockwood, who observed an excessive production of β-amyloid after the ingestion of large amounts of cholesterol, whereas treatment with statins reduced the risk. Although the action mechanisms of statins are very different, and probably well beyond the classical inhibition of cholesterol metabolism, some clinical experiments with these inhibitors have suggested a beneficial effect in preventing cognitive decline. But no study has yet been able to ascertain a direct causal relationship between cholesterol and the incidence of Alzheimer's disease or other types of dementia. It is hoped that the development of research in this area could quickly provide valuable insights for simple preventive treatments of these serious diseases.

Many older studies have suggested that lowering cholesterol has been associated not only with a decrease in the incidence of cardiovascular disease but also with increased mortality due to suicide, homicide, or accident. These results have been disputed but more often by experiments using diets with very different fatty acid compositions. An analysis of the literature in 1995 by J.R. Hibbeln has helped to highlight the interference of the n-6 to n-3 fatty acid ratio in the overall effect observed. This example illustrates very well the interconnection between dietary lipids and the physiology or the behavior of individuals; a beneficial action for a system may disorganize one or more other systems.

4.3.1.3   Cancers

The relationships between cholesterol and cancer have resulted in numerous epidemiological studies. Whereas some investigations have shown an association between the ingestion of large amounts of cholesterol and an increased risk of lung cancer, others did not support these conclusions. It is the same for the risk of breast, colon, and prostate cancers. As for cognitive disorders, some investigations have suggested a beneficial effect of statins on various cancers. It will take many more studies before prescribing these inhibitors of cholesterol biosynthesis in preventing one or several forms of cancer, especially after the implication by J.S. Thomson in 2010 of statins in the development of certain forms of cancer.

4.3.2 Phytosterols

In the early 1950s, the hypothesis of a close relationship between cholesterol and atherosclerosis prompted research on the effects of plant sterols on plasma cholesterol. Thus, the cholesterol-lowering effect of β-sitosterol (Section 2.6) was observed in 1951 in the chicken and confirmed shortly after in humans, but with high doses (9 g/day of β-sitosterol). Later, it was shown that a daily dose of 2.6 g of sitostanol ester was sufficient to lower serum cholesterol by 10% after 6 months in slightly hypercholesterolemic subjects. In 2001, the Adult Treatment Panel from the National Cholesterol Education Program of the United States recommended the use of phytosterols to a maximum dose of 2 g/day in the case of elevated cholesterolemia. This treatment allowed the lowering of LDL-cholesterol by about 25% compared with a usual American diet. For an optimal effect, the program advised a parallel reduction of the intake of saturated lipids (<7%) and cholesterol (<200 mg/day). Although the underlying mechanisms of this inhibition are not yet fully understood, the major effect might lie in reducing the intestinal absorption of cholesterol. The validity of this hypothesis is reinforced by the very low intestinal absorption of phytosterols (maximum 2%) and especially stanols (<0.1%). Other targets may be envisaged if one considers that the endogenous synthesis of cholesterol would offset any reduction of the exogenous source.

According to the consumption of plant products, the natural daily sterol intake is between 150 and 400 mg, stanols (saturated sterols) not exceeding 25 mg. Despite the dominance of statins in the treatment of hypercholesterolemia, many tests were conducted to demonstrate the efficiency of food supplementation with plant sterols or stanols. In most cases, margarines have been used as commercial vectors and sterols or stanols were supplied in the form of fatty acid esters (Section 3.2.3.2). A meta-analysis of 59 studies involving over 4500 subjects, carried out in 2008 by S.S. Abumweis, has shown that an intake of approximately 2 g of phytosterols per day allowed a decrease in LDL-cholesterol of about 0.3 mmol/L (0.12 g/L), the effect being all the more important as the initial rate was higher (Abumweis et al. 2008). That analysis has also shown that the effect was dependent on the nature of the fortified food, the number, and the meal timing. If it could be shown that stanols were more effective over long periods and very poorly absorbed, a meta-analysis of 14 studies published in 2010 by R. Talati, the United States, showed that sterols and stanols were also effective in reducing total cholesterol, LDL-cholesterol, and also HDL-cholesterol. All of these observations led to the marketing of various products added to some common foodstuffs. A French study by C. Weidner in 2008, on the effects of the consumption of milk supplemented with soybean phytosterols, clearly showed that the ingestion of 1.6 g/day of plant sterols decreased after 8 weeks by 7% the LDL-cholesterol and by almost 5% the total cholesterol. Despite the popularity of these foods enriched with sterols (nutraceuticals) and various supplements in the market, it should be noted that their ingestion may be recommended mainly for subjects with moderately elevated cholesterol levels (5.2–6.2 mmol/L or 2–2.4 g/L), but it must be accompanied by a controlled intake of fats and especially of unsaturated fatty acids.

Numerous studies have led the EFSA to decide on the daily intake of phytosterols that should not exceed 3 g as adverse effects are expected for higher doses, such as a deficiency in carotene and vitamin E. Indeed, a consumption of 4 g of stanol ester may induce a decrease in serum carotene and vitamin E, the other fat-soluble vitamins remaining unchanged. An intake of phytosterols may also be combined with a statin therapy to achieve greater efficiency. To this day, it is unfortunate that no scientific study has come to confirm that this treatment may have any effect on morbidity or mortality caused by cardiovascular diseases. A supplementation with phytosterols is obviously not indicated in patients with sitosterolemia (or phytosterolemia), a rare genetic disorder characterized by the appearance of xanthomas and a premature coronary artery disease. This situation justifies caution for prolonged absorption of phytosterols, their anti-atherogenic role being far from demonstrated.

The use of phytosterols as dietary supplements is still debated among those in the medical profession, because in addition to the absence of clinical evidence of a positive effect on cardiovascular disease the consequences of the increase in their plasma levels are unknown. Despite these reservations, all health authorities are currently favorable to the use of plant sterols in hypercholesterolemic subjects. A very detailed review of the latest knowledge on the use and involvement of phytosterols in the prevention of cardiovascular disease may be consulted in the literature (Marangoni and Poli 2010).

Some epidemiological studies have suggested that phytosterols exert a protective effect against some cancers (lung, breast, and stomach), but no causal link has been established. If a greater consumption of phytosterols could not be associated with a reduction in colorectal cancer risk, there would be some evidence for a curative effect of phytosterols with respect to cancers of lung, stomach, ovary, and breast (Woyengo et al. 2009). β-Sitosterol has been used for the treatment of benign prostatic hypertrophy. Although the mechanism of action is not yet elucidated, this sterol is believed to inhibit 5 α-reductase, an enzyme that plays an important role in the development of the disease.

To date, the AFSSA in France has authorized a sterol enrichment of margarines and dairy specialties, saying that "there is no need to multiply these products intended exclusively for persons with hypercholesterolemia. In addition, clinical studies have shown that a daily intake of more than 3 g sterol did not further lower cholesterol." The advice given by the various official health authorities are reported in Section 3.2.3.2.

4.4   VITAMINS AND HEALTH

The knowledge of the molecular action of vitamins that may be considered as full lipids (vitamins A, D, E, and K) has grown enormously in the last decade. These vitamins, formerly confined in a purely catalytic action, have become micronutrients with a status of hormone or protection of cells against various toxic agents. New experiments or inquiries have extended the scope of these vitamins by making them play a major role as nutritional agents essential to health.

4.4.1 VITAMIN A AND CAROTENOIDS

Vitamin A was the first "vitamin type" factor firstly identified as a "fat-soluble factor A" in the field of nutrition by E.V. MacCollum (1879–1967), a famous biochemist at the University of Wisconsin.

Since its discovery, vitamin A has gradually emerged as a regulatory agent of many biological processes such as reproduction and development. The knowledge of nuclear retinoid receptors stems from the work of Prof. Pierre Chambon in Strasbourg, France. Retinoic acid thus behaves like a hormone capable of altering gene expression and directing mainly cell differentiation, which explains its diverse roles in the immune system and carcinogenesis. Like other factors essential to organism functions, our knowledge, although incomplete, is mainly due to the study of disorders observed in deficiency situations, natural or experimental. This method has quickly updated a large number of clinical signs, ranging from mechanisms of vision to growth control and defense against infections. More recently, many epidemiological studies and few intervention studies have allowed the investigation of the role of vitamin A in the cancer process.

If among the carotenoids some are acting as provitamins A, most are powerful antioxidants, thus completing the main function of vitamin E (Section 4.4.3).

4.4.1.1 Vision

4.4.1.1.1 Vitamin A

The American biochemist G. Wald (1906–1997) discovered that vitamin A is a component of the retina. In 1935, he showed that rhodopsin, a protein pigment, was split after light exposure into a protein, opsin, and an aldehyde derivative of retinol he named retinene (*cis*- 11-retinal), a chromophore essential for vision (Section 2.7.1). He received Nobel Prize in 1967 for his work on "the molecular basis of visual excitation."

The vision troubles named xerophthalmia (dryness of the conjunctiva and cornea) and night blindness (decreased night vision) were known since ancient times and were gradually related to a poor nutritional status. Since the early twentieth century, numerous studies have clarified the role of vitamin A in the process of vision.

In children, vitamin A deficiency is the leading cause of blindness in the third world. The first symptom of this deficiency is loss of night vision, followed by damage to the cornea leading to xerophthalmia, ulcer formation, and loss of all vision. The WHO has estimated that more than 250 million children under 5 years are vitamin A deficient in the world, with 3 million suffering xerophthalmia and nearly 500,000 cases of blindness occurring due to this deficiency. A large meta-analysis of 43 large trials (total 215,633 children under 5 years) has shown that the administration of vitamin A reduced the prevalence of vision problems, mainly night blindness (–68%) and xerophthalmia (–69%) (Mayo-Wilson et al. 2011). This supplementation should be applied in all children at risk, particularly in developing countries. Similarly, the development of retinitis pigmentosa, a hereditary disease causing blindness in young people, can be slowed down after administration of vitamin A. Campaigns of vitamin A administration (60 mg every 4–6 months in children 1–4

years old) are regularly organized in Africa and Asia by several international organizations (http://www.who.int/vaccines/en/vitamina.shtml).

4.4.1.1.2 Carotenoids

Carotenoids are present in the retina and two isomeric compounds belonging to the
subgroup of xanthophylls, lutein, and zeaxanthin (Section 2.7.1); they are especially
concentrated in the macula (yellow spot) where they constitute nearly 70% of the
total carotenoids. The presence of these compounds has attracted a large number
of works related to retinal diseases. It appears therefore that macular pigment could
protect the retina against blue light (the most energetic) and neutralize the singlet
oxygen, which is one of its highly oxidizing reaction products, thus fighting against
the harmful effects of light exposure.

This biochemical feature of the macula rapidly directed the interest of clinicians
to its potential role in a retinal pathology yet poorly controlled, age-related macular degeneration (AMD). This pathology indeed affected an area extremely rich in
retinal photoreceptor cells, which results in severe deterioration of the central vision
and may even lead to blindness. The prevalence of this disease, in the form of two
types (dry form and wet form), increased with age especially after 50 years, and in
more than a quarter of people over 75 years. As a result of the aging population,
this proportion should be multiplied by three during the next 25 years. The WHO
has estimated that 8.7% of the blind people in the world are suffering from AMD.
Macular degeneration is affecting as many as 15 million Americans, and it has been
estimated that 14%–24% of the U.S. population aged 65–74 years and 35% of people
aged 75 years or more have the disease. In Europe, the number of people with AMD
is estimated to be around 20 million. In France, more than a million people, including a significant number of blinds, are suffering from AMD.

Numerous studies have been devoted to the relationship between AMD and retinal pigments.

The U.S. study of the Eye Disease Case-Control Study Group was published
in 1993 and has explored in nearly a thousand subjects the potential relationships
between xanthophylls and the risk of AMD. These research efforts were among the
first demonstrating an inverse relationship between plasma concentrations of lutein
and zeaxanthin and AMD risk. They also revealed that the amount of ingested xanthophylls was closely associated with a reduced risk of AMD, subjects consuming the
most having a 60% lower risk than those consuming the least. It has also been demonstrated in 2010 by S.Y. Cohen that the concentration of xanthophylls in the macula
itself, as assessed by direct measurement of optical density, was well correlated with
food intake. This last relationship was confirmed by the results obtained through the
French study PIMAVOSA (macular pigment in healthy volunteers) reported by M.N.
Delyfer in 2012. This study further suggested that n-3 fatty acids (22:5 n-3 and EPA)
may act synergistically with the xanthophylls in the macular pigment accumulation.
As the author points out, despite the ignorance of the involved mechanisms, this
association is an additional reason for a combined intake of carotenoids and n-3 fatty
acids to protect against AMD. This joint effect was also evaluated more precisely in
the context of American AREDS 2. These relationships between AMD and n-3 fatty
acids are discussed in Section 4.2.4.5.2.4.

The first epidemiological survey carried out in France by INSERM (Survey Ocular Diseases Linked to Age, POLA) with 2600 people focused on the close association between plasma levels of lutein and zeaxanthin and the risk of age-related eye diseases (AMD and cataract). Thus, it was shown that the AMD risk decreased by 93% and the nuclear cataract risk by 75% among the subjects with the highest levels of zeaxanthin compared to subjects with the lowest levels. Meanwhile, the AMD risk was reduced by 69% for high levels of lutein and 79% for high levels of both lutein and zeaxanthin. The study suggested that these xanthophylls were playing an important protective role toward AMD and cataract. Further studies are needed to confirm and extend these results, some showing a beneficial effect (the U.S. Male Health Professionals study) and others (the Beaver Dam Study) giving inconsistent results.

What about the role of diet? Little evidence has been accumulated so far; but, with few exceptions, they go in the same direction as the previous studies. It should be noted that epidemiological investigations have determined either the number of consumed fruits and vegetables or the amount of carotenoids in these foods. It has been thus verified that only fruit consumption reduced the risk of AMD, whereas vegetables; vitamins A, C, and E; and xanthophylls had no effect. Complex calculations from poorly documented data tables cast doubt on such results while reviving the debate on the most available forms of bioactive lipids. In addition, clinicians have emphasized the importance of confounding factors such as smoking and obesity, which disturb the statistical evaluation of possible relationships between the two main parameters. The complexity of the mechanisms involved suggests caution when based on consumer surveys, even if the study involves a large number of subjects for long periods.

With respect to intervention studies, there is, for the moment, some direct evidence of a protective effect of a supplementation with lutein and zeaxanthin against AMD. Thus, in 2009 F. Carpentier, Canada, showed that a supplementation with 30 mg/day of each of these carotenoids increased the pigment density in the macula, these changes enabling therefore a protection of the retina against any sunlight excess. Although a few studies found no effect of a lutein supplement on vision, others have shown improved visual performance in subjects with AMD. Thus, after a supplementation of 10 mg/day of lutein in subjects with a dry form of AMD, S. Richer of the Chicago Medical Center Eye Clinic, the United States, measured a 50% increase in the density of the macular pigment. At the same time, he observed an improved visual performance such as a glare recovery and an improved contrast vision and visual acuity. The importance of the topic and the results already obtained motivated the National Eye Institute, under the control of the NIH to launch in 2006 a large 5-year study (AREDS 2) with 4000 people from 50 to 85 years with AMD (www.areds2.org).

Participants took one of four additional supplements or combinations: These included lutein/zeaxanthin (10 /2 mg) combined or not with n-3 fatty acids (EPA/ DHA 650/350 mg) or placebo. Progression to advanced AMD was established by examination of retina photographs or treatment for advanced AMD. The final report in 2013 described that participants who took lutein and zeaxanthin during the initial 5-year trial were 25%–30% less likely to develop advanced AMD, mostly due to a reduction in the number of neovascular, or wet, AMD cases, over the next 5 years, compared with participants who took a placebo. *n*-3 Fatty acids but not carotene may further improve the formulation. None of the formulations helped to reduce the risk of progression to cataract surgery.

The analysis of many observations and experiments has now led to the conclusion of a beneficial effect of lutein and astaxanthin on retinal function. Not only did these carotenoids improve visual acuity and contrast sensitivity but they may also prevent or slow the development of some types of AMD, especially in their advanced forms. Many other works should be done on a large scale before finally deciding on the regular intake of lutein alone or associated with zeaxanthin for the prevention and treatment of the two known AMD forms. The same conclusions can be drawn for the prevention of cataracts. In both cases, many ophthalmologists do not hesitate to advise a supplementation with lutein (10 mg/day) in patients with early disease of the retina or the eye lens.

Apart from serious eye diseases, it is interesting to note that in 2004 A. Nakamura has shown that the ingestion of astaxanthin (Section 2.7.1) can improve visual acuity and reduce the accommodation time in healthy subjects.

In the current state of knowledge on this subject, it seems wise to be in agreement with the authors of a large meta-analysis of six studies involving large cohorts of 1,700 to more than 71,000 people who were followed for 5–18 years (Ma et al. 2012). This analysis has revealed a significantly reduced prevalence (–26%) of late AMD for the highest consumption of lutein and zeaxanthin; in contrast, no association was observed for the early form. In the absence of compelling results, it is still relevant to recommend to people aged 50 years and older with at least one other risk factor (smoking, heredity, and intense illumination) to eat at least three fruits per day. Any supplemental lutein and zeaxanthin should be taken only after medical advice.

4.4.1.2   Immune System and Infections

4.4.1.2.1   Vitamin A

In 1925, vitamin A was known to be involved in the development of lymphoid organs. About 3 years later, it was seen as an anti-infective factor and, shortly after, it was shown to be able to halve the mortality of patients with measles. In 1933, A.F. Hess, already a leading American specialist of vitamins C and D, stated that young children did not require more vitamin A than the amount contained in 750 mL of milk. Moreover, he added that an excess of vitamin A was a therapeutic absurdity.

After these successes, spectacular trials were conducted in the United States to solve the problem of absenteeism in factories caused by multiple respiratory diseases. Again, the administration of vitamin A, provided through a cod-liver oil, was able to reduce absenteeism and save millions of dollars in workdays. The real effects attributable to vitamin A have not yet been separated from those attributable to vitamin D and *n*-3 fatty acids present in fish oil. In the 1940s, these clinical and economic outcomes caused U.S. and English policy makers to offer an extra ration of butter to workers living in difficult environments and a ration of milk to schoolchildren. This measure was taken in France in 1954 by the former President Mendes-France, but it was under the cover of fighting against alcoholism.

The gradual improvement of dietary conditions and the discovery of antibiotics in the 1940s relegated to the background the use of vitamin A in antimicrobial control. Renewed interest came when doctors practicing in the third world again made the discovery of a bacterial protection in children supplemented with vitamin A for treatment of xerophthalmia. Thus, in 1983 in Indonesia, A. Sommer observed a relationship between vitamin A deficiency and respiratory diseases, diarrhea, and blindness. These works received the Lasker Award in 1994 for the "contribution and demonstration that supplementation with small doses of vitamin A to millions of children in the third world can prevent blindness and mortality due to illness." From these observations, several clinical trials in various countries of the third world have shown that vitamin A supplementation could reduce by a third infant mortality. To achieve this aim, a special session of the United Nations (WHO/UNICEF) established in 2002 the goal of eliminating before 2010 all vitamin A deficiency and its consequences (http://www.who.int/vaccines/en/vitamina.shtml). The strategy most commonly adopted was that of a supplementation combined with vaccination against measles and polio. The recommended dose was 100,000 International Unit (IU) (30 mg) of retinol for children less than 1 year and 200,000 IU (60 mg) for up to 4 years and for lactating women. This preventive treatment could effectively reduce child mortality by about 60%. The WHO has estimated that around 1.25 million lives have been saved for 10 years through programs overcoming vitamin A deficiency in 40 countries. Thus, every dollar invested in this action has according to the WHO saved over $17 that have been invested in treatments and economic compensation. This international program focused on the prevention of vitamin A (led by the Micronutrient Initiative agency) has been rated the "best investment for development in the world" (Copenhagen Consensus 2008).

A vitamin A deficiency has been observed in HIV-positive patients, the infection being facilitated by skin changes. This observation has led to supplementation trials in the infected mother, leading to the suggestion that vitamin A may reduce the transmission of the virus to the child. Further works will be needed before concluding the usefulness of a systematic preventive treatment.

Although the mechanisms involved in the action of vitamin A on the immune system are not well known, its role is not discussed especially since the discovery of an enhancer effect on immunoglobulins and nonspecific resistance to infection. It is now well established that vitamin A is involved in the regeneration of damaged mucosal infection, stimulates white blood cells such as T and NK lymphocytes and phagocytic cells, and increases the antibody activity. Research in animals in 2006 by Y. Nozaki, Japan, has helped to show that retinoic acid reduced the formation of inflammatory cytokines, chemokines, and immunoglobulins. These functions are related to the defense against infections and healing of infected tissues. Several studies have reported that vitamin A could be part of a comprehensive approach to the treatment of rheumatoid arthritis.

4.4.1.2.2   Carotenoids

Experiments in animals have also shown that carotenoids, including those without provitamin A activity, may strengthen the immune system. Their antioxidant properties might participate in these properties. Apart from experiments in rodents or in cultured cells, one study in humans confirms these initial encouraging results (Park 2010). In fact, the ingestion of 2–8 mg of astaxanthin for 2 months was capable of modulating immune response, estimated by a tuberculin test and by the concentration of T cells, and reducing inflammation, as assessed by C-reactive protein plasma levels. These positive reactions must only encourage the consumption of fish with pink flesh, such as farmed salmon. A 200-g serving of these fish may supply 2–8 mg of astaxanthin, besides *n*-3 fatty acids with a high biological value (Section 3.2.4.1).

In 2012, S.H. Jang, Korea, showed that another carotenoid, lycopene, is very efficient in inhibiting the aggressive responses of gastric epithelium generated by *Helicobacter pylori*, a major factor in the onset of gastritis, ulcers, and stomach cancer. These effects might arise from the powerful antioxidant properties of lycopene (Section 3.2.4.1), suppressing apoptosis and damages to DNA by an excess of free radicals. These results obtained in cell cultures will certainly be exploited in future clinical investigations.

4.4.1.3   Skin

4.4.1.3.1   Vitamin A

For a long time, vitamin A has been related to homeostasis of the epidermis, characterized by maintenance of the balance between cell proliferation and differentiation. Also, early on, it appeared that hyperkeratinization resulted from a vitamin A deficiency. In humans, this deficiency is accompanied by dry skin (xerosis) with the appearance of areas with scales (ichthyosis), followed by corneal keratinization, which is a source of blindness.

Retinol and retinoic acid, as well as a number of active derivatives, have been used in the treatment of some skin pathologies, such as acne, acne rosacea, keratosis, eczema, and psoriasis. Conversely, an excess of vitamin A–inhibited keratinization may even induce metaplasia with local irritation. The use of new natural or synthetic derivatives helped to minimize these side effects. Several of the compounds forming the complex of vitamin A are included in the cosmetic products offered to fight against premature aging of the skin under the influence of ultraviolet (UV) light.

4.4.1.3.2   Carotenoids

Despite the lack of rigorous evidence, a diet supplemented with or enriched in carotenoids (β- carotene and lycopene from fruits and orange, yellow, or red vegetables) has often been proclaimed to be beneficial for skin protection against UV aggression. Several studies on skin fibroblasts in culture have suggested that oral administration of astaxanthin could minimize the harmful effects of UV-A on wrinkles.

4.4.1.4   Cancers

4.4.1.4.1   Vitamin A and β-Carotene

In 1925, the first suggestion of a link between cancer disease and vitamin A deficiency was made. Since that first event, issued about the protective effect of β-carotene, several epidemiological studies have shown some benefits for consuming plenty of fruits and green or yellow vegetables. These findings have naturally led to many investigations on the protective effects of carotenoids and vitamin A, compounds known to be abundant in these foodstuffs. Thus, *in vitro* or animal models allowed the conclusion that different retinoids have protective effects against the development of several types of cancers (skin, prostate, liver, lung, breast, and colon). In addition, extensive research, mainly conducted in 2012 by Y. Sharoni, Israel, has led to the attribution of these effects not to carotenoids themselves but to their oxidation products, apocarotenoids. The action mechanisms of these derivatives are unfortunately still little explored. Despite this encouraging research and an indication that the consumption of fruits and vegetables helps to reduce the risk of cancer, the analysis of numerous studies in humans has not provided clear answers that could lead to preventive and effective treatments. Thus, a review of about 10 works on the potential role of lutein and zeaxanthin in the prevention of colorectal cancer has not concluded positively on the usefulness of these carotenoids. A recent study by S.Y. Park, the United States, in 2009, of the association between the intake of carotenoids and the incidence of colorectal cancer in a large multiethnic cohort has shown a significant result only for β-cryptoxanthin, which seemed protective in humans. According to several epidemiological studies, carotenoids used as food additives (E161c) might act as a chemopreventive agent against lung cancer.

In the United States, the vast multicenter Carotene and Retinol Efficacy Trial study led by G.S. Omenn, the United States, tested the supplementation of more than 18,000 people at risk (smokers and asbestos victims) with 8 mg of retinol and 30 mg of β-carotene per day. This test had to be stopped prematurely in 1996 due to the observation of an excess of cases of lung cancer (+28%) and mortality (+17%) in the treated group compared to the control group. In the absence of additional data, it seemed important to discourage smokers and other people at risk from taking any supplementation of carotenoids and retinol, except from a diet that should anyway be rich in fruits and vegetables. Several other studies have confirmed that supplementation with β-carotene was not associated with a lower risk of lung cancer.

Similarly, in the case of breast cancer there is little evidence that vitamin A supplementation may reduce the risk, except a prospective study with premenopausal women that was carried out in 1999 by S. Zhang, the United States.

It seems likely that the beneficial effect of β-carotene occurs only at doses found currently in food, and it would be rather detrimental at pharmacological doses.

Indeed, the use of blood tests has reopened the debate as it has been shown that an increased amount of 50 μg/100 mL in the β-carotene blood level was associated with a 26% decrease in the risk of cancer (Aune et al. 2012). Future studies should benefit from the demonstration in 2009 by R.M. Tamimi, the United States, that there was a clearer inverse relationship between circulating carotenoids and breast cancer among women with high mammographic density.

Many intervention studies have been realized with mixtures of β-carotene and retinol, sometimes with other vitamins, that unfortunately did not allow any rigorous conclusions on the effect of each component. To complicate matters, it is possible that each carotenoid has a protective effect *per se* without necessarily being converted into retinol. It has been shown that various natural carotenoids could be more effective than β-carotene for cancer prevention.

An extensive meta-analysis of various surveys, carried out in 2009 by S.C. Larsson, Sweden, involving more than one million people followed for a maximum of 26 years has demonstrated a beneficial effect of carotene on the incidence of estrogen-dependent breast cancer, but, as the author points out, without being able to completely eliminate the influence of other dietary factors.

4.4.1.4.2   Other Carotenoids

Regarding lycopene, which is not a provitamin A, many studies have demonstrated that an increase in consumption is accompanied by a lower risk of prostate cancer. These observations could be related to the demonstration of a protective effect of DNA against free radicals, carotene being one of the most powerful antioxidants that humans can find in food. Moreover, the study of patients with prostate cancer by A.V. Rao, Canada, in 1999 has shown that their blood and some tissues contained less lycopene and more oxidized lipids than healthy men, but the link between dietary lycopene and the reduction of the risk of prostate cancer remains poorly defined. One of the largest epidemiological studies focusing on the relationship between prostate cancer and carotenoids in nearly 48,000 people was conducted in 1995 in the United States by E. Giovannucci. The study found that among dietary carotenoids only a large consumption of lycopene could induce a 21% reduction in the incidence of prostate cancer. In addition, better effects (35% reduction) were observed in subjects consuming weekly at least 10 servings of tomatoes accommodated in different ways (raw, sauce, and pizza).

Given the numerous results confirming this protective effect, the Global Research Fund against Cancer recommended in 2007 lycopene as a protective agent against prostate cancer. Emphasis was placed on the advantage of cooked foodstuffs with tomato and mixed with vegetal oil to ensure better availability of lycopene in these types of preparations. It would appear that the use of thick tomato sauce is 20 times more efficient than raw tomatoes with an equivalent amount of lycopene. Despite these results, the FDA of the United States concluded in 2007 that the evidence supporting an association between tomato consumption and reduced risk of various cancers, including prostate cancer, was not yet sufficiently substantiated.

Although it has been shown that among the three carotenoids astaxanthin, canthaxanthin, and β-carotene astaxanthin had *in vitro* the most potent antitumor activity in mouse or human cancer cells, no clinical study has focused on this carotenoid. This example shows that there is a wide field of research on the potential anticancer activity of each of the carotenoids present in the plant world.

Although no clinical trials have examined the role of lycopene in the prevention of prostate cancer, a few trials have examined the effect of a supplementation on disease progression. In a study conducted in 2001 by O. Kucuk, Turkey, with 36 men suffering a prostate cancer that was clinically diagnosed, the ingestion of 30 mg of lycopene per day significantly reduced the tumor growth after 3 weeks. A meta-analysis conducted in 2009, led by F. Haseen, England, brought some interesting conclusions, although not definitive on this subject, from eight rigorous studies. This review has highlighted that in six studies an inverse relationship was observed between the amounts of lycopene intake and the PSA level. Treatment with lycopene reduced the symptoms associated with the disease (pain and urinary disorders). Despite these observations, the authors concluded that a lack of results did not allow definitive conclusions. Other controlled trials with more individuals will be required before establishing clinical or nutritional recommendations relevant to patients.

On a global level, clinicians and nutritionists have a great interest on the potential protective effects of carotenoids from a diet rich in fruits and vegetables against cancers. Given the difficulty repeatedly emphasized to assess the quantities of carotenoids actually ingested or assimilated in epidemiological surveys, determination of carotenoid plasma levels seems much more rational. This is the approach that has been adopted by many clinicians and has been the subject of a meta-analysis from 62 controlled studies in various countries (Donaldson 2011). Not only have 52 of the 62 reviewed studies concluded a beneficial effect of carotenoids on cancer incidence, but they have also allowed to define a plasma concentration equal to or greater than 2.5 μmol/L as the most favorable while major risks appeared to a concentration less than 1 μmol/L. These results only confirm the public health messages advocating the consumption of fruits and vegetables rich in various carotenoids, thus ensuring protection against various pathologies often considered to be associated with advanced civilizations.

After analyzing the large trials conducted worldwide, a group of experts of the International Agency for Research on Cancer (IARC) concluded that supplementation with one of the most common carotenoids should not be recommended. In contrast, it seems clear that any protection against the development of cancer may be exercised by a diet rich in carotenoids; its efficiency will come more certainly from the mixture contained in the foodstuffs themselves than from supplementation with a single compound.

4.4.1.5   Cardiovascular System

4.4.1.5.1   Vitamin A

Several cellular responses to retinoic acid, like other retinoids, such as growth and differentiation are also at the center of vascular pathologies such as atherosclerosis. Trials in humans have shown that vitamin A was much involved in angiogenesis and cell growth, but its effects on cardiovascular disease and mortality remained difficult to interpret. Some epidemiological studies have yet shown an inverse relationship between blood retinol levels and the incidence of myocardial infarction. This is the case for the vast Prospective Epidemiological Study of Myocardial Infarction carried out in France and Ireland with nearly 10,000 people for 5 years and completed in 2012.

Many studies have shown that vitamin A was involved in the mechanisms of blood coagulation, inflammation, and vascular calcification, proving that this vitamin is an essential partner in the understanding of vascular pathologies. The complexity and diversity of the mechanisms involved could only delay clinical trials, the results being obtained with retinoids as well as carotenoids.

4.4.1.5.2 Carotenoids

As soon as coronary heart disease related to atherosclerosis was largely explained by the phenomena of lipoprotein oxidation, clinicians have focused their efforts toward a detailed examination of the possible beneficial effects of foods rich in fat-soluble antioxidants, especially carotenoids and vitamin E. Clinicians have attempted early to establish links between the supply of these compounds by the consumption of plants and the incidence of cardiovascular disease, most often estimated by the mortality of persons included in larger cohorts followed for several years. In 1995, the follow-up during 5 years of approximately 1300 seniors living in institutions in Boston, Massachusetts, by J.M. Gaziano has shown that the risk of death by cardiovascular disease was 46% lower in subjects ingesting the highest amount of carotenoids compared to those ingesting the least. The benefit on myocardial infarction was more evident since the risk was reduced by 75%. An analysis of similar data made by B. Buijsse in 2008, with the Dutch Zutphen study based on monitoring 1000 adults of average age 72 years for 15 years, has confirmed these initial findings.

The effect was found again in a European study (SENECA study) taking into account the plasma levels of α- and β-carotene, markers of the ingestion of these compounds, and the mortality from cardiovascular disease in nearly 1200 elderly followed for 10 years. More recently, these blood parameters were positively associated with markers of endothelial function and with a decrease in inflammation and oxidative stress.

Whereas virtually all epidemiological studies have shown undeniable consistency, intervention studies have shown no effect of the supplementation of carotene on the incidence of cardiovascular disease. Thus, several types of treatments by antioxidants, including β-carotene, had no effect on the progression of atherosclerosis. A large meta-analysis conducted by the Cochrane Institute concluded that β-carotene supplementation has no interest in primary and secondary prevention (Bielakovic et al. 2012). Furthermore, of the 78 clinical trials examined, 38 even demonstrated a significant increase in mortality. What is the origin of these differences in findings between observational and intervention studies? The most likely hypothesis is based on the complexity of our foodstuffs and the possible interactions between the components in the ingested plant tissues. Another possibility may be the difference in lifestyles and diets among people accustomed to a natural diet rich in antioxidants and those who prefer to absorb vitamin supplements.

Few works have been concerned with the potential effect of carotenoids other than carotene on the cardiovascular system. Among these, especially astaxanthin was investigated in animals, their results being in favor of a very beneficial effect against oxidative stress and inflammation. Clinicians were able to verify the validity of these findings in humans, but no study to date has been devoted to any effect in individuals with cardiovascular disease. Despite this lack of data in the field of medicine, astaxanthin remains a well-placed candidate in the search for an active and safe therapeutic agent for cardiovascular disease (G. Hussein, International Research

Center for Traditional Medicine, Japan). It is the same for lycopene, based on surveys assigning to tomatoes a protection power against cardiovascular disease (all accidents combined) or more particularly against ischemic stroke (Karppi et al. 2012).

The problem is far from being resolved since an experiment reported in 2012 by F. Thies, England, did not allow the detection of a beneficial effect of a supplementation with lycopene in the form of capsules (10 mg/day) or tomatoes (32–50 mg/day) in improving the traditional indicators of cardiovascular disease.

If epidemiological studies, as those conducted *in vitro*, appeared promising paths for future research (antioxidation and anti-inflammation), V. Böhm, Germany, admitted in 2012 that the experimental studies in humans have failed to provide evidence of any beneficial cardiovascular effect. As for other carotenoids, other experiments in the long term and in subjects with varying degrees of cardiovascular injury will be needed to reach more convincing conclusions on the therapeutic effectiveness of lycopene.

In any case, the recommendation to eat five fruits and vegetables per day, claimed by all national agencies, fits well with all the conclusions of scientific and clinical works concerning the possible relationships between carotenoids and cardiovascular system. The acuteness of the situation is underlined by the WHO, which assigned some three million annual deaths to an inadequate consumption of fruits and vegetables, a risk factor almost as high as tobacco. For its part, the director of the Division of Food and Nutrition of the FAO stated that "to increase the consumption of fruits and vegetables is a fundamental public health issue at the moment."

4.4.1.6 Alzheimer's Disease

Laboratory experiments have allowed for many years to know that retinoic acid, a form of vitamin A, plays a key role in the development of the nervous system. Thus, F. Mingaud, University of Bordeaux, France, showed in 2008 that vitamin A was involved in the regulation of synaptic plasticity, which is a fundamental parameter for learning and the functioning of memory, mainly through the expression of target genes.

Many data converged on the concept of maintaining memory performance through the regulatory functions of retinoids in the brain. On the one hand, several studies have helped to demonstrate that dysregulations of neuronal transport of vitamin A and cell signaling were related to the onset of Alzheimer's disease (Goodman et al. 2003). Furthermore, the serum retinol content and the efficiency of retinoic acid biosynthesis were very often decreased in patients with Alzheimer's disease; this appears to be mainly involved in the etiology of the "late-onset" form of this disease. This working hypothesis seems to have been confirmed by J.P. Corcoran, England, in 2004 by the demonstration in animals of a deposition of β-amyloid peptide, an extracellular component of senile plaques, when the experimenter inhibited the retinoic acid signaling pathway with a vitamin A deficiency.

Many studies have shown that the formation of senile plaques was associated with local oxidative stress, the latter contributing to the development of pathogenesis of Alzheimer's disease. Several treatments have also tried, but without much success, to use antioxidant molecules on these potential oxidative reactions. However, epidemiological studies realized in 1999 by F.J. Jiménez-Jiménez, Spain, have verified that the serum concentrations of several antioxidants, such as β-carotene and vitamins A and E, were lower in subjects with Alzheimer's disease than in control subjects. Several studies, such as that carried out by E.J. Johnson (2012), tended to prove that lutein and zeaxanthin may indeed play a role in maintaining cognitive function in older people.

In an animal model of the disease, Y. Ding, the United States, demonstrated in 2008 that a treatment with retinoic acid, which itself has antioxidant properties, was able to reduce the accumulation of the β-amyloid peptide and the neurone loss and also to improve the memory performances of the animals. These results are obviously encouraging for a future therapeutic use already considered by several authors, perhaps in synergy with the current treatments. In future research, it will be necessary to better understand the implications of vitamin A and carotenoids in the chain of oxidative reactions and in the slowing of signaling pathways. These pathways are indeed known to be involved in the genesis of pathological lesions in the brains of patients with Alzheimer's disease.

No clinical trials on this topic are currently reported in the international databases.

4.4.2 Vitamin D

Vitamin D, especially vitamin D_3 (Section 2.7.2), is now regarded as a compound playing in different areas an essential role in our health. In addition to controlling calcium metabolism, it seems to be involved in metabolic disorders, diabetes, cardiovascular disease, cancers, brain function, and immune system (Cherniack et al. 2008). Recent data on the genome expression of white blood cells collected from healthy adults have revealed that the expression of 66 genes was altered in deficient subjects. The recovery of their expression after vitamin D treatment supports the view that more than 160 pathways linked to several chronic diseases are associated with a deficiency state (Hossein-nezhad et al. 2013). These results support the use of gene expression analysis in future research on the effects of nutritional factors in deficiency situations.

To confirm the importance of vitamin D, the EFSA officially approved in 2010 that vitamin D not only contributed to phosphocalcic metabolism but also ensured the normal functioning of muscles, immune and inflammatory reactions, and the cardiovascular system and contributed to the mechanism of cell division [Article 13 of Regulation (EC) No. 1924/2006] (http://www.efsa.europa.eu/en/scdocs/doc/s1468.pdf).

The existence of an endogenous origin and its mode of action have caused vitamin D to be considered as a hormone rather than as a vitamin. It has even been nicknamed "solar hormone" (Dupont 2011) and even "soltriol."

The main physiological systems affected by a deficiency of vitamin D from the diet or from an underexposure to solar radiation are summarized in Figure 4.1.

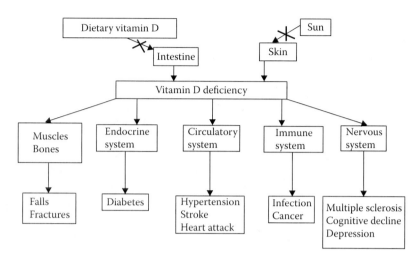

FIGURE 4.1 Physiological systems affected by a vitamin D deficiency: X indicates the essential origins of vitamin D deficiency.

The multiplicity and the physiological importance of the areas of influence of vitamin D explained the growing interest of investigators and medicine doctors for this vitamin-hormone, the interest resulting in over 6000 scientific articles in 2010, that production increasing by nearly 300 every year (PUBMED database). Thirty-six clinical trials on vitamin D were declared in 2012 on the official website http://www.clinicaltrials.gov.

4.4.2.1 Phosphocalcic Metabolism

The regulation of phosphate and calcium homeostasis by increasing their intestinal absorption was the first main role assigned to vitamin D. In the absence of this vitamin and compared to normal levels, the human body does absorb a maximum of 15% calcium and 60% phosphorus. Physiologically, the increase in this absorption is generally a response to a phosphocalcic deficit in body fluids during certain physiological (growth and pregnancy) and pathological (hyperparathyroidism) conditions. Any deficiency in vitamin D may lead to bone disease, starting with a disorder of mineralization that may worsen into rickets (stunted growth) and, if left untreated, into osteomalacia (progressive bone softening). The initial failure of the mineralization is amplified by a concomitant increase in plasma parathyroid hormone that may contribute in the long term to osteoporosis. A dietary deficiency of vitamin D has obviously more acute consequences when accompanied by intestinal malabsorption (celiac disease, Crohn's disease, etc.). In this case, the intake of vitamin D as all the fat-soluble vitamins is lessened and therefore must be compensated by a significant nutritional supplementation.

A supplementation with calcium and vitamin D is commonly used in aged or osteoporotic subjects to consolidate their bones. It is now accepted that this simple treatment reduces their risk of falls and fractures, through an action on the skeleton as well on the muscles and the central nervous system. The beneficial effect of vitamin D on muscle function and physical condition of women over 65 years was demonstrated in 2011 by S.R. Mastaglia, Argentina. Several epidemiological studies have shown that a daily intake of 20 μg of vitamin D reduced the risk of falls and fractures. It has been shown that a significant proportion of adults may be deficient in vitamin D, while the supplies were sometimes higher than the recommended doses. An assay of serum vitamin D is essential to better detect a deficiency state and especially the risk of fractures. A serum concentration of this compound below 30 μg/L is now clearly considered as inadequate. Several clinical studies have evaluated the effects of a supplementation with vitamin D (associated or not with calcium) on the risk of limb fractures. Thus, a meta-analysis, published in 2011 and realized by B. Tang, Australia, has concluded that the efficacy of a daily intake of at least 800 IU (20 μg) was indeed effective especially in patients over 70 years. The association with about 1 g of calcium per day improved the benefits of treatment for the prevention of fractures and bone mineral loss. A recent review of the Cochrane Institute analyzing eight studies has shown that supplementation with vitamin D combined with calcium significantly reduced the risk of hip fracture in elderly people in institutions, whereas vitamin D alone had little effect (Avenell et al. 2009).

4.4.2.2   Metabolic Diseases

Diabetes is one of the most important chronic diseases in all populations and inducing severe and often fatal complications (8% of world mortality). Worldwide, nearly 380 million people had diabetes in 2013 and by 2035 this will rise to 592 million. The prevalence of diabetes has been estimated in 2013 to 37 million people in North America, 56 million in Europe, and 72 million in Southeast Asia (http://www.idf.org/diabetesatlas). It is estimated that over 3.5 million people living in France are suffering from diabetes (prevalence is steadily rising), with most patients having type 2 diabetes (about 90% of the total).

For many years, the incidence of type 1 diabetes has been related to geographical locations, the season and the sunshine, all these observations suggesting the involvement of vitamin D in the etiology of the disease. Based on these findings, many epidemiological studies have shown an inverse relationship between blood levels of vitamin D and diabetes risk. A recent review of eight investigations has shown that the absorption of more than 500 IU/day (12 μg/day) of vitamin D reduced the risk of type 2 diabetes by 13% compared to the lowest intakes and that a blood level higher than 25 μg/L reduced that risk by 43% compared to the lowest levels (Mitri et al. 2011).

Various experiments in animals, and especially the discovery of vitamin D receptors in the pancreatic insulin-secreting cells, have rapidly directed research toward an increasingly evident relationship between vitamin D and glucose metabolism. From a clinical point of view, the problem was rapidly expanded toward a role of vitamin D in the genesis of both types of diabetes. Thus, a reduction of insulin secretion in animals deficient in vitamin D and the observation of a higher incidence of type 2 diabetes in individuals deficient in vitamin D clearly suggested a participation of this vitamin in the genesis of both types of diabetes.

For type 1 diabetes (insulin dependent), it was often observed that children with rickets had a diabetes risk nearly three times higher than children with normal growth. Besides numerous observations going in the same direction, a striking example of intervention was given in 2001 by E. Hyppönen, England, who followed more than 10,000 newborns until their first birthday. In these young children, a supply of 50 µg/day (2000 IU/day) of vitamin D allowed to reduce by 80% the risk of developing childhood diabetes. In this area, the exact role of vitamin D at the molecular level is still poorly known, but it seems more probable that it involves a protection of pancreatic β cells against damages induced by cytokines and other inflammatory agents. A recent meta-analysis on more than 20,000 participants has demonstrated that a vitamin D supplementation during early life was more efficient in reducing the risk of type 1 diabetes in the childhood period than a maternal intake of vitamin D during pregnancy (Dong et al. 2013).

Clinical use of vitamin D in the prevention of diabetes is still limited due to the hypercalcemic effects of the required doses. Currently, much research is directed toward the use of vitamin D structural analogues showing especially immunomodulatory effects.

For type 2 diabetes, disease characterized by a loss of cellular insulin sensitivity, the investigations are less numerous but could lead to promising clinical applications. Thus, a major study on the subject, led in 2008 by N.G. Farouhi under the Medical Research Council Ely Prospective Study, confirmed earlier epidemiologic evidence that vitamin D deficiency is associated with the development of hyperglycemia and insulin resistance. The study followed more than 500 nondiabetic subjects for 10 years. Similarly, the monitoring of more than 4000 Finns for 17 years allowed C. Mattila in 2007 to show that the risk of developing type 2 diabetes was lower for those having a high plasma vitamin D level. An extensive study by A.G. Pittas in 2010, the United States, as part of the Nurses' Health Study has confirmed this association. In 2009, as a conclusion of an epidemiological study of more than 3200 people, L. Lu, China, proposed that the blood level of vitamin D was predictive of the onset of type 2 diabetes, characterizing the metabolic syndrome.

Obesity has been related to a deficiency state of vitamin D for a long time. This deficiency could be the result of storage of the vitamin in adipose tissue, and also of a dietary deficiency or of its impaired metabolism. It is significant that an Italian study led by G. Muscogiuri in 2010 has shown that the plasma levels of vitamin D were inversely correlated with body mass index values in obese patients.

In terms of intervention studies, the former observation of an improvement after treatment with vitamin D of the glucose tolerance and the insulin secretion in deficient subjects has not been confirmed, except, perhaps, in nondiabetic subjects.

Diabetes and vitamin D are the subject of much fundamental and clinical research. All have confirmed the close relationships between these two themes, but progress is still needed to identify metabolic and molecular targets that may help treatment and prevention of this complex disease. As pointed out in 2012 by C.E. Chagas, Brazil, future studies should be focused on the role of vitamin D in developing type 2 diabetes but favoring the participation of inflammation mechanisms.

Based on the results already achieved, it seems urgent to plan rigorous clinical trials to determine whether vitamin D can really help the prevention and treatment of both types of diabetes. As emphasized by some authors, it is imperative to raise doubts about the well-known possible interferences between circulating levels of vitamin D, sun exposure, physical activity, diet, and cellular sensitivity to insulin.

4.4.2.3 Cardiovascular Diseases

The beneficial effects of high plasma vitamin D levels or vitamin D supplementation on overall mortality have been the object of many epidemiological studies. To illustrate this approach, one only has to examine the results of a large meta-analysis of 18 studies performed on a total of nearly 57,000 participants followed for nearly 6 years (Autier and Gandini 2007). This work came to the main conclusion that a major 7% reduction in mortality was obtained from the participants who received vitamin D (between 7.5 and 50 µg/day, 300 and 2000 IU/day). These observations were verified in 2012 by A. Zittermann, Germany, after performing a meta-analysis suggesting an inverse but nonlinear relationship between mortality and circulating levels of vitamin D with an optimal level of 33 µg/L. This result probably explained why some investigators have observed no effect of a regular intake of moderate doses of vitamin D (15 µg/day) on markers of cardiovascular risk. Confusing results were released in 2012 by E.D. Michos, the United States, after conducting a large epidemiological survey based on monitoring 5000 white subjects and nearly 3000 Afro-Americans for 14 years. Like his predecessors, this author has verified that vitamin D deficiency is closely associated with the incidence of myocardial infarction, but only in white subjects. The lack of relationship among Afro-Americans shows the complexity of the physiological responses, revealing a possible influence of the natural resistance of these subjects to vitamin D deficiency, a condition frequently reported in this population (Section 3.2.4.2). It has already been suggested in 2011 by O.M. Gutierrez that this resistance could be related to their lower susceptibility to bone fractures.

The close relationships between circulating levels of vitamin D and cardiovascular disease mortality have been extensively studied in various countries. All agreed to fix the lower limit at about 15 µg/L, the risk of heart attack at lower concentrations being about twice as large as that observed for concentrations greater than or equal to 30 µg/L. Moreover, the risks seem to be largely more important in subjects with a metabolic syndrome at the level of the sudden death and congestive heart failure, but curiously not for myocardial infarction (Thomas et al. 2012).

In addition to heart, similar observations were made for cerebral circulation. After following in Finland nearly 6200 people (the Mini-Finland Health Survey) for 26 years, A. Kilkkinen showed in 2009 that the incidence of stroke is more directly correlated to the vitamin D status than to vascular events at the cardiac level. These observations on a Finnish population deserve to be replicated in other populations in various countries to eliminate possible bias induced by the age of the subjects, the particular modes of life, and food preferences. The influence of diet is more difficult to measure, probably because of the difficulty of assessing the daily intake of vitamin D (imprecision in the composition tables and the consumed rations) and the participation of sun exposure. However, through the study of a large part of a population of health professionals (more than 118,000 Americans were followed for

22 years), Q. Sun could highlight in 2011 that among men ingestion of more than 600 IU (15 µg) of vitamin D per day was associated with a risk of cardiovascular disease 16% lower than that with an ingestion of less than 100 IU (2.5 µg) per day. Curiously, women showed no difference in risk.

It is obviously necessary to emphasize that any direct causal link between vitamin D and cardiovascular disease cannot be definitively established on the basis of the aforementioned results, despite the elimination of many factors interfering in the genesis of these diseases. Indeed, some of these factors cannot be quantified accurately (sun exposure), and others are too complex to be taken into account (nutrition and lifestyle).

So far, any intervention study in humans has failed to show a positive effect on the development of cardiovascular diseases.

Vitamin D deficiency has also been implicated in the phenomenon of hypertension since the discovery that sunlight exposure may induce a decrease in blood pressure. After that ecological data, several studies have been conducted to clarify the efficacy of a very inexpensive treatment of hypertension. Indeed, this simple treatment could be applied to 14 million French hypertensive patients in whom 7.6 million are treated pharmacologically for an estimated €2.6 billion cost (IRDES). One of the most documented clinical studies, conducted by J.P. Forman in 2007, the United States, was to monitor blood pressure parameters and circulating levels of vitamin D in nearly 1800 subjects over 4 years. A highly significant correlation was found between risk of hypertension and plasma levels of vitamin D. All subjects combined, the risk appeared to be more than three times higher among those with vitamin D deficiency (levels < 15 µg/L) than in those with a subnormal level (≥30 µg/L). These differences appeared to be more significant in the male cohort (risk 6 times higher) than in women (risk 2.7 times higher). This phenomenon has been observed consistently in many other studies, with slightly different values for the risk as for the critical levels of vitamin D (Burgaz et al. 2011). Faced with the evidence of these findings, it seems already recommended to include the determination of circulating levels of vitamin D in any clinical investigation of hypertension. One can also imagine the therapeutic value of these findings for the treatment of hypertension in low-income countries. This seems even more evident in people with highly pigmented skin since S.G. Rostang, the United States, in 2010, correlated the frequently high hypertension among Afro-Americans to their frequently inadequate vitamin D blood levels. What might be the mechanisms underlying the protective effect of vitamin D on the cardiovascular system? Many hypotheses have been proposed; one of the most likely seemed to be that involving the activation of the renin–angiotensin–aldosterone system by a too low level of vitamin D (V.C. Garcia, Brazil, 2010). This hypothesis has been reinforced by the findings of J.P. Forman, an American hypertension specialist. For the first time in 2010, this investigator demonstrated in humans the association between vitamin D deficiency, increased angiotensin II circulating levels, and a decrease in renal perfusion. Recent genetic research in the same laboratory has verified that the regulation of plasma renin was at the level of the vitamin D receptor. Another promising possibility arose from the demonstration in 2012 by A.E. Rick, the United States, of the genesis of a type of proatherogenic monocytes/macrophages under the effect of vitamin D. These cells involved in the formation of atherosclerotic plaques more intensely express adhesion molecules such as integrins and selectins, promoting their association with the vascular endothelium.

Curiously, J.L. Perez-Castrillon, Spain, observed in 2010 that high levels of vitamin D promoted the action of statins, which are powerful cholesterol-lowering chemicals. This observation must be seriously taken into account before doing large surveys on the physiological effects of vitamin D because of the widespread use of statins.

A large study (Vitamin D and Omega-3 Trial) was initiated in 2010 (Harvard Medical School) to specify the terms of prevention of cardiovascular disease (and cancer) by vitamin D (and *n*-3 fatty acids) in 20,000 subjects over the age of 50 who will be followed for 5 years (www.vitalstudy.org). The protocol planned inter alia the consideration of medications and supplements absorbed by the participants during the study period.

As for diabetes, it is desirable that the effects of vitamin D on the prevention of all cardiovascular diseases, and mainly hypertension, have the benefit of other intervention trials, precise and extended, to define the exact therapeutic gain of an inexpensive treatment reduced to a simple vitamin supplementation. Once the causal link between cardiovascular disease and vitamin D is established, clinicians will be able to prescribe an upgrading of the vitamin status and ensure an effective and inexpensive primary or secondary prevention.

Given the number of studies finding a beneficial effect of vitamin D on heart disease and hypertension, it now seems important to determine circulating levels of vitamin D in any clinical investigation carried out to explore the risk of cardiovascular disease. Restoring a vitamin level consistent with official recommendations may be provided by natural foods (fatty fish) and more conveniently by prescribing regular doses of vitamin D.

4.4.2.4 Cancers

Several environmental investigations have shown a close relationship between the prevalence of many cancers and the habitat latitude, the distance from the equator generally increasing the incidence of certain cancers (colon, breast, prostate, brain, pancreas, etc.). Given the wide variety of the sampled locations, the correlation seems to be explained by the decrease in the cutaneous synthesis of vitamin D at increasing latitudes.

First suggested in 1937 by the famous American epidemiologist S. Peller, the hypothesis of an inverse relationship between sun exposure and cancer mortality has been clearly reported after the detailed analysis of official data from the United States and Canada. Epidemiological work of C. Garland, the United States, published in 1980 has clarified the beneficial effects of sun exposure on the development of colorectal cancer. This author showed that colon cancer was less common in the southern cities (very sunny) than in the northern ones (less sunny cities) in the

United States. The breakthrough of this study was the inverse relationship between the frequency of the disease and the intensity of vitamin D biosynthesis among the inhabitants of these cities. A similar relationship was subsequently confirmed by the same author between vitamin D intake and the same type of cancer. For Europe, the *Vitamin D and Cancer* report published in 2008 by the IARC (IARC–WHO) concluded that epidemiological studies provided convincing evidence for an inverse association between plasma vitamin D and the incidence of colorectal adenoma and sporadic colorectal cancer. It is clear that no causal link can yet be established. Results on colorectal cancer, obtained from the EPIC survey in 10 European countries involving 520,000 participants, have been published in 2010 by Jenab, France. The survey has shown a clear advantage to having high plasma vitamin D levels, as for the highest values (>40 µg/L) the risk of cancer is 40% lower than for the lowest values (<10 µg/L).

So far, it should be noted that no vitamin D supplementation could lower the risk of this type of cancer.

For breast cancer, seven major studies have been conducted since 2005 in different countries, some revealing a prominent inverse relationship between serum vitamin D and the risk of developing this type of cancer and others detecting no connection.

The French INSERM study, conducted by P. Engel, verified in 2010 an inverse relationship, as a previous study from Denmark, but it was more significant in younger women (<53 years) than in older women. As the author pointed out, these data argued in favor of maintaining adequate biological levels of vitamin D especially in populations with low serum levels. Indeed, to increase from 10 to 30 µg/L the concentration of vitamin D considered as adequate a daily intake of 50 µg of vitamin D would be necessary in women with low sun exposure, which corresponds to 10 times the current recommendations of the AFSSA (Section 3.2.4.2). In the face of the wide range of epidemiologic and laboratory studies providing compelling evidence of a beneficial role of vitamin D in the field of breast cancer, it becomes evident that a supplementation is urgently needed. As requested by some clinicians, a vitamin D supplementation remains a low-cost and safe strategy for breast cancer prevention that should be implemented without delay (Mohr et al. 2012).

These epidemiological studies highlight the complexity of the involved phenomena and the need for further investigations especially in the field of vitamin D intakes required to effectively reduce the risk of breast cancer. Much research will certainly be needed before answering this important question.

Among all types of cancer, prostate cancer has initiated the largest research efforts. The first assumption of a protection against prostate cancer with vitamin D was clearly made in 1990 by G.G. Schwartz, the United States. Since this work, more than 500 works, experimental or clinical, have clarified this role in the onset of the disease, its progression, angiogenesis, and metastasis genesis. To be convinced of the close association between this vitamin and this type of cancer, it is enough to browse one of the more recent studies conducted on 1822 patients with prostate cancer who were followed for nearly a decade (Fang et al. 2011).

The authors have shown that individuals with the lowest vitamin D levels (mean 20.4 µg/L) have a mortality risk 60% higher than those with the highest levels (mean 49 µg/L). In view of these data, it is surprising that the *Vitamin D and Cancer* report published in 2008 by the IARC (IARC–WHO) concluded that epidemiological studies provided little or no evidence of any effect of vitamin D on the incidence of prostate cancer.

Research published since this report seemed to tip the balance toward a clearer association between vitamin D and prostate cancer. This association gave rise to numerous experimental studies that have explored on cell cultures the effects of vitamin D on cell proliferation. All confirmed that this vitamin, and synthetic analogues, inhibited growth and accelerated the differentiation of many types of cancer cells, limiting their uncontrolled growth. Its molecular mode of action includes a cross talk between its own signaling pathways and that of other growth factors regulating cell proliferation, differentiation, and survival (Moukayed and Grant 2013). The recent demonstration by two American teams of several molecular mechanisms involved in the control of cell division has shed new light on the anticancer potential of this vitamin. Indeed, in 2012 R. Salechi-Tabar showed that vitamin D blocked carcinogenesis by inhibiting the production and function of a specific protein required for cell division, c-Myc, and later R. Kasiappan discovered a novel mechanism of inhibition by vitamin D of the expression of an essential enzyme for cell proliferation, telomerase.

Despite all the uncertainties related to "ecological" and clinical investigations, it has been estimated that nearly 60,000 individuals in the United States and nearly 25,000 in the United Kingdom die each year from a cancer caused by a vitamin D deficiency. In the United States, the expenses induced by the diseases that might be linked to this deficiency were assessed in 2004 to be between $40 and $56 billion. The potential benefits are such that an increase in plasma vitamin D of 25–40 µg/L could reduce in western Europe a health expenditure of about €190 million per year. Similarly, C.F. Garland estimated in 2009 that the daily intake of 50 µg of vitamin D could help prevent 58,000 new cases of breast cancer and 49,000 cases of colon cancer and cause a 75% reduction in mortality associated with these diseases. Thus, people with high circulating levels of vitamin D are twice less likely to contract breast cancer or colon cancer than those with low levels and their chances of survival would be better when they are affected. All these works have led experts to unanimously suggest that daily intake in the form of a dietary supplement should now be increased for all people. In addition to these nutritional recommendations, it must not be overlooked that vitamin D is induced by sun exposure, although the supply is irregular and difficult to control.

Patients on anticancer chemotherapy or radiotherapy may also take advantage of the benefits of vitamin D, even when metastases are present. These benefits might result from the potentiation of cytotoxic anticancer agents by vitamin D, as demonstrated in 2010 by D.L. Trump, the United States.

The importance of the subject and the large number of results justified the claim by the IARC in 2008 to support more research in this area in accordance with the classical experimental rules (*Vitamin D and Cancer*: http://www.iarc.fr/en/publications /pdfs-online/wrk/wrk5/Report_VitD.pdf).

In conclusion, the only cancers that may be linked to vitamin D are, for the moment, colorectal, breast, and prostate cancers. It is increasingly clear that high plasma levels of vitamin D effectively inhibit the development of colorectal cancer. The results are also promising for breast cancer but have yet to be confirmed. For prostate cancer, the evidence is less strong but many biological data suggest that the clinical use will be quickly implemented. Pending rigorous facts and clear explanations, it seems wise to maintain circulating levels of vitamin D at least as high as those recommended by medical authorities.

4.4.2.5 Nervous Diseases

Many studies, in both humans and animals, have made it increasingly clear that vitamin D could be involved in the synthesis of neurotransmitters, the expression of neurotrophic factors, neurogenesis, brain function via cytokine production, and protection against oxidative stress (McCann and Ames 2008). As vitamin D, through its nuclear receptor, modulates the transcription of 500–1000 genes and these receptors are present in the hippocampus and the cortex of human brain, research interest has naturally focused on the possible links between this vitamin and brain dysfunction. Thus, gradually, vitamin D has achieved the status of neurohormone, similar to some steroids (pregnenolone, progesterone, and dehydroepiandrosterone).

4.4.2.5.1 Cognitive Decline

According to a European prospective study conducted by D.M. Lee, England, and published in 2009, vitamin D deficiency was frequently associated with cognitive decline in the elderly. The results have indeed shown that men with the lowest plasma levels of vitamin D were affected by high cognitive impairment and, conversely, those with higher levels have better mental performances. A recent multicenter study (Ireland, France, and Italy), directed and published in 2010 by K.M. Seamans, Great Britain, has confirmed these results in subjects 55–87 years old and tested with the psychophysiological procedure CANTAB. All research in this area has found that individuals deficient in vitamin D (plasma levels < 10 µg/L) have an increased risk of deterioration of the MMSE score that could reach about 60%. This comprehensive test explores various memory abilities, such as temporospatial orientation, learning, memory, attention, calculation, reasoning, language, and constructive praxis. A large meta-analysis of 37 studies conducted in 2012 by C. Balion has verified that the lowest vitamin D blood levels (<20 µg/L) were associated with the lowest cognitive abilities and the highest risk of further developing Alzheimer's disease.

Following many other specialists, D.J. Llewellyn, England, issued in 2009 the principle that "deficits in vitamin D are a very promising therapeutic target for the prevention of dementia. Furthermore, that supplementation is inexpensive, safe, and has already shown its benefits in reducing the risk of falls, fractures and death." Recent research has added new insights to the treatment or prevention of Alzheimer's disease in showing that a combination of vitamin D and an inhibitor of N-methyl-D-aspartate receptors (memantine) was the most potent in protecting cortical neurons against degeneration (Annweiler et al. 2014).

In the absence of clinical trials, it is premature to conclude that vitamin D supplementation can prevent the onset of cognitive impairment in persons in good health or improve the cognitive status of people already suffering from these deficiencies. It is hoped that health authorities of all countries would consider that there is urgency to drive further research in this area, given the predicted epidemic of Alzheimer's disease, the ultimate in cognitive decline. Future research should address the pressing questions of the population by verifying the usefulness of a therapy with vitamin D and establishing the dose required to prevent or treat diseases while avoiding harmful side reactions. Future research should also identify the roles of the vitamin D receptor in the pathogenesis of cognitive decline and possibly dementia and its close relationship with the plasma levels of vitamin D and *n*-3 fatty acids.

In any event, even in the absence of compelling evidence of a causal link between vitamin D and intellectual capabilities, emphasis should be placed on the observance of a daily intake of vitamin D sufficient and consistent with the recommendations of competent medical organizations. This compliance must be especially strict for the elderly and even more for those with highly pigmented skin. In case of a deficiency detected in the blood, an oral intake of vitamin D is recommended after medical advice.

4.4.2.5.2 Schizophrenia

A possible link between vitamin D deficiency and schizophrenia was raised in 1999 by J. McGrath, England. Many observations have tried to confirm this possibility based on winter births, a state of dark-skinned subjects who migrated to the Nordic countries and the incidence of the disease at high latitudes. An original approach to this problem was addressed later by the same investigator considering the well-known importance of nutrition and lifestyle during the first year of life on adulthood health. Thus, the accumulated data on vitamin D supplementation during the first year of life and the diagnosis of schizophrenia established in the thirty-first year in 9114 Finnish subjects enabled clinicians to verify in 2004 the validity of an association between vitamin D intake and a reduced risk of disease. Although nearly 10% of cases of schizophrenia may be blamed on a winter date of birth, the physiological causes of this prevalence are far from being understood. It is more than likely that vitamin D deficiency during pregnancy and in infants could affect brain development and brain's future functions, thus inducing disorders such as schizophrenia, and even autism.

In 2010, R. Amato, Italy, demonstrated a close relationship between geographical areas and genes associated with neuropsychiatric diseases, such as schizophrenia, and those associated with vitamin D. This work confirmed at the molecular scale the relationships between psychiatric illness and vitamin D. The author further suggested that schizophrenia may result from a trouble in the adaptation of vitamin D metabolism as a function of geographic latitude. The importance of this parameter suggests that this mechanism could occur for other diseases in the field of vitamin D. Much research will be needed to determine the validity of these assumptions, but all experts stress the need to detect and compensate rapidly any vitamin D deficiency in patients with psychiatric disorders including schizophrenia.

4.4.2.5.3 Parkinson's Disease

Since 1997, the observation of a vitamin D deficiency has been reported by Y. Sato, Japan, in patients with Parkinson's disease, which was later often confirmed. Obviously, it could not be determined whether the presence of such a chronic debilitating disease was the cause of this deficiency or if, conversely, the latter is the cause of the disease, or at least its aggravation. On the other hand, it is clear that this disease is often accompanied by osteoporosis, a bone condition causing frequent fractures.

Among new research, it must be recalled that E. Evatt, the United States, proposed in 2011 that Parkinson's disease, and to a lesser degree Alzheimer's disease, could be related to vitamin D deficiency. Moreover, he said that hypovitaminosis does not seem related to lifestyle. A large study conducted in 2010 by P. Knekt in Finland, where sun exposure is reduced and vitamin D deficiency is common, tended to confirm the relevance of the previous hypothesis. Indeed, taking into account all other factors for a period of 30 years, the number of patients who reported the disease in the group of people with the lowest plasma vitamin D levels was three times greater than in the group with the highest levels.

Research in this sensitive area will certainly accelerate in the near future following the discovery of an association between Parkinson's disease and a polymorphism of the gene controlling the synthesis of vitamin D receptors, an association likely related to the pathogenesis of that neurological disease. A recent genomic study published in 2011 by W. Butler, the United States, has reinforced the latter working hypothesis.

Three trials involving vitamin D in the treatment of Parkinson's disease are reported on the official website http://www.clinicaltrials.gov.

In conclusion, besides protecting the nervous system for which an optimal level has not yet been determined, it seems appropriate to advise people diagnosed as suffering from Parkinson's disease to increase their sun exposure and take all measures to fight against osteoporosis, including vitamin D and calcium supplementation. For a still hypothetical prevention against this disease, it seems prudent to recommend for all seniors, regardless of the contribution of their skin photosynthesis, a daily intake of vitamin D from 25 to 50 µg (1000–2000 IU), an intake required to struggle against bone demineralization (Section 3.2.4.2).

4.4.2.5.4 Depression

Besides effects on cognitive impairments, it seems more and more likely that vitamin D is involved in the development and severity of depression.

The correlation between depression and neurotransmitter activity (serotonin, norepinephrine, and dopamine) is now well established. These neurotransmitters participate in the regulation of emotional activity, reaction to stress, regulation of sleep cycles, appetite, and many other functions. It therefore seems logical that vitamin D as an antidepressant controls the regulation of the balance of neurotransmitters involved in depression. This sun hormone may well be, as it was hypothesized in 1991 by W.E. Stumpf, a link in a pathway, skin–brain, similar to the well-known neuroendocrine pathway retina–diencephalon–epiphysis. It was not until 1984 that the use of light therapy in clinical psychiatry made its first appearance for treating seasonal affective disorder (SAD), also known as winter depression. This discovery was published in 1984 by N.E. Rosenthal, the United States. The idea that low serum levels of vitamin D may be related to depression stemmed initially from the observation of a high incidence of SAD when sun exposure was reduced, in autumn and winter. An experiment conducted in 1999 by F.M. Gloth, the United States, led to believe that the administration of vitamin D (100,000 IU) is more effective than light therapy to treat SAD. The relationship between vitamin D and melatonin are nevertheless quite obscure, the latter being the hormone produced by the pineal gland primarily in response to the absence of light. Many more investigations are needed to explore this complex area of the relationship between light, neurohormones, and vitamin D.

Many clinicians are invested in the relationships between vitamin D and depression, pushed toward that direction by the high incidence (up to 45%) of this disease among people living in nursing homes, and so with little sun exposure. To get an idea of the magnitude of this disease, one only has to look at the results of the French study PAQUID conducted since 1989 on a cohort of approximately 2800 people over 65 years living in the Aquitaine region. The study has shown that the prevalence of depressive symptoms identified from responses to an adapted questionnaire (CES-D) was about 16%, similar results being reported in several other countries (United States, United Kingdom, north Europe, and South Africa). Moreover, experts agreed that screening and treating depression is the best prevention of suicide risk in the elderly. Despite this, some works remain unconvincing because they rely on a limited number of subjects who are uncontrolled in their lifestyle or diet, unlike other studies conducted in nursing institutions with better control. The difficulty of comparing the results to assess the intensity of the involved depressions often arises from the type of population and the diversity of tests used in different countries. Thus, some epidemiological studies have shown no association between plasma vitamin D and depressive symptoms. There is no data to clarify the reason for these negative results. The simple analysis of food sources of vitamin D compared to the presence of depressive symptoms in large populations of elderly has generally shown a close relationship between these two parameters. This was the case for the study published in 2011 by E.R. Bertone-Johnson, the United States, who showed that subjects ingesting more than 800 IU/day (20 µg/day) of vitamin D had a risk of depression 21% lower than the subjects ingesting only 100 IU/day (2.5 µg/day).

If one considers vitamin D blood levels instead of vitamin food intake, many well-conducted studies have still shown a close relationship between depression and vitamin D in the young as in the elderly. This improvement in the significance of these relationships emphasized the importance of collecting the blood vitamin D levels to establish a reliable vitamin status. An example of this demonstration was given in the study published in 2010 by V. Ganji, the United States, under the Third

National Health and Nutrition Examination Survey. The survey, conducted on 7970 people aged 15–39 years, found that subjects with serum vitamin D levels greater than 30 µg/L (75 nmol/L) had significantly fewer depressive episodes than those with a level lower than 20 µg/L (50 nmol/L). A cross-sectional and prospective study of a cohort of about 6000 subjects followed from age 45 to 50 years provided a precise support for a close association of low vitamin D plasma levels with current and subsequent risk of depression in midadulthood (Maddock et al. 2013). Many other epidemiological surveys carried out in different countries with older people led to the same results; the association between depression and inadequate serum vitamin D was always observed.

These descriptive epidemiological studies are informative; but they cannot help determine whether depression is a direct consequence of vitamin D deficiency or if the deficiency is secondary to depression, resulting, for example, from an unbalanced diet, a slowing of vitamin D metabolism, or limited sun exposure. Obviously, only intervention studies can demonstrate whether high levels of vitamin D have antidepressant effects and may lead to a vitamin treatment. One such study, published in 2004 by R. Vieth, Canada, has investigated the effect of administration of vitamin D for 6 months on serum vitamin D levels, and also on mood and well-being. The latter has been estimated by assessing energy level, sleep, interest, pleasure, attention, and weight loss. The author showed that significant and beneficial results were already obtained with a dose of 600 IU/day (15 µg/day), but they were even clearer with a dose of 4000 IU/day (100 µg/day). It seemed that a regular intake of vitamin D was much more effective than the intake of occasional high doses.

It must be recognized that all of the most recent works on this subject allowed the conclusion of a correlation between vitamin D and depression, but without a possibility to define the mechanisms. To support this demonstration, some studies using a supplementation seem to prove that the symptoms of depression or mood disorders are influenced by vitamin D deficiency, the latter resulting from a too low dietary intake and/or insufficient sun exposure. In the current state of knowledge, experts suggested that vitamin D supplementation may be beneficial in depressed subjects, especially if they are in a state of true deficiency or if their lifestyle and their location suggest a potential risk of permanent deficiency. In the absence of further studies, no formal medical recommendation has been proposed.

4.4.2.6 Immunity

4.4.2.6.1 Bacterial and Viral Infections

Since 1840, sunlight and cod-liver oil were known to prevent rickets and tuberculosis, but it took until the 1980s to understand the nature of the possible links between this disease and vitamin D, found in large amounts in fish oil. In 1985, the relationship between the development of tuberculosis and vitamin D deficiency was finally raised with a serious argument by the English specialist P.D. Davies. A great development of "solarium" (or sanatorium) was initiated in France for the treatment of rickets and tuberculosis in 1920 by J. Saidman following the discovery by the Danish physician Niels Ryberg Finsen (1860–1904). The latter was awarded the Nobel Prize in 1903 "in recognition of his contribution in the treatment of infectious diseases with concentrated light radiation, thus opening a new avenue for medical science." It would be interesting to revisit these initiator works in the light of current knowledge. More than a century later, this vision seems to be definitively established considering the latest research published in 2011 by A. Martineau, England. This author has shown that throughout the year the incidence of tuberculosis in South Africa varied inversely with serum vitamin D. On a purely experimental design, it is important to know that vitamin D is able to enhance the *in vitro* inhibition of mycobacteria growth by leukocytes in association with the production of cathelicidin, a powerful antimicrobial peptide. In 2012, it was demonstrated by A.K. Coussens, England, that supplementation with vitamin D potently inhibited inflammatory responses during the traditional treatment of tuberculosis, responses related to a significant risk of mortality. A causal relationship between sunlight, vitamin D, and the immune system thus seems to be finally established. These results even pushed W.B. Grant, recognized American expert on vitamin D, to formulate in 2011 the hypothesis of a significant contribution of vitamin D deficiency in the premature death of Mozart, a famous musician who lived and worked mainly at night.

Besides this battle against tuberculosis, other research routes were opened in areas as diverse as influenza and AIDS. The increase in mortality due to influenza, as well as other respiratory infections, also seems directly related to a deficiency of vitamin D. Numerous epidemiological studies have supported this hypothesis, including one of the most recent historical studies published by W.B. Grant in 2009 showing a correlation between sun exposure and reduction of mortality during the great flu epidemic of 1918 in the United States. The likely involvement of vitamin D in the etiology of this disease has been widely analyzed, and a new clinical study has reinforced the initial arguments. One Japanese experiment, published in 2010 by M. Urashima, was performed on 430 children aged 6–15 years followed for 4 months. The results have shown that vitamin D supplementation (30 µg/day or 1200 IU/day) was associated with a significant decrease (–42%) of influenza A cases.

Apart from a few rigorous studies, it is currently difficult to assess whether vitamin D supplementation is of public utility to mitigate the impact of influenza.

Again, the variety of treatments and the almost general absence of measurement of blood vitamin D levels have prevented an accurate and indisputable statistical analysis. More rigorous research with larger populations and biochemical analyses are required before possibly recommending a preventive therapy of influenza using vitamin D. The epidemiological study conducted in 2010 by J.R. Sabetta, the United States, has shown that a serum vitamin D concentration equal to or greater than 38 µg/L reduced by 50% the risk of developing an acute viral respiratory tract infection. In 2012, D.R. Murdoch, New Zealand, determined that above 30 µg/L a supplementation is ineffective in modifying the risk of infection of the upper respiratory tract.

Apart from flu, other viral respiratory infections could be liable to a rigorous determination of vitamin D status in patients. Encouraging results obtained in 2010 by J.R. Sabetta seemed to confirm this. They may only encourage the medical profession to intervene with the general population and especially those at risk (pregnant women, people with pigmented skin, and obese) in proposing if necessary an appropriate vitamin D supplementation during periods of low sunshine (autumn and winter) or when professional activity is out of direct sunlight. Ultimately, this approach could improve during winter the population's health by reducing the incidence of related diseases, thus reducing the high cost of care at the expense of individuals and society.

This protection against respiratory infections was confirmed in 2011 by M.E. Belderbos, Netherlands, by highlighting a prevention of bronchiolitis caused by respiratory syncytial virus in infants by supplementing vitamin D in pregnant women. Thus, infants who had the lowest vitamin D levels at birth had six times more risk of bronchiolitis in their first year than those who had the highest levels. Several similar works have clearly shown the importance of a preventive action in the mother in restoring a normal vitamin D status. The treatment was thus able to develop in the child the struggle against lung infection that affected more than 460,000 children (30% of infants) in France each winter and whose number is steadily increasing each year. It is hoped that larger clinical studies will soon be undertaken to consider preventive measures that are simple and nationally inexpensive.

The worsening of disease symptoms due to AIDS, characterized primarily by infections, anemia, and death, also seemed to be associated with the vitamin D status. The results from the large EuroSIDA study in 20 European countries (http://ec.europa.eu/research/success/fr/med/0349f.html) confirmed the diagnosis as 83% of infected people were vitamin D deficient and had higher levels of complications and mortality than others. An emerging information is a probable link between the mother-to-child transmission of HIV and child mortality to vitamin D deficiency (Fitchett 2013).

If these epidemiological studies are confirmed by controlled clinical trials, intake of vitamin D could represent a simple and inexpensive way to postpone or alleviate antiretroviral therapy and may thereby reduce the severity of bacterial complications induced by the treatment of AIDS. It is possible that vitamin D deficiency in people is due to a direct effect of antiretrovirals on their metabolism. EuroSIDA investigators are currently pursuing a study of observing the effects of vitamin D supplementation in HIV-positive people.

By 1939, a correlation between the incidence of dental caries in young children (94,000 children aged 12–14 years) and frequency of sunshine was established in 24 U.S. states. This relationship was verified shortly later by H. Kaiser in 500,000 young American children. These facts were naturally forgotten after the generalization in the 1950s of fluoride treatments. The debate was revived when the development of caries was limited in children by exposure to light containing UV-B. Similarly, in the elderly E.A. Krall, the United States, showed in 2001 that the loss of teeth can be effectively combated by the administration of vitamin D and calcium. The explanation of this effect certainly lies in the direct relationship between periodontal inflammation, a risk factor for tooth loss, and hypovitaminosis D. This work, which did not have too many echoes in the world of dental health, nevertheless has important implications in the prevention of a costly and widespread disease in Western countries.

4.4.2.6.2 Autoimmune Diseases

The immunomodulatory role of vitamin D has been discussed for over 25 years, but it was not until 2012 that the physiological aspect was strongly emphasized by Y. Schoindre. This evolution was due to our knowledge of many epidemiological investigations establishing correlations between vitamin D and various autoimmune diseases and especially to the identification of vitamin D synthesis in immune cells. Thus, epidemiological studies have shown that vitamin D deficiencies were often associated with autoimmune diseases such as type 1 diabetes and multiple sclerosis and inflammatory diseases such as rheumatoid arthritis, Crohn's disease, or systemic lupus erythematosus.

As shown by C. Pierrot-Deseilligny, France, in 2010, it seemed that hypovitaminosis D was a risk factor for the development of multiple sclerosis. As with other diseases, one of its features is its geographical distribution with an increase in frequency in areas with low sunlight, thus with a reduced cutaneous synthesis of vitamin D. The relationship between prevalence and sunshine was recorded in various countries, including North America and even France. In that country, S. Vukusic observed in 2007 that multiple sclerosis was twice as common among farmers living in the north compared to those living in the south. A similar relationship with dietary vitamin D has also been observed in Nordic countries with little sunshine. Even more surprising was the relationship highlighted by R. Robson, United Kingdom, in 2012, who observed a greater risk of developing multiple sclerosis in adulthood when pregnancy elapsed during the fall and winter, periods of low sun exposure. Whereas the connection between vitamin D and multiple sclerosis remains poorly elucidated, it is noteworthy that vitamin D insufficiency is a strong risk factor for long-term multiple sclerosis activity and progression (Ascherio et al. 2014).

Despite these evidences, the studies undertaken to demonstrate a therapeutic effect of vitamin D on multiple sclerosis are still unconvincing. Further studies should establish the precise roles of vitamin D and the doses required to slow down the inflammation and demyelination observed during that severe nerve pathology. As pointed out in 2011 by C. Pierrot-Deseilligny, "rather than waiting for the results of phase III trials that still take several years, it is wise in a preventive perspective to supplement patients with vitamin D, especially those who are deficient and if they have factors that tend to aggravate that condition."

The identification of vitamin D receptors, in intrinsic form in macrophages and dendritic cells and under inducible form in the cells, highlighted the important role of this vitamin in the regulation of immune and inflammatory responses (Borges et al. 2011). These findings open the way to new treatments for autoimmune diseases and original struggle strategies against allograft rejection. The mechanisms involved in the development of the immune system have been partially elucidated by the discovery of vitamin D receptors in circulating mononuclear leukocytes and the expression of specific peptides (cathelicidin and β-defensin) controlled by vitamin D. These peptides may also be directly involved in the inactivation of viruses. As yet, the precise relationship between vitamin D and native immunity has been explored only in animals, but the very encouraging results and the diversity of vitamin D analogues suggest an interest on the development of prevention and new treatments in clinical practice in the near future.

4.4.2.6.3 Psoriasis

Psoriasis, a disease with a high autoimmune component, is a chronic inflammatory disease of the skin and joints that could also be related to the vitamin D status. In fact, dermatologists treat this condition more and more often with topical applications of vitamin D (calcitriol) or with an analogue (calcipotriol) and sometimes with phototherapy (UV radiation or light therapy). Although these treatments induce an increase in the circulating levels of vitamin D, the causal relationships between this vitamin and its effects on psoriasis have not yet been identified. Despite this ignorance, the treatment of psoriasis with vitamin D remains an inexpensive and efficient option that is the subject of ongoing research and a significant clinical development that may rapidly change the approach to this disease and its complications.

4.4.2.6.4 Asthma

The relationships between asthma and vitamin D became the object of much research when immune defense mechanisms involved in the onset of this disease were identified. Several studies, including those by G. Devereux, Great Britain, published in 2007 pointed out clearly the association between vitamin D deficiency and frequency of lung disease in adults and the association between maternal deficiency and asthma frequency in young children. A large survey in the United States and Australia has provided additional arguments for a close link between the prevalence of asthma, as for other allergic diseases, and the vitamin D status in populations variously exposed to solar radiation. Thus, G. Krstic showed in 2011 that a change of $10°$ in latitude northward in both countries led to a significant increase, of 2%, in the prevalence of this disease. Besides other epidemiological studies, this investigation has suggested that vitamin D plays an important role in the development of allergies and asthma, diseases known to be linked to the reduced capacity of the immune system. A supplementation at an early age could be part of the prevention of these respiratory disorders that are increasing in all countries. In France, nearly three million people, one-third aged under 15 years, are affected by this disease and treatment costs amount to nearly €1 billion. Again, a prevention campaign linked to a control of the vitamin D status and a long-term supplementation could contribute to reducing these health expenditures.

The recent increase in the incidence of diseases involving immune responses has caused U. Christen to propose the "hygienist" hypothesis in 2005; but, according to the explanations given in 2008 by M.T. Cantorna, the United States, it seems more logical to relate these diseases to the vitamin D status directly affecting these diseases. Indeed, fluctuations in circulating levels of vitamin D are directly related to outdoor activities, seasons, pollution, and vitamin-depleted diets. These factors could, according to this author, influence particularly the individuals genetically susceptible to the development of autoimmune responses.

This evocation of the wide implications of vitamin D in the immune process shows that there is much to learn in this area. It seems necessary to call for the creation of new research efforts using appropriate doses of vitamin D in the struggle against bacteria, viruses, and autoimmune reactions to effectively reduce the morbidity and mortality associated with them.

4.4.3 Vitamin E

Vitamin E is the most abundant lipophilic antioxidant in our body; this probably explains its protective effects for biological membranes and lipoproteins rich in n-3 and n-6 polyunsaturated fatty acids. These effects are also potentiated in the presence of vitamin C regenerating the α-tocopherol inactivated after the reaction with free radicals. An adapted consumption of vitamin E would increase immunostimulation and reduce the incidence of chronic degenerative diseases (atherosclerosis, myocardial infarction, cancer, neurodegenerative diseases, etc.), where oxidative stress plays a major role. Only the best documented effects of the components of the vitamin E complex on some important physiological systems or pathological situations are reported here.

First, it should be noted that for many scientists the term "vitamin E" is still equivalent to α-tocopherol, ignoring other forms whose various properties are gradually being discovered (Section 2.7.3). It is significant, for example, that less than 1% of the literature on vitamin E is concerned with tocotrienols.

4.4.3.1 Cardiovascular Diseases

A therapy by a vitamin E supplementation in the context of prevention of cardiovascular diseases is feasible, consistent with the importance of protection of the coronary vasomotor function in patients with atherosclerosis, as shown in 1999 by S. Kinlay, the United States. Similar effects were observed in retinal vessels in patients with type 1 diabetes, which may require a vitamin E supplementation in the context of diabetic retinopathy prevention. An additional advantage of this vitamin treatment stems from the observation of an inhibitory effect of vitamin E on platelet adhesion, the first step in the development of arterial thrombosis. The largest and best known intervention study was the Cambridge Heart Antioxidant Study conducted in 1996 among nearly 2000 coronary patients who received up to 800 mg of α-tocopherol per day. The most significant result of this test was the dramatic 47% decrease in the risk of nonfatal myocardial infarction in the group supplemented with vitamin E. Although these studies have suggested treatment in a secondary prevention coronary disease by high doses of vitamin E, the supplementation in a population with no a priori atherosclerotic risk currently remains irrelevant.

Numerous other studies have also been carried out in the context of primary prevention (studies in healthy subjects) and secondary prevention (study of patients with various cardiovascular diseases). The complexity of the problem is such that a recent meta-analysis, done in 2010 by Z. Cordero, Germany, could not reach any conclusion on the undeniable relationship between vitamin E supplementation and cardiovascular disease. This lack of conclusive evidence is probably based on the diversity of the ingested tocopherols (in the diet or as capsules), different biological markers used, and clinical investigations. Future research should take into account the discovery made in 2008 by S. Das, the United States, on the cardioprotective properties of tocotrienols in animals. Thus, it seems actually imperative to consider the impact of each component of the vitamin E complex on one of the most important causes of mortality in Western countries. Investigations of the same author published in 2012 supported this hypothesis by showing that some tocotrienols may give the heart a new resistance to ischemia.

4.4.3.2   Biosynthesis of Cholesterol

In 1986, research in animals showed that tocotrienols have cholesterol-lowering properties that they do not share with α-tocopherol. Soon, the same effects were verified in humans. Later, it was shown that δ-tocotrienol and γ-tocotrienol displayed the strongest effects. Their metabolic target is HMG-CoA reductase, the key enzyme in the biosynthesis of cholesterol, known to be inhibited by statins, the drugs widely used in hypercholesterolemia. A synergy between tocotrienols and a statin (lovastatin) was even observed in 2001 by A.A. Qureshi, the United States. In the near future, great expectations can be formulated for the use of tocotrienols, allowing the prescription of lower doses of statins or even their removal. An experiment by K.H. Yuen, Malaysia, in 2011 in hypercholesterolemic subjects has shown that, compared with placebo, a treatment with a mixture of tocotrienols lowered blood cholesterol by 9% and LDL-cholesterol by 13% after 4 months. It should be noted that this effect was accompanied by a 20 times increase in the plasma concentration of tocotrienols. The long-term safety of such a supplementation is also not yet proved.

4.4.3.3   Cancers

From an epidemiological point of view, results regarding the effects of vitamin E on cancer were not always consistent and seemed to vary depending especially on the studied organs but remained under the influence of other nutritional factors. Nevertheless, it should be noted that an inverse relationship was often observed between blood vitamin E levels and the risk of cancer in various organs (lung, colon, stomach, breast, and cervix). These results argued for a larger dietary vitamin E intake than that recommended by official agencies (Section 3.2.4.3).

From a clinical point of view, recent studies have emphasized the beneficial effects of a combination of vitamin E and vitamin C in the case of various tumors (esophagus, intestine, and lung), but the interpretation of widely dispersed results remained difficult. In no less than three intervention studies (the Linxian Nutrition Intervention Trial, Polyp Prevention study, and ATBC cancer prevention study) involving up to nearly 30,000 subjects, it was observed that a vitamin E supplementation could significantly reduce the risk of certain cancers, the effects being a function of the prescribed doses, affected organs, and habits of the populations studied.

A large Finnish study, involving more than 29,000 smokers followed for 19 years, has shown that subjects with very low serum tocopherol levels (<9.3 mg/L) had a risk of pancreatic cancer twice as large as those with the highest levels (>14.2 mg/L) (Stolzenberg-Solomon et al. 2009). However, the findings of this probative study could not be recommended to other populations such as women or nonsmoking men. These results help to appreciate the difficulty in analyzing and especially comparing studies involving subjects with various behaviors and sometimes in using different biological criteria.

The results of another study in smokers, published in 2007 by E. Wright, the United States, showed that vitamin E supplementation had no positive effect on the risk of cancer of oral cavity, larynx, and esophagus.

In 2009, a prevention study (Selenium and Vitamin E Cancer Prevention Trial [SELECT]), including 35,500 men followed for more than 5 years, showed that neither selenium (200 μg/day) nor vitamin E (dose equivalent to 270 mg α-tocopherol per day), singly or together, affected the risk of prostate cancer among healthy individuals. However, a significant beneficial effect was highlighted, but again this was only in smokers. After an extended follow-up for 7 years with the same participants of the SELECT study, it was able to detect a significant increase (+17%) in cases of prostate cancer, with only 270 mg/day of α-tocopherol. It is therefore likely that the high dose of α-tocopherol used was the cause of this increased incidence. In fact, this amount was much higher than the recommended dietary allowance for vitamin E, which is only up to 12 mg/day in adults (Section 3.2.4.3). Further clinical investigations, including a monitoring of blood vitamin E levels (and its various vitamers), must be undertaken over long periods and with multiple subjects before recommending the use of this vitamin for the prevention or control of prostate cancer.

Since 1989, an inhibitory effect of tocotrienols on the formation of mammary tumors was described by K. Sundram in animals. This effect was found again with human mammary cells and explained by the inhibition of estrogen-dependent mechanisms or by phenomena of apoptosis. Many other *in vitro* studies have suggested that components of the vitamin E complex could effectively control the development of many cancers by their antiproliferative or anti-angiogenic action. Several studies analyzed by T. Miyazawa, Japan, in 2009 were carried out on cell cultures or in animals and have shown that this anti-angiogenic effect was mainly based on the control of endothelial cells by growth factor signaling. The demonstration in 2011 by Y. Li, China, of a strong angiogenesis inhibition by γ-tocotrienol in a human cell culture model has confirmed the anticancer therapeutic properties of these compounds.

All existing research suggested that the tocotrienol fraction of vitamin E contributed to the implementation of new therapies combining this vitamin to the conventional treatment of some cancers.

4.4.3.4 Nervous Diseases

The frequent demonstration of a close link between oxidative stress; aging; and some nerve disorders, such as Alzheimer's disease or depression, has focused on the possible use of antioxidants in the prevention and the treatment of these pathological conditions. Since antioxidants have proved beneficial for the cardiovascular system, itself involved in cognitive disorders, they could also be useful to fight against memory loss.

4.4.3.4.1 Cognitive Decline

Numerous studies based on nutritional surveys have attempted to reveal a possible association between antioxidants and memory. The responsibility of vitamin E in the diet is difficult to estimate as it is ingested with many other natural antioxidants such as vitamin A, vitamin C, carotenoids, and many polyphenols.

Among the epidemiological studies open to debate, there was the U.S. survey conducted in 2002 by M.C. Morris, who followed nearly 2900 people aged 65–102 years for over 3 years. A statement of the amounts of vitamin E intake (in all its forms) and the use of four different psychological tests have correlated high doses of ingested foodstuff vitamin with a reduced cognitive decline.

In contrast, the use of the same methods in a large Dutch study in 1996 (the Rotterdam Study) did not reveal any correlation between vitamin E intake and cognitive function; only low levels of carotene were apparently associated with a worsening of this function. The extension of this study for 6 years has revealed a modest but significant beneficial effect of vitamin E on the risk of dementia and Alzheimer's disease, the effect being more pronounced in smokers. Monitoring the same study for another 10 years confirmed these results, eliminating the combined effects of β-carotene and vitamin C (Devore et al. 2010).

In addition to these studies based on dietary surveys, it is interesting to note that measurements of serum vitamin E made in 1999 by A.J. Perkins, the United States, have confirmed the association between poor memory and low vitamin E levels, excluding other natural antioxidants. The study was conducted for 6 years in more than 4800 multiethnic subjects 60 years and over in the southern United States (Indianapolis).

Conversely, several studies based on the administration of vitamin E have failed to demonstrate any effect on cognitive performance. Intervention studies could not detect any clear relationship between vitamin E consumption and the prevalence or progression of Parkinson's disease or Alzheimer's. Yet, the study by M. Sano, Japan, published in 1997 and carried out by treating patients with high doses of vitamin E, showed a slowing of the progression of Alzheimer's disease at an early stage. This has given rise to many other research efforts, opening prospects for new perspectives in the treatment of this disease by establishing a direct link between vitamin E and protection against the oxidation of brain lipids and cerebrospinal fluid lipoproteins.

The vast French SUVIMAX study consisted of a daily supplementation for eight years of vitamin E (30 mg), β-carotene (6 mg), vitamin C (120 mg), and trace elements (Zn and Se). The report published in 2011 by E. Kesse-Guyot has shown a beneficial effect of these vitamin intakes on the cognitive performance of subjects aged 45–60 years. What part is due to vitamin E is yet difficult to establish, but the role of vitamins and natural antioxidant substances in the prevention of cognitive decline now seems indisputable. A recent intervention study with 140 patients with mild to moderate Alzheimer's disease supplemented daily with 2000 IU of α-tocopherol for 5 years allowed significant slowing of the functional decline (Dysken et al. 2014).

As with other physiological effects, it should be noted that evaluation of the vitamin E intake should also take into account the different tocopherols and tocotrienols, components of this vitamin complex. Thus, the comparison between studies performed in different countries is made difficult by the specificity of the ingested forms: the Europeans absorb mainly α-tocopherol from sunflower and olive oils, whereas the Americans consume mostly γ-tocopherol from soybean and corn oils. These experimental precautions should also apply to the products administered in intervention studies. Given the biochemical activity of γ-tocopherol, it now appears necessary to distinguish the various molecular forms of vitamin E in clinical trials before considering the merits of chronic vitamin supplementation. The situation is the same with tocotrienols. A study considering these parameters was conducted in Sweden and published in 2010 by F. Mangialasche, Stockholm. This author collected 232 subjects aged 80 years and followed them for 6 years. Analysis of the different vitamin E vitamers present in plasma revealed that the risk of developing Alzheimer's disease was reduced only in subjects with elevated plasma levels of β-tocopherol. This investigation emphasized that the neuroprotective effect of vitamin E is based more on the combination of various molecular forms than on a particular form as α-tocopherol. Another study by the same group has established that in using brain MRI and measuring the plasma vitamin E forms it is possible to differentiate Alzheimer's disease and MCI from control healthy subjects (Mangialasche et al. 2013). Thus, it now seems possible to use vitamin E forms as nutritional markers of specific brain diseases.

The discovery in animals in 2011 by N. Crouzin, France, of an interaction between α-tocopherol and the brain cannabinoid system opened up new perspectives on a central action of that vitamin, which clearly has a neuromodulatory action, besides any antioxidant effect.

α-Tocotrienol has also shown specific neuroprotective properties at nanomolar concentrations, where its antioxidant activities are not commonly expressed. Although they have been the subject of only fundamental research on the control of glutamate neurotoxicity in cultured nerve cells, these properties are remarkable and are already the subject of much research. Examples are the studies of F. Hosakada, Japan, in 2004, showing that rat striatum neurons are protected against apoptosis by very low concentrations of α-tocotrienol in the absence of any oxidative stress. Many molecular mechanisms have been invoked to explain these spectacular specific effects, but it seems likely that a major mechanism, highlighted by S. Khanna, the United States, in 2010, could be the inhibition by α-tocotrienol of the activation of phospholipase A_2 by glutamate.

4.4.3.4.2 Depression

Many studies have suggested that the generation of free radicals, the main actors of membrane lipid oxidation, may play a role in neuropsychiatric disorders such as depression. These free radicals may in fact be produced during inflammatory episodes, immune responses, and monoamine catabolism. Some research seemed to confirm this hypothesis. Thus, an increase in lipid oxidation accompanied by intense antioxidant enzyme activities has been observed in patients with major depression. Conversely and surprisingly, M. Bilici, Turkey, noted in 2001 that a treatment with an antidepressant (inhibitor of the serotonin reuptake) may reduce lipid oxidation. This close relationship between depression and lipid oxidation appeared to be confirmed by the presence of lower serum vitamin E levels in patients with major depression compared to healthy subjects.

The vitamin E dietary intake did not appear to be directly involved as A.J. Owen, Australia, observed in 2005 that the relationship between depression and low serum α-tocopherol levels is present even with an adequate vitamin intake. This condition would result in an increased consumption of vitamin E along with an increased oxidative stress probably due to depression. The aforementioned results were still in disagreement with a Japanese study, but differences in age and sex of the persons studied were probably at the origin of this apparent discrepancy.

From all these works, one may conclude that, in the absence of controlled clinical trials with a large number of subjects, any significant supplementation with vitamin E or other antioxidants cannot be recommended, either preventive or curative within nerve pathologies. Future trials should take into account the equipment of individual antioxidants, plus a possible toxicity of the vitamin E doses required to achieve a satisfactory effect. Maintaining a diversified food intake in accordance with official recommendations (intake of about 15 mg of vitamin E per day) even in the elderly remains the only advice that can yet be delivered to the population.

4.4.3.5 Immune System Disorders

The relationships between vitamin E and immune responses were particularly studied in aged animals, as in the elderly, representing a population frequently suffering from a progressive weakening of these responses. It has been well demonstrated in 1997 by S.N. Meydani, the United States, that a high vitamin E supplementation (200 mg daily for 4 months) reinforced effectively the functional clinical indices of T lymphocytes. In 2011, a rigorous study by D. Mahalingam, Malaysia, examined the effect of a daily supplementation with 400 mg of tocotrienols on the immune response induced by vaccination against tetanus. The conclusions were clear: tocotrienols enhanced the production of interferon-γ and interleukin (IL-4) by leukocytes and especially the production of the tetanus immunoglobulin. Therefore, these vitamins have an obvious clinical interest in the strengthening of the immune response following administration of vaccines.

4.4.3.6 Reproduction

Infertility affects 15% of couples and in 30% of these couples, the cause of infertility is associated with aberrations found in the male partner, termed male infertility. Recent research has suggested that male infertility found its source in the oxidative attack of lipids from spermatozoa, a cell poorly equipped in antioxidants and therefore very sensitive to free radicals. Therefore, it seemed natural to fight against this infertility by a dietary supplementation with vitamin E.

These mechanisms could be partially verified *in vitro* when it was shown clearly that an excess of free radicals peroxidized spermatozoa membrane lipids, fragmented DNA, and weakened their mobility and ability to fusion with oocytes. Meanwhile, it has been determined in 1996 by P. Therond, France, that concentration in normal spermatozoa was closely correlated with the α-tocopherol concentration in the spermatozoa membranes. Following the work by F. Diafouka, Ivory Coast, in 2009, it seemed significant that impaired fertility (asthenospermia and oligoasthenospermia) was related to a very low concentration of α-tocopherol in the seminal fluid.

If one considers that almost half of the French population has an inadequate vitamin E intake (Section 3.2.4.3), it is surprising that medical authorities do not focus more on this disturbing nutritional status, knowing the steady increase in male infertility in France as in many other countries.

Some clinical trials have been conducted to estimate the potential benefits of a vitamin E supplementation on fertility. Despite the paucity of studies and the small number of subjects per study, the investigation of E. Kessopoulou, England, reported in 1995, allowed the conclusion that a vitamin E supplementation (twice 300 mg of vitamin E per day for 3 months) had a significant effect in humans. A practical confirmation of these results may be found in the publication of E. Geva, Israel, reporting in 1996 an increase in fertilization rate when attempting to medically assist procreation after supplementation of men for 3 months with 200 mg of vitamin E per day. The publication of some negative results in this area showed that a long research effort will still be needed to unravel the mechanisms involved and lead to reliable results that will make recommendations in cases of male infertility. Pending these findings, it seems not unreasonable to advise infertile men to increase their dietary intake of vitamin E to prevent a deficiency status. It is therefore recommended to maintain a daily intake of at least 12 mg of α-tocopherol equivalents in adults (Section 3.2.4.3) or increase it even up to 50 mg, the recommended dose in elderly people using vitamin supplements. Only one clinical research program on this subject is recorded on the official website http://www.clinicaltrials.gov.

4.4.4 Vitamin K

Since its discovery in 1935 and for nearly 40 years, the functions of vitamin K were limited to the control of blood coagulation via the hepatic synthesis of several coagulation factors (prothrombin and factors VII, IX, and X). The discovery in 1974 of the role of vitamin K in the synthesis of coagulation factors by carboxylation of glutamic acid residues has broadened the scope of extrahepatic proteins also dependent on vitamin K (Gla-protein) but without any coagulation activity. The discovery of these proteins, primarily in the bone matrix (osteocalcin) and later in the wall of arteries (matrix Gla-protein) and in the nervous system, greatly expanded the physiological importance of vitamin K.

4.4.4.1 Blood Coagulation

A vitamin K deficiency is extremely rare in humans, a consequence of the high concentration of vitamin K_1 in dietary green vegetables (Section 3.2.4.4). It is also possible to add the production of vitamin K_2 (menaquinones) by intestinal flora. A deficiency status may still be found in conditions of intestinal malabsorption or in poor nutritional conditions combined with antibiotherapy.

Unlike adults, the risk of vitamin K deficiency is greater in newborns, due to their low hepatic stocks and the low vitamin K concentration in milk. Again, this condition is worsened by a lack of intestinal flora at birth. This deficiency status may lead to hemorrhagic disease of the newborn, a disease that is often avoided by the systematic administration of vitamin K at birth. Although the WHO did not recommend the routine administration to all infants in good health, the practice of an injection of 1 mg (or 2 mg *per os*) of vitamin K_1 is common in many countries (the United States, Canada, Great Britain, Belgium, Switzerland, and Germany) for the prevention of hemorrhagic disease. In France, this practice also seems very well observed.

The observation of a hemorrhagic syndrome in cattle fed with poorly preserved and fermented hay led to the discovery in 1941 by H.A. Campbell, the United States, of an anticoagulant, dicumarol, produced by fungi. Used first as rat poison, dicumarol was rapidly used in the treatment of myocardial infarction. Since 1953, several derivatives, including warfarin, have been used as anticoagulants along with heparin in the treatment of deep vein thrombosis. During these treatments with anti-vitamin K, it is prudent to avoid eating vegetables containing the highest vitamin K concentrations, such as kale, broccoli, spinach, and salad (Section 3.2.4.4). Other vegetables containing less vitamin K should be eaten in moderation, and the situation is the same for rapeseed and soybean oils. As even a moderate intake of vitamin K_2 may interfere with oral anticoagulant treatment, it is best to also limit the consumption of fermented milks and cheeses.

It has been experimentally shown that administration of high doses of vitamin E lengthened the clotting time (prothrombin time), a disorder corrected by the administration of vitamin K.

4.4.4.2 Calcification (Bones and Arteries)

The demonstration of a vitamin K–dependent protein (osteocalcin) in chicken bones was done at the same time when P.V. Hauschka discovered in 1978 the glutamic acid residues in that protein. Since then, considerable research efforts have been devoted to the role of vitamin K in calcium homeostasis.

Two major areas have been the targets of fundamental and clinical research: bone metabolism and arterial calcification. These works will certainly give rise to nutritional recommendations and to therapeutic developments related to those observed for vitamin D in the near future.

4.4.4.2.1 Bone Calcification

Among the many proteins involved in bone formation, several groups have glutamic acid whose carboxylation is vitamin K dependent. Among them, osteocalcin is synthesized in bone and directly involved in the calcification of bone tissue by binding to hydroxyapatite, this binding to the mineral being very strong and dependent on the protein carboxylation level. The exact role of osteocalcin in bone metabolic balance is not yet known exactly; but its activation by vitamin K, especially vitamin K_2, is becoming increasingly involved besides vitamin D in maintaining bone health. Thus, although the bone density appeared not to be directly connected to vitamin K absorption, the rate of hip fracture has been described as being associated with reduced circulating levels of vitamins K_1 and K_2, or with an insufficient vitamin K_1 dietary intake. Thus, the studies in 1997 by L.J. Sokoll, the United States, led to providing an estimate of the risk of hip fracture by measuring circulating levels of undercarboxylated osteocalcin. An investigation reported in 2007 by M.H. Knapen, the Netherlands, and performed in postmenopausal women followed for 3 years showed that a daily administration of vitamin K_2 (menaquinone-4) enabled stable bone strength at the neck of femur by increasing its calcium content. These results confirmed those obtained in 2006 by J. Iwamoto, Japan, showing the interest of an osteoporosis treatment with a combination of bisphosphonates or raloxifene and menaquinone-4. Also in Japan, Y. Ikeda established in 2006 a close relationship between fracture of the neck of femur and intake of menaquinone-7 by comparing two cohorts of women, one in the east of the country consuming traditionally natto, a preparation rich in menaquinone-7, and the other in the west only rarely consuming natto. Subsequently, this author has verified that bone mineral density was higher in women consuming more natto. Ongoing experiments have suggested that similar effects may be obtained after the administration of menaquinone-7 over the long term (2 to 3 years). It seemed that menaquinone-7 was more bioavailable than menaquinone-4 and also had a half-life longer than vitamin K_1 in the circulation.

Several other epidemiological studies have pointed in the same direction, but it remains difficult to eliminate the influence of general nutritional conditions. This problem should be solved by performing perfectly controlled intervention studies. Although still rare, clinical and experimental observations in animals tended to direct research toward the positive effects of vitamin K_2 on bone quality and a possible application in patients with type 2 diabetes, who are subjects naturally exposed to bone fractures. It was observed in 2008 by M. Ferroon, the United States, that osteocalcin could affect peripheral insulin sensitivity and also its production by pancreatic cells. Possible relationships between vitamin K, osteocalcin, and insulin sensitivity have found an important clinical development when N. Ibarrola-Jurado, Spain, highlighted in 2012 the strong decrease in the risk of developing type 2 diabetes by maintaining a high dietary vitamin K intake.

Since a beneficial effect of vitamin D on bone mineralization induced by menaquinones was discovered in 1996 by Y. Koshihara, Japan, it became essential to thoroughly explore the joint action of these two vitamins in chronic diseases such as osteoporosis and in the treatment with an antivitamin K. The potential superiority of these vitamins compared to conventional treatments (alendronate, risedronate, or strontium ranelate) should also be defined.

Given all these convincing research on the relationship between vitamin K, bone density, and fracture risk, clinicians should take into account this knowledge in chronic anticoagulant therapy capable of generating osteoporosis based on coumarin and coagulation inhibitors. Further research is needed to establish recommendations for a specific vitamin K intake for the treatment of osteoporosis and perhaps soon to improve the prevention of type 2 diabetes.

4.4.4.2.2 Vascular Calcification

Based on extensive research in circulation pathology, it is now accepted that the presence of arterial calcification, mainly in coronary arteries, is a marker of a well-established atherosclerotic disease. Many studies have shown a relationship between the calcification of vessel walls and that of bone. A protein closely related to osteocalcin, the matrix Gla-protein, is indeed present in the vessel wall and also in the cartilage. This protein plays a role in regulating the vitamin K–dependent calcification.

Early in 1995, K. Jie showed that a high dietary intake of vitamin K_1 was accompanied by a reduction of aortic calcification, an inevitable aging phenomenon. The opposite behavior of the two sites of calcification, bone and arterial walls, was highlighted by the analysis of the "calcification paradox," which was often observed in postmenopausal women. In this condition, osteoporosis is frequently observed with a concurrent arterial calcification, but it is also accompanied by very low vitamin K levels. The possible consequences were clarified by the demonstration of an inverse correlation between vitamin K_1 and the incidence of myocardial infarction and cardiovascular disease mortality. To complete the picture, a large study (the Rotterdam

Study), in about 4800 Dutch followed for 10 years, has related the dietary intake of vitamin K_2 (as several menaquinones), but not that of vitamin K_1, with a small aortic calcification and with a reduction in mortality from cardiovascular disease (Geleijnse et al. 2004). This work was confirmed 5 years later by a larger Dutch study (Prospect-EPIC cohort). The results have indicated that long-chain menaquinones (from MK-7), present in fermented cheeses, were most beneficial in reducing mortality by 9% for each menaquinone intake of 10 µg/day. No optimal intake value of menaquinones is defined to date.

All these results suggested a protective effect of vitamin K_2 against cardiovascular disease, partially induced by an arterial calcification. As the coronary calcification is recognized as an important risk factor for coronary heart disease, it is not surprising that research in this area is increasing despite questionable results. Presently, as emphasized by the large epidemiological study reported in 2009 by J. Beulens, Netherlands, vitamin K_2 could quickly become a valuable therapeutic agent in the prevention of cardiovascular disease.

As requested by C. Vermeer, Netherlands, in 2012, new recommendations for the intake of vitamins K_1 and K_2 should be available after serious experimental studies for the prevention of cardiovascular disease in healthy persons as well as for the treatment of patients under antivitamin K.

Despite all this evidence, the nutrition committee of the EFSA concluded in 2012, at the request of the European Commission, that on the basis of current data no causal link has been established between ingestion of vitamin K_2 and proper cardiovascular system function.

Six studies on the relationships between vitamin K and arterial calcification are declared on the official website http://www.clinicaltrials.gov.

4.4.4.3 Nervous Diseases

The presence of vitamin K (particularly menaquinone-4) in the brain has been demonstrated in rats since 1994 by H. H. Thijssen, Netherlands. A role of vitamin K in the metabolism of sphingolipids was mentioned first in bacteria and in rat brain in 1988 by K.S. Sundaram. Later, many works have clarified the metabolic controls of several types of sphingolipids (gangliosides, sphingomyelin, and sulfatides) by vitamin K. Another aspect of the role of vitamin K in the functioning of the nervous system was highlighted in 1999 when A.L. Prieto discovered in the central nervous system a protein (Gas6) close to the vitamin K–dependent coagulation factors and ligand for tyrosine kinase receptors. Since then, the functions of this protein have been linked to the mechanisms of myelination by oligodendrocytes and of stem cell proliferation in certain brain areas. Meanwhile, some works, such as those by D.M. Cochetto in 1985, have shown in rats that vitamin K deficiency induced hypoactivity and decreased exploratory activity of the animal. These results have not raised awareness among clinicians probably because of the scarcity of vitamin K deficiency in humans, at least in terms of clotting factors. Although a low serum vitamin K characterized patients with Alzheimer's disease, currently no conclusion can be drawn from the small number of epidemiological and clinical studies in this field (Ferland 2012). However, the demonstration for the first time of an association of serum phylloquinone concentration with a better performance in verbal episodic memory in healthy old subjects (70–85 years) adds evidence to a role of vitamin K in cognitive functions (Presse and Belleville 2013). Future studies will be needed to strengthen our understanding of the various roles of vitamin K in cognition without forgetting the possible participation of menaquinones and vitamin K antagonists.

4.5 PHOSPHOLIPIDS

The high concentration of some phospholipids (phosphatidylcholine and phosphatidylserine) in the central nervous system and the discovery of their involvement in the mechanisms of cellular function have suggested to investigators that they could be used as food supplements. The possible intake of some components of these phospholipids (choline and fatty acids) may also be considered. Intestinal absorption of these lipids, even in very little amounts, could then compensate or enhance a previously established lipid deficiency, helping the formation of membranes and new cell mediators. Tests on cell cultures, animals, and sometimes humans have suggested that some advantages could be expected from a diet enriched with phospholipids.

4.5.1 PHOSPHATIDYLCHOLINE

Phosphatidylcholine (Section 2.4) ingested by humans from either meat or vegetables or as commercial lecithin supplement is a good source of choline, an essential compound that helps in the biosynthesis of cell membranes (Section 3.2.5.1). Choline is also required for methylation reactions and the synthesis of acetylcholine, a fundamental neurotransmitter. Besides choline, phosphatidylcholine may be a carrier of specific fatty acids, mostly long-chain n-3 fatty acids of marine origin (krill oil). This source has been used for the treatment of inflammation and cancer. These properties have been described in Section 4.2.4.

4.5.1.1 Cardiovascular Disease

Although several early studies have reported conflicting results on the effects of lecithin on well-known risk factors such as hypertriglyceridemia and hypercholesterolemia, some encouraging results were recently obtained in patients with type 2 diabetes. Thus, D. Ristic Medic, Serbia, reported in 2006 that a daily supplementation with 15 g of soy lecithin induced a lipid-lowering effect accompanied by a decrease in LDL-cholesterol and an increase in HDL-cholesterol. An interesting property of lecithin could be its antioxidant power with respect to serum lipoproteins, primarily LDL, thus helping a reduction of the atherogenesis risk. This mechanism has already been demonstrated in 1999 by K.P. Navder, the United States, in baboon monkeys treated with alcohol.

Despite these positive results, it seems too early to draw any conclusion on the protective effects of phosphatidylcholine on cardiovascular disease. The main source of discrepancies is the variability at the level of participants, the quality and doses of lecithin used, and the duration of treatments. The animal studies still remain a good data source on the mechanisms of action of lecithin and should be used for future clinical research in the field of cardiovascular disease (Chanussot 2008).

4.5.1.2 Nervous Diseases

Thanks to the research by B.H. Peters, phosphatidylcholine was considered in 1979 as a potential donor of choline to improve cholinergic systems weakened in patients with dementia. In this initiatory work, some encouraging results on memory improvement have been described in patients with Alzheimer's disease and treated with phosphatidylcholine. Subsequently, many clinicians have explored the possible effects of the administration of this phospholipid mainly on memory and also on other cognitive functions.

Many of these studies have been performed in rats with some success on brain development as well as on behavior or memory. Unfortunately, few works has been done in humans mainly in the elderly or subjects with memory loss or dementia. Some positive results were reported in 2003 in a review by M.A. McDaniel, the United States, but when screening many other published studies no definitive conclusions may be drawn because of the diversity of the populations studied. A meta-analysis conducted in 2003 by the Cochrane Institute from 12 clinical trials involving patients with Alzheimer's disease, Parkinson's disease, and various memory disorders failed to demonstrate any benefit of the administration of phosphatidylcholine (Higgins and Flicker 2003).

Some recent work done in Japan by T. Nagata may open a new approach. After doing basic research on acetylcholine receptors in the hippocampus of rat brain, the author obtained in 2011 surprisingly positive results on moderate memory loss and dementia in humans after the oral administration of two molecular species of phosphatidylcholine, one containing palmitic acid and oleic acid and the other containing two linoleic acid molecules. This study highlights the difficulty of concluding from effects observed using complex lipids whose origin and composition may be diversified, even if they originate from plants.

Despite the uncertainty about the validity of many results, dietary supplements containing large amounts of phosphatidylcholine (soybean lecithin) are available in the market claiming beneficial effects for memory, concentration, and sleep. The allegations concerning these neurological disorders are often based on confusion between phosphatidylcholine and phosphatidylserine (Section 4.5.2), as these two phospholipids are present together but in various proportions in all natural extracts.

4.5.1.3 Liver Diseases

Recently, phosphatidylcholine has been associated with the protection of the liver against toxic substances, such as pharmacological molecules, alcohol, or damages generated by bacteria or viruses. Thus, one of the most interesting applications of lecithin could be in the clinical treatment of steatosis (fat accumulation in hepatocytes). Indeed, several animal studies have shown that lecithin can prevent steatosis induced by a high-fat diet, probably by increasing the elimination of excess fat in bile. In humans, a test conducted by A.L. Buchman, the United States, in 1992 has shown that steatosis observed in patients undergoing total parenteral nutrition could be improved with a lecithin supplementation. These beneficial effects are likely related to a large and rapid rise in choline blood levels, emphasizing the essential role of this molecule and the value of a contribution of phosphatidylcholine in choline deficiency.

Commercially, phosphatidylcholine was sometimes given in food supplements to alleviate liver diseases (hepatitis) and nausea. It is also proposed for increasing appetite and reducing plasma cholesterol levels. In some European countries and following some studies, phosphatidylcholine was used as an adjuvant in the treatment of hepatitis using interferons. In Germany, phosphatidylcholine is widely marketed to fight against liver diseases: chronic hepatitis, cirrhosis, and poisoning. In the United States, the FDA has decided to ignore the production and sale of supplements containing this phospholipid.

4.5.1.4 Physical Performances

Because cholinergic nerves induced muscle contraction, it is understandable that the level of free choline may affect the mechanism by modulating the synthesis of acetylcholine, a neurotransmitter at the basis of this contraction. Thus, it can be assumed that a decrease in the concentration of choline during intense physical activity may affect muscle performance and therefore sport performance. Extensive measurements in the blood of athletes have confirmed the decrease in choline, the first evidence being provided in 1986 by L.A. Conlay, the United States, who was studying the participants in the Boston Marathon in 1985 and 1986. This decrease in plasma choline can also be counteracted by prior administration of lecithin, a compound known to be much more efficient than choline chloride. Several attempts have been made but without much success to study simultaneously the recovery of blood choline levels and the improvement of physical performances.

It is still too early to recommend to sportsmen a supplementation of lecithin to improve their performance. In this complex area, too little research has been focused on the nature of the sport, importance of the environment in achieving good results, and quality of the ingested phosphatidylcholine. Given the low intakes of choline in the modern diet, the best advice to sportsmen seeking better performances is to increase their consumption of foods naturally rich in free choline and phosphatidylcholine before exercise (Section 3.2.5.1).

4.5.2 Phosphatidylserine

Phosphatidylserine (Section 2.4) is present in all membranes of plants as in those of animals. Its particularity is to be located in the inner leaflet of the cell membrane where it exerts multiple functions, including the control of receptors, enzymes, and ion channels. Through these functions, phosphatidylserine could modulate cognitive, endocrine, and possibly muscle functions and thus athletic performance. It has been suggested in 2003 by M.A. McDaniel, the United States, that phosphatidylserine taken in large quantities was able to improve the status of neuronal membranes, increase the number of receptors and dendritic spines, and stimulate the neurotransmitter release. Despite the lack of accurate data, the European Commission decided in 2011 that phosphatidylserine prepared from soybean phospholipids could be used in foods for special medical purposes (2011/513/EU Decision of August 19, 2011).

4.5.2.1 Nervous Diseases

In the brain, many studies on isolated cells and even in small mammals have shown that phosphatidylserine was involved in maintaining the structure of neuronal membranes, development of dendritic branches, and proliferation of receptors. Its intervention in the phenomena of apoptosis and secretion of neurotransmitters (acetylcholine and catecholamines) is well known. The brain function of phosphatidylserine has been related with its richness in DHA, a fatty acid considered as a powerful neuroprotective agent (Section 4.2.4.5). As all these functions alter gradually with age, investigators have often tried to use this phospholipid to improve certain brain performances. Thus, considering the encouraging results obtained in animals, tests to improve memory by supplementing young children or the elderly with phosphatidylserine were undertaken. The first tests carried out in 1990 by M. Maggioni, Italy, consisting of an oral treatment with phosphatidylserine isolated from bovine brain were able to improve depressive symptoms in elderly people as well as a number of behavioral and cognitive parameters. One of the most interesting studies was conducted in 1991 by T. Crook, the United States, in patients suffering from probably Alzheimer's disease. He has shown that a marked improvement in cognitive performances was obtained with the daily administration of 100 mg of phosphatidylserine of bovine origin. Based on some of the changes recorded, T. Crook calculated that the cognitive clock of the participants was turned back about 12 years. It should be recognized that the results were less convincing in several later human trials compared with the results obtained in animals. A moderate improvement of short-term memory performances was nevertheless repeatedly observed in elderly people with some cognitive deficits. Thus, a recent study of 78 Japanese 50–69 years old, supplemented daily with 300 mg of soybean phosphatidylserine, has shown a significant improvement in memory performance (Kato-Kataoka et al. 2010). It should be noted that in general phosphatidylserine brought about no improvement in patients with early degenerative dementia or Alzheimer's disease. An experience realized in 2011 by A.G. Parker on 18 American students 22 years old showed an unambiguous improvement of cognitive function (subtraction test) after a daily intake of 400 mg of soybean phosphatidylserine for 2 weeks, each subject being compared to himself after 2 weeks on placebo. A similar supplementation experiment was conducted in 2008 by J. Baumeister, Germany. It showed no effect on cognitive function but revealed specific changes in EEG, changes in accordance with a new state of relaxation following a period of stress. This new state may be useful when preparing for a sporting or intellectual event requiring great concentration.

Experiments reported in 2010 by V. Vakhapova, Israel, with phosphatidylserine-containing long-chain n-3 fatty acids (EPA and DHA, 100 mg/day) have shown after 15 weeks of treatment a substantial improvement in working and long-term memory in the elderly. This improvement was considered to be sharper when the cognitive status of the treated person was slightly altered. This finding would argue for taking into account therapeutically early memory complaints. Other studies have confirmed these results and even reported an inhibition of stress accompanied by a decrease in plasma cortisol.

It is obvious that these experiments do not attribute the results to the phospholipid itself rather than to one of its components (serine and fatty acids).

Although the effects on memory are often modest, this medicine area would benefit greatly from trials over long periods (on the order of years) with products that are chemically defined and with patients who are psychologically well controlled and analyzed with several types of memory tests. It would be important to define the efficiency of phosphatidylserine in selected individuals as a preventive or curative medication. The abundance in the market of food supplements enriched with phosphatidylserine from various sources should motivate laboratories to conduct serious behavioral and psychological studies. These would clearly inform consumers about the potential benefits of such a supplementation.

Fortunately, the tolerance of the body with respect to this phospholipid seems very good, since B.L. Jorissen showed in 2002 that supplementation with 600 mg/day of soybean phosphatidylserine for 12 weeks in elderly patients did not induce any biochemical or physiological change.

Five studies on the relationships between phosphatidylserine and psychiatric disorders are recorded in the official website http://www.clinicaltrials.gov.

4.5.2.2   Physical Performances

Phosphatidylserine supplementation has become very popular among athletes who wish to increase their muscle mass and reduce the effects of fatigue after a prolonged exercise. The commonly advanced assertions are muscle protection in efforts, improvement of performances, muscle recovery, and increased concentration and precision.

These assertions are based primarily on the demonstration in 1990 by P. Monteleone, Italy, of a slowing of cortisol secretion by phosphatidylserine during intensive sport events, while the importance of this effect on the improvement of physical performances remains yet unknown. It is true that an excessive production of cortisol during intense exercise proved to be detrimental to muscle fibers by contributing to fatigue and longer recovery periods. This assumption has led to some research that provided early verifications involving several types of sports, unfortunately with a too small number of subjects. Thus, R. Jäger, the United States, found in 2007 that the oral supplementation with 750 mg of soybean phosphatidylserine per day for 10 days improved performance during cycling or running exercises. A similar experiment was reported in 2008 by M.A. Starks, the United States, after a study of students absorbing 600 mg of phosphatidylserine for 10 days and using a bicycle ergometer, blood tests, and a control of oxygen consumption. The results have shown that this treatment is able to effectively combat effort-induced stress by slowing the rise in blood cortisol. A 6-week treatment of a dozen golfers aged 30 years with 200 mg of soybean phosphatidylserine per day has been able to significantly improve their performance. The products used have been the subject of a patent covering the use of a "food, preferably in the form of chocolate bar containing 100 to 300 mg of phosphatidylserine and a high amount of sugar. This combination ensures an increase in cognitive abilities in individuals about 40-year old" (Giventis GMBH Patent No. EP1335732A2).

This kind of study has often been renewed in very different sport conditions but has provided very similar conclusions. However, only one trial conducted in 2005 by M.I. Kingsley, the United States, did not demonstrate any effect on cortisol, sensitivity, lipid peroxidation, or markers of muscle condition. Nevertheless, this author found a positive effect on the possible effort limits just before exhaustion.

On the basis of these few works, manufacturers and sports trainers recommend phosphatidylserine supplementation without direct evidence to support these claims. The mode of action of this phospholipid is still too little known to encourage its use. Other better controlled research efforts with placebo with a larger number of laboratories and subjects will be required to verify the possible effects of phosphatidylserine on endurance, muscle mass, and the psychological state of athletes before and after exercise. Placing in the market certain products and their eventual acceptance by competent authorities would be able to accelerate their dissemination and encourage the development of new research.

4.6   SPHINGOLIPIDS

An abundant literature has shown that sphingolipids, complex lipids represented mainly by sphingomyelin (Section 2.4) and glycosphingolipids (Section 2.5), are highly functional molecules even without taking into account about structural role in cell membranes. Indeed, they may act as bioactive molecules in the intestinal lumen with their intact structure. Furthermore, it has been known for several years that their hydrolysis products (ceramide and sphingosine) or derivatives (ceramide phosphate and sphingosine phosphate) are absorbed by the intestine and have important functions in cell signaling processes. The functions of these derivatives are sometimes complex and confusing as some, such as ceramide and sphingosine, are antimitotic, inhibiting growth and inducing cell death, whereas sphingosine phosphate is mitogenic and inhibits apoptosis. One can imagine the importance of the balance between these metabolites to control cell life. Basically, these sphingolipids play multiple roles, the most important being the modulation of G-protein-coupled receptors, generation of intracellular signals especially when they are glycosylated, and participation in the formation of complex membrane structures ("rafts") gathering many receptors.

As the dietary ceramide intake was estimated to around 50 mg/day in infants and between 100 and 300 mg/day in adults (Section 3.2.6), it must be natural to address the problem of their potential effects directly on the intestine, and also on the consumer physiology. Nevertheless, it must not be forgotten that the cells are capable of ensuring the *de novo* biosynthesis of all sphingolipids, glycosylated or not. This situation suggests that there is no need to consume these lipids included in foodstuffs or to absorb supplements, at least under normal nutritional conditions.

4.6.1   NERVOUS DISEASES

The discovery of gangliosides in nervous tissues in 1942 by E. Klenk, Germany, prompted rapid investigations on their possible relationships with the development and functions of the brain. These complex glycolipids are based on a glucosylceramide core (Section 2.5) where a glucose molecule is bound to a more or less complex glycan chain but containing at least one characteristic amino sugar, sialic acid (or *N*-acetylneuraminic acid, NeuAc). GM3 is one of the simplest forms and the best represented ganglioside in breast milk (Figure 4.2). Another ganglioside, GD3,

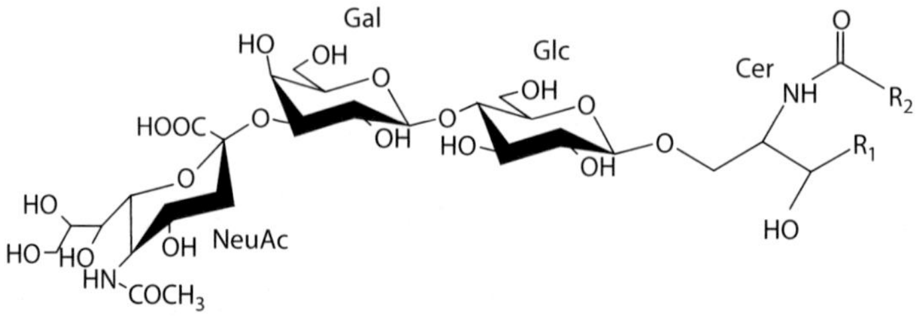

FIGURE 4.2 GM3 ganglioside (NeuAcα2-3Galβ-4Glc-Cer): R_1, amino alcohol carbon chain; R_2, fatty acid carbon chain.

similar to the previous one but with two molecules of NeuAc on the terminal galactose, is the most abundant in cow's milk lipid. In the brain, gangliosides are now recognized to be involved in the development process (neurotrophic effect), repair mechanisms, and neuronal signaling. They improve the stability of myelin.

It has been well demonstrated in 1975 by A.J. Dunn, England, that the biosynthesis or transformation of these compounds was increased in mice subjected to fitness exercises. Naturally, the theories of the neural basis of learning have rapidly called on these glycosphingolipids. Several studies carried out in animals have shown that gangliosides, and especially their residue, sialic acid, may be a significant supplement through milk during the development. Early research by B.L. Morgan, the United States, in 1980 highlighted the incorporation of sialic acid in the brain of young rats and the concomitant improvement in their learning ability. Later, in 2009 it was shown by P. McJarrow, New Zealand, that the presence of gangliosides in human milk could be a guarantee of beneficial effects on the maturation and function of the nervous system. Other works done in 2007 by B. Wang, Australia, have confirmed that the enrichment of milk with natural sialic acid could improve the performance of learning and memory in young pigs.

Few works in this area have been undertaken in humans. B. Wang, Australia, still showed in 2003 by analyzing the brains of young children who died suddenly that the ganglioside concentration was higher in breast-fed infants than in those fed infant formula. So far, the only trial to improve cognitive development by the administration of gangliosides was carried out in 2012 by D.A. Gurnida, Indonesia. This author has shown that, compared to a control group, supplementation of infants with gangliosides prepared from milk increased their concentration in the blood and improved after 22 weeks scores of hand–eye coordination and IQ in treated children. It is unfortunate that the relationships between gangliosides and cognitive or behavioral performances have not been further explored in adult men. As pointed by T. Ariga, the United States, in 2008, the close relationships between gangliosides and β-amyloid peptide in the brain of patients with Alzheimer's disease should strongly encourage clinicians to develop the therapeutic potential of these lipids after encouraging results in animals.

In patients with Parkinson's disease, J.S. Schneider, the United States, described in 1998 the possibility of improving the motor function by administering the GM1 ganglioside. The author was able to obtain a significant improvement in motor performance (hands, feet, and walking) with an injection of 200 mg twice daily for 16 weeks. In 2010, the same author showed that long-term use (5 years) of treatment was safe and that motor performances remained improved and stable throughout the treatment period.

No trial of dietary supplementation appears to have been experienced to date.

4.6.2 Intestinal Diseases

Bacteria and their toxins, and viruses, often adhere to cells via sphingolipids. K. Hanada, Japan, considered in 2005 that rafts, microdomains enriched in sphingolipids, could be the structures responsible for this adhesion. Thus, dietary sphingolipids may have an important role in the elimination of pathogenic organisms in intestinal lumen. This possibility was verified in 1997 by J. Fantini, France, with synthetic sphingolipids that were shown to be effective in inhibiting the binding of the AIDS virus to membrane receptors. This defensive action could play the role of soluble receptors for microorganisms or their toxins.

Biologists have especially made all efforts to demonstrate the direct effect of dietary sphingolipids on intestinal mucosa, its development, and on the occurrence of cancer disease. It is likely that milk sphingolipids play an important role in the intestinal mucosa of children as these lipids are the equivalent of membrane receptors for bacteria, viruses, and even toxins. The best known example is cholera toxin, whose only known receptor on the surface of intestinal cells is the GM1ganglioside. It is therefore conceivable that milk sphingolipids are "false receptors" for intestinal pathogenous bacteria or their toxins. The dietary sphingolipids could compete with the bacterial attachment sites, making their translocation to the blood more difficult. Thus, it has been observed for the first time in humans in 1998 by R. Rueda, Spain, that the addition of gangliosides to an infant formula induced a depletion of *Escherichia coli* in the infant gut, thereby facilitating the installation of other beneficial bacteria (Bifidobacteriae). These effects underline the potential probiotic role of these glycolipids, a role that can be reinforced by the growth inhibition of several pathogenic bacteria by the digestion products of sphingolipids (ceramide and sphingosine). It is significant that breast milk contains a glycolipid (globotriaosylceramide) known to bind to the Shiga toxin secreted by a variety of *E. coli*. I. Herrera-Insua, the United States, showed in 2001 that infants are protected by the ingested glycolipids besides other factors against diarrhea associated with the Shiga-toxin or related toxins. Another important aspect is the development of a modulation of immune cells by gangliosides. This property, demonstrated in animals in 2007 by R. Rueda, Spain, suggested that gangliosides in colostrum and milk (native or added) may influence the proliferation or activation of immune cells in the infant intestine.

Milk sphingolipids thus appear to be essential factors of a still naive immune system that helps to protect infants against bacterial infections. In this area, the properties of different milks used in infant feeding have not yet been established. Many research efforts will be needed before considering a preventive use of sphingolipids as an anti-infection agent in young children or adults.

4.6.3 CANCERS

Although not yet explored in humans, sphingolipids, glycosylated or not, may well become rapidly an effective weapon in the treatment of intestinal cancers, as suggested in 2004 by E.M. Schmelz, the United States. Indeed, since ceramides and sphingosine are regulators of growth and cell differentiation, what are the effects of the metabolites formed in the gut from dietary sphingolipids? Are these metabolites able to modulate the function of cells of the intestinal epithelium, normal cells or cells engaged in a carcinogenesis process? Although the digestion of sphingolipids remains limited, their presence in the gut may be a means to influence the risk of local carcinogenesis, or at least to limit the inflammatory processes.

Some answers have been given to these questions in 1991 after the demonstration by C. Borek, the United States, of an anticancer effect of sphingosine in cultured epithelial cells of mice intestine. These promising results were confirmed soon after *in vivo* in the colon using a diet containing 25–100 mg of sphingomyelin per 100 g of food. This sphingolipid effectively reduced the number of aberrant intestinal crypts and may even promote the transformation of malignant adenocarcinomas in benign adenomas. It is interesting to note that the effective doses in mice are not very far from the estimated consumption in humans. Also in 1998, E.M. Schmelz, the United States, revealed that the ceramides and sphingosine were able to induce apoptosis in cultured carcinoma cells. In 2004, H. Symolon, the United States, demonstrated that soybean glucosylceramides produced the same effect in suppressing colon tumorigenesis in mice. These results suggest that ingestion of sphingolipids may contribute to a reduced risk of colon cancer. Although the underlying mechanisms are poorly understood, these data support the allegations of anticancer effects attributed to many food sources rich in sphingolipids such as soybeans.

In 2001, an epidemiological study by R. Jarvinen, Finland, showed that the consumption of dairy products was associated with a decreased risk of colon cancer. This effect could probably be partly related to the role of sphingomyelin and its metabolites in cell signaling and apoptosis. A. Moschetta, Italy, showed in 2003 that sphingomyelin ingested with foodstuffs may also protect against intestinal carcinogenesis after mixing with bile salts. In fact, the author demonstrated that sphingomyelin reduced the deoxycholate-induced hyperproliferation of cells derived from the colon. This protective effect may therefore have important implications for the prevention of colon cancer simply by changing the diet (egg, calf liver, and fish) (Section 3.2.5.2). Currently, despite a large number of registered patents, no supplement enriched with sphingomyelin is commercially available.

Gradually, through much fundamental research, the mechanisms of action of sphingolipid metabolites have been explored, and the latest developments offer hope for effective therapeutic outcomes of cancer treatment. The ability of this group of molecules to be involved in basic cellular processes such as proliferation, apoptosis, transformation, differentiation, and motility has provided novel therapeutic strategies based on the effect of synthetic molecules derived from certain natural sphingolipids. The emerging role of sphingolipids in the genesis of diseases was exposed in 2007 by Y.H. Zeidan, the United States.

As pointed out by B. Ogretmen, the United States, in 2004, the observation of a parallelism between the effectiveness of drugs used in chemotherapy and their influence on the metabolism of ceramides lends credence to the existence of successful developments. Although the administration of these metabolites, or synthetic derivatives, is a part of the future strategies of therapeutic treatments, the specific application of dietary sphingolipids in the prevention or treatment of intestinal cancer in humans has not yet been initiated. To date, no clinical intervention trial has been published.

4.6.4 Biosynthesis of Cholesterol

At the molecular level, the specific association between cholesterol and sphingomyelin was established in 1980 by Y. Barenholz, Israel. The nutritional effect of sphingolipids was examined later. In 1997, T. Kobayashi, Japan, showed in the rat that the addition of sphingolipids to diet lowered serum cholesterol by 30%. These preliminary studies have been confirmed in mice in 2006 by I. Duivenvoorden, Netherlands, using a diet supplemented with sphingolipids causing a decrease of more than half of the cholesterol and triacylglycerol blood contents. The most significant effect was observed for equimolar concentrations of sphingomyelin and cholesterol in the diet, the composition also being observed in the human intestinal lumen. The interaction between the two lipids was favored by the presence of a saturated fatty acid in sphingomyelin, thus explaining the more pronounced effect of milk sphingomyelin compared with that of egg yolk.

A recent study limited to 20 healthy subjects, undertaken in 2010 by L. Ohlsson, Sweden, has shown that the addition of 975 mg of glycolipids, including 700 mg of sphingomyelin to a fat-rich breakfast, had no effect on postprandial plasma triacylglycerols levels. Only a slight decrease of cholesterol linked to lipoproteins was detected. A study with a larger number of subjects and with larger ingested amounts could perhaps reveal more convincing effects.

Although the mechanisms responsible for these effects observed in animals are far from being elucidated, some sphingolipids or their metabolites could be protective agents against the risk of atherosclerosis. Further epidemiological studies will be needed to determine the true role of dietary sphingolipids, glycosylated or not, in the protection against cardiovascular disease.

4.7 REPLACEMENT LIPIDS

Dietary fat substitutes are poorly metabolized molecules or have fewer calories than conventional lipids. They are created in the laboratory to replace a part of the usual fat intake, thus reducing the calories from the ingested lipids. These molecules were placed in the market in response to consumer demand in the 1990s for low-calorie foods to help lose weight.

The main quality requirements of these products are as follows:

- Lack of toxicity
- Resistance to digestive enzymes or low caloric intake

- Preservation of the organoleptic qualities of foods. These lipids currently on the market may be divided into two categories:

1. Artificial lipids derived from natural lipids but designed to make them less easily hydrolyzed in the intestinal lumen and thus hypoenergetic (structured triacylglycerols and glycolipids) (Section 2.8.1)
2. Natural lipids (diacylglycerols [DAGs]) (Section 2.8.2) virtually absent from the diet but with some physicochemical aspects of vegetal oils

4.7.1　Hypoenergetic Fats

4.7.1.1　Structured Triacylglycerols

Since the energy supplied by a fatty acid is proportional to its chain length, it soon appeared interesting to synthesize triacylglycerols with mixed composition, containing one long-chain fatty acid (18:0 or 22:0) and two medium-chain (8:0 and 10:0) or short-chain (2:0, 3:0, and 4:0) fatty acids.

Salatrim® from Danisco (Benefat® in the United States) is the best known low-calorie substitute in the group of triglycerides obtained by interesterification (Section 3.2.1.4). These molecules are formed by incubating triacetin (two carbons), tripropionin (three carbons), or tributyrin (four carbons) isolated or in mixture with hydrogenated vegetable oil (rapeseed, cotton, and sunflower) providing mainly stearic acid (18:0) (Section 2.8.1). As a result of the low absorption of long-chain fatty acids and the low energy supplied by butyric acid, Salatrim brings only about 5 kcal/g (instead of 9 kcal/g for the natural lipids) and can therefore be useful as part of a hypoenergetic diet. The product was approved in Europe in 2003 as a fat substitute in baking and confectionery (chocolate). Salatrim does not reduce the absorption of fat-soluble vitamins (vitamins A, D, E, and K) and does not affect cholesterol. It is recommended to consume no more than 30 g of Salatrim a day.

Salatrim obtained in 2003 an approval from the European Parliament to be present on the market. In France, the AFSSA was in favor of this product in 2004. Its use was restricted to professionals and should be labeled with a quantitative limit of consumption. It should be stated that the products are not intended for consumption by children and that it can cause gastrointestinal disturbances.

A product close to the previous one, Caprenin®, has been marketed for some time in the United States. This substitute contained a long-chain fatty acid (22:0) and two with short chains (8:0 and 10:0) (Section 2.8.1). However, due to its hypercholesterolemic effect it was withdrawn from the market.

Neobee® is a product marketed by the Stepan Company (http://www.stepan.com /en/) (Section 2.8.1). It is a real mixture of MCTs (with six or eight carbons). Available in the United States, it has been proposed as an additional energetic source for athletes, adults, and children, and as a substitute of traditional lipids for patients with intestinal malabsorption. Its lower energetic value compared with vegetal oils (6.8 instead of 9 kcal/g) made it useful as a substitute for natural fats in energy-restricted diets. The manufacturer claimed that the product is eight times more rapidly metabolized in the liver than conventional triacylglycerols without being accumulated in adipose tissue. This product was approved in 2008 for use as safe in some foodstuffs by the FDA in the United States, but it is not yet approved in Europe.

4.7.1.2 Glycolipids

Many attempts have been made to place indigestible products on a very promising market but with organoleptic qualities comparable to those of culinary fats. Due to a lack of toxicological tests, the majority of products were not marketed. The best known and most common glycolipids are sucrose polyesters. Only the hexa- and octo-esters are used with some reservations as substitute lipids.

Olestra (Olean® in the United States), manufactured by Procter & Gamble in 1971, consists of a sucrose molecule esterified with six to eight fatty acids (Section 2.8.1). This sucrose polyester is the only representative of that group to have obtained a marketing authorization. Olestra is the first substitute fat to obtain FDA approval, in 1996. Although allowed in the United States, it is still illegal in Canada and Europe. It is not hydrolyzed or absorbed in the intestine due to the lack of accessibility of lipases for ester bonds, and therefore it has no nutritional value (no calorigenic power). This lack of absorption gives it a great attraction for low-calorie diets but results in some intestinal disorders (diarrhea and abdominal pain). Another disadvantage of this type of lipid substitute is the reduction of absorption of fat-soluble vitamins. From a culinary point of view, it gives to food the same taste and texture as oils and fats usually used for frying. It is used for making chips, crackers, and fried products. Fat-soluble vitamins must be added to foods to compensate for their loss by elimination in feces with Olestra. There are regular contestations in the press, and the dietary interest of Olestra is far from being unanimous among nutritionists. It is likely that Europe will be reluctant to give an authorization for placing this product on the market, as no acceptance procedure for the moment seems to have been filed.

Olestra is not to be confused with monoesters of sucrose (sucrose ester, E473), being a part of the compounds known as the "Ryoto sugar ester" molecules. Their use is permitted in Europe as a food additive for a maximum concentration of 20 g/ kg of food, and 5 g/L of drink, the consumption having to be less than 40 mg per kilogram of body weight per day. This product, containing one fatty acid for one carbohydrate molecule, is hydrolyzed in the intestine and is therefore not a true fat substitute. It is mainly used as a texturing agent promoting the manufacture of processed foods, supplement in beverages, and surfactant in fruit preservation.

4.7.2 Diacylglycerols

Following research carried out in the rat, it has been shown in humans that postprandial lipemia could be reduced by replacing dietary triacylglycerols by DAG (mainly 1,3-DAG) (Taguchi et al. 2000) (Section 2.8.2). These metabolic works led nutritionists to look for possible effects of the consumption of these DAGs on the development of fat deposits. It has been shown in 2000 by T. Nagao, Japan, that replacing a fourth of the lipid ration by DAG significantly reduced the accumulation of visceral and subcutaneous fat after 16 weeks in overweight subjects. That diet enriched in DAG was also efficient in reducing weight gain in obese individuals. H. Yanai, Japan, suggested in 2007 that the effects of DAG on lipid and glucose metabolism could be the basis for improving the condition of patients with metabolic syndrome and possibly for preventing that disease. These effects are not directly related to the caloric content of DAG, as it is very similar (approximately 98%) to that of triacylglycerols. However, a recent meta-analysis conducted in 2010 by W. Wang, China, has shown that replacing vegetal oil with an identical amount of DAG did not reduce fasting triglyceridemia in healthy subjects, some effects being only suggested in diabetic patients. A study in 2012 by H. Yanai, Japan, on a small number of young men, has clearly shown that DAG (30 g/day) suppressed the postprandial increase of VLDL and insulin. In addition, the author has found in the same conditions a plasma serotonin increase, an effect related to thermogenesis and therefore to increasing energy expenditure after the meal.

With these encouraging results, a DAG-based oil (about 80% from rapeseed oil and soybean oil) was placed in the Japanese market under the name Healthy Econa (Kao Society). After checking the safety of consumption of this oil with respect to the absorption of fat-soluble vitamins, its marketing was authorized in 2003 in the United States under the name Enova® and in 2004 in Canada. In 2009, the sale of this oil was temporarily suspended in the United States, due to the presence of a probable carcinogen, glycidol.

In Europe, a favorable opinion for its use was given in 2004 by the EFSA and an authorization to market this new food was given by the European Commission in 2006. The European decision stated that "this oil must be used in cooking oils, fat spreads, salad dressings, mayonnaise, drinks presented as a replacement for one or more meals of the daily intake, bakery products and yoghurt-type products. The designation 'vegetal oil with a high content in diacylglycerols' (at least 80% diacylglycerol) must appear on the label or in the list of ingredients of foodstuffs containing it" (Decision 2006/720/CE).

It should be noted that although the EFSA accepted the marketing of DAG-based oils in 2006, it gave an unfavorable opinion in November 2011 for their use in losing weight (http://www.efsa.europa.eu/en/efsajournal/pub/2469.htm). The working group (EFSA Panel on Dietetic Products, Nutrition and Allergies) concluded that a causal connection was not established between consumption of DAG oil (replacing triglyceride oils) and weight loss (http://www.efsa.europa.eu/fr/efsajournal/doc/2469.pdf). This decision was taken after a review of six published trials that provided inconsistent results unrelated to the doses used and the duration of the study. The experts found no evidence of any mechanism by which DAG would act on the weight of the subjects.

REFERENCES

Abumweis, S.S., Barake, R., et al., 2008. Plant sterols/stanols as cholesterol lowering agents: A meta-analysis of randomized controlled trials. *Food Nutr. Res.* 52:10.3402/fnr.v52i0.1811.

Ameur, A., Enroth, S., et al., 2012. Genetic adaptation of fatty-acid metabolism: A human-specific haplotype increasing the biosynthesis of long-chain omega-3 and omega-6 fatty acids. *Am. J. Human Genetics* 90:809–20.

Amminger, G.P., Schäfer, M.R., et al., 2010. Long-chain omega-3 fatty acids for indicated prevention of psychotic disorders. *Arch. Gen. Psychiatry* 67:146–54.

Annweiler, C., Brugg, B., et al., 2014. Combination of memantine and vitamine D prevent axon degeneration induced by amyloid-beta and glutamate. *Neurobiol. Aging* 35:331–5.

Armstrong, B., Doll, R., 1975. Environmental factors and cancer incidence and mortality in different countries, with special reference to dietary practices. *Int. J. Cancer* 15:617–31.

Ascherio, A., Kassandra, L., et al., 2014. Vitamin D as an early predictor of multiple sclerosis activity and progression. *JAMA Neurol.* 71:306–14.

Aune, D., Chan, D.S., et al., 2012. Dietary compared with blood concentrations of carotenoids and breast cancer risk: A systematic review and meta-analysis of prospective studies. *Am. J. Clin. Nutr.* 96:356–73.

Autier, P., Gandini, S., 2007. Vitamin D supplementation and total mortality: A meta-analysis of randomized controlled trials. *Arch. Intern. Med.* 167:1730–7.

Avenell, A., Gillespie, W.J., et al., 2009. Vitamin D and vitamin D analogues for preventing fracture associated with involutional and post-menopausal osteoporosis. *Cochrane Database Syst. Rev.* 2:CD000227.

Bang, H.O., Dyerberg, J., et al., 1976. The composition of food consumed by Greenland Eskimos. *Acta Med. Scand.* 200:69–73.

Baracos, V.E., Mazurak, V.C., et al., 2004. n-3 Polyunsaturated fatty acids throughout the cancer trajectory: Influence on disease incidence, progression, response to therapy and cancer-associated cachexia. *Nutr. Res. Rev.* 17:177–92.

Bauer, I., Hughes, M., et al., 2014. Omega-3 supplementation improves cognition and modifies brain activation in young adults. *Human Psychomarmacol.* 29:133–44.

Bent, S., Bertoglio, K., et al., 2009. Omega-3 fatty acids for autistic spectrum disorder: A systematic review. *J. Autism Dev. Disord.* 39:1145–54.

Bernard, J.Y., De Agostini, M., et al., 2013. The dietary n6:n3 fatty acid ratio during pregnancy is inversely associated with child neurodevelopment in the EDEN mother-child cohort. *J. Nutr.* 143:1481–8.

Beydoun, M.A., Fanelli Kuczmarski, M.T., et al., 2013. Omega-3 fatty acid intakes are inversely related to elevated depressive symptoms among United States women. *J. Nutr.* 143:1743–52.

Bhatia, H.S., Agrawal, R., et al., 2011. Omega-3 fatty acid deficiency during brain maturation reduces neuronal and behavioral plasticity in adulthood. *PloS ONE* 6(12):e28451.

Bielakovic, G., Mikolova, D, et al., 2012. Antioxidant supplements for prevention of mortality in healthy participants and patients with various diseases. *Cochrane Database Syst. Rev.* 3:CD007176.

Bloch, M.H., Hannestad, J., 2012. Omega-3 fatty acids for the treatment of depression: Systematic review and meta-analysis. *Mol. Psychiatry* 17:1272–82.

Bonanone, A., Grundy, S.M., 1988. Effect of dietary stearic acid on plasma cholesterol and lipoprotein levels. *N. Eng. J. Med.* 318:1244–8.

Borges, M.C., Martini, L.A., et al., 2011. Current perspectives on vitamin D, immune system, and chronic diseases. *Nutrition* 27:399–404.

Bougnoux, P., Hajjaji, N., et al., 2009. Improving outcome of chemotherapy of metastatic breast cancer by docosahexaenoic acid: A phase II trial. *Br. J. Cancer* 101:1978–85.

Bourre, J.-M., 1990. *La diététique du cerveau de l'intelligence et du plaisir.* Odile Jacob, Paris.

Boyd, N.F., Stone, J., et al., 2003. Dietary fat and breast cancer risk revisited: A meta-analysis of the published literature. *Br. J. Cancer* 89:1672–85.

Bradsky, T.M., Till, C., et al., 2011. Serum phospholipid fatty acids and prostate cancer risk: Results from the prostate cancer prevention trial. *Am. J. Epidemiol.* 173:1429–39.

Brenna, J.T., Salem, N., et al., 2009. α-Linolenic acid supplementation and conversion to n-3 long-chain polyunsaturated fatty acids in humans. *Prost. Leukotr. Essential Fatty acids* 80:85–91.

Breslow, J.L., 2006. n-3 Fatty acids and cardiovascular disease. *Am. J. Clin. Nutr.* 83:1477S–82S.

Broadhurst, C.L., Stephen, C., et al., 1998. Rift Valley lake fish and shellfish provided brain-specific nutrition for early *Homo. Br. J. Nutr.* 79:3–21.

Burgaz, A., Orsini, N., et al., 2011. Blood 25-hydroxyvitamin D concentration and hypertension: A meta-analysis. *J. Hypertens.* 29:636–45.

Butler, L.M., Wang, R., et al., 2009. Marine n-3 and saturated fatty acids in relation to risk of colorectal cancer in Singapore Chinese: A prospective study. *Int. J. Cancer* 124:678–86.

Chanussot, F., 2008. *Lécithine, métabolisme et nutrition*. Lavoisier, Paris.

Chavarro, J.E., Stampfer, M.J., et al., 2007. A prospective study of polyunsaturated fatty acid levels in blood and prostate cancer risk. *Cancer Epidemiol. Biomarkers Prev.* 16:1364–70.

Cherniack, E.P., Florez, H., et al., 2008. Hypovitaminosis D in the elderly: From bone to brain. *J. Nutr. Health Aging* 12:366–73.

Clarke, R., Frost, C., et al., 1997. Dietary lipids and blood cholesterol: Quantitative meta-analysis of metabolic war studies. *Brit. Med. J.* 314:112–7.

Cole, G.M., Frautschy, S.A., 2010. DHA may prevent age-related dementia. *J. Nutr.* 140:869–74.

Cunnane, S.C., Nugent, S., et al., 2011. Brain fuel metabolism, aging, and Alzheimer's disease. *Nutrition* 27:3–20.

Cunnane, S.C., Plourde, M., et al., 2009. Fish, docosahexaenoic acid and Alzheimer's disease. *Prog. Lipid Res.* 48:239–56.

Denis, I., Heberden, C., et al., 2011. Acides gras polyinsaturés n-3 (omega 3) et cerveau. *Med. Nutr.* 47:17–28.

Derosa, G., Cicero, A.F.G., et al., 2011. Effects of n-3 PUFA on insulin resistance after an oral fat load. *Eur. J. Lipid Sci. Technol.* 113:950–60.

Devore, E.E., Grodstein, F., et al., 2010. Dietary antioxidants and long-term risk of dementia. *Arch. Neurol.* 67:819–25.

Donahue, S., Rifas-Shiman, S. L., et al., 2011. Prenatal fatty acid status and child adiposity at age 3 y: Results from a US pregnancy cohort. *Am. J. Clin. Nutr.* 93:780–8.

Donaldson, M.S., 2011. A carotenoid health index based on plasma carotenoids and health outcomes. *Nutrients* 3:1003–22.

Dong, J.Y., Zhang, W., et al., 2013. Vitamin D intake and risk of type 1 diabetes: A meta-analysis of observational studies. *Nutrients* 5:3551–62.

Dupont, P., 2011. *Vitamine D, Hormone solaire source d'éternelle jeunesse?* Clara Fama Ed., Plaisance du Touch.

Dysken, M.W., Sano, M., et al., 2014. Effect of vitamin E and memantine on functional decline in Alzheimer's disease. *JAMA* 311:33–44.

Epstein, M.M., Edgren, G., et al., 2012. Temporal trends in cause of death among Swedish and U.S. men with prostate cancer. *J. Natl. Cancer Inst.* 104:1335–42.

Erridge, C., Attina, T., et al., 2007. A high-fat meal induces low-grade endotoxemia: Evidence of a novel mechanism of postprandial inflammation. *Am. J. Clin. Nutr.* 86:1286–92.

Fang, F., Kasperzyk, J.L., et al., 2011. Prediagnostic plasma vitamin D metabolites and mortality among patients with prostate cancer. *PloS ONE* 6:e18625.

Féart, C., Peuchant, E., et al., 2008. Plasma eicosapentaenoic acid is inversely associated with severity of depressive symptomatology in the elderly: Data from the Bordeaux sample of the Three-City Study. *Am. J. Clin. Nutr.* 87:1156–62.

Ferland, G., 2012. Vitamin K and the nervous system: An overview of its actions. *Adv. Nutr.* 3:204–12.

Fitchett, J.R., 2013. Placental HIV transmission and vitamin D: Nutritional and immunological implications. *Nutr. Bull.* 38:410–3.

Francois, C.A., Connor, S.L., et al., 2003. Supplementing lactating women with flaxseed oil does not increase docosahexaenoic acid in their milk. *Am. J. Clin. Nutr.* 77:226–33.

Geelen, A., 2007. Fish consumption, n-3 fatty acids, and colorectal cancer: A meta-analysis of prospective cohort studies. *Am. J. Epidemiol.* 166:1116–25.

Geleijnse, J.M., Vermeer, C., et al., 2004. Dietary intake of menaquinone is associated with a reduced risk of coronary heart disease: The Rotterdam study. *J. Nutr.* 134:3100–5.

Gerber, M., 2009. Background review paper on total fat, fatty acid intake and cancers. *Ann. Nutr. Metabol.* 55:140–61.

Gillies, D., Sinn, J.K.H., et al., 2012. Polyunsaturated fatty acids (PUFA) for attention deficit hyperactivity disorder (ADHD) in children and adolescents. *Cochrane Database Syst. Rev.* 7:CD007986.

Gillingham, L.G., Harris-Janz S., et al., 2011. Dietary monounsaturated fatty acids are protective against metabolic syndrome and cardiovascular disease risk factors. *Lipids* 46:209–28.

Goldberg, R.J., Katz, J., 2007. A meta-analysis of the analgesic effects of omega-3 polyunsaturated fatty acid supplementation for inflammatory joint pain. *Pain* 129:210–23.

Goodman, A.B., Pardee, A.B., et al., 2003. Evidence for defective retinoid transport and function in late onset alzheimer's disease. *Proc. Natl. Acad. Sci. USA* 100:2901–5.

Gopinath, B., Flood, V.M., et al., 2010. Consumption of omega-3 fatty acids and fish and risk of age-related hearing loss. *Am. J. Clin. Nutr.* 92:416–21.

Gow, R.V., Hibbeln, J.R., 2014. Omega-3 and treatment implications in attention deficit hyperactivity disorder (ADHD) and associated behavioral symptoms. *Lipid Technol.* 26/1:7–10.

Gow, R.V., Sumich, A., et al., 2013. Omega-3 fatty acids are related to abnormal emotion processing in adolescent boys with attention deficit hyperactivity disorder. *Prostaglandins Leukot. Essent. Fatty Acids* 88:419–29.

Greer, J.B., O'Keefe, S.J., 2011. Microbial induction of immunity, inflammation and cancer. *Front. Physiol.* 1:168.

Griel, A.E., Cao, Y., et al., 2008. A macadamia nut-rich diet reduces total and LDL-cholesterol in midly hypercholesterolemic men and women. *J Nutr.* 138:761–7.

Gu, Y., Cosentino, S.A., et al., 2012. Nutrient intake and plasma β-amyloid. *Neurology* 78:1832–40.

Hall, M.N., 2007. Blood levels of long-chain polyunsaturated fatty acids, aspirin, and the risk of colorectal cancer. *Cancer Epidemiol. Biomarkers Prev.* 16:314–21.

Halton, T.L., Willett, W.C., et al., 2006. Low-carbohydrate-diet score and the risk of coronary heart disease in women. *New Engl. J. Med.* 355:1991–2002.

Harbild, H.L., Harlof, L.B.S., et al., 2013. Fish-oil supplementation from 9 to 12 months of age affects infant attention in a free-play test and is related to change in blood pressure. *Prost. Leukotrienes Essent. Fatty acids* 89:327–33.

Harris, W.S., Von Schacky, C., 2004. The Omega-3 Index: A new risk factor for death from coronary heart disease? *Prev. Med.* 39:212–20.

He, K., Rimm, E.B., et al., 2002. Fish consumption and risk of stroke in men. *JAMA* 288:3130–6.

He, K., Song, Y., et al., 2004. Accumulated evidence on fish consumption and coronary heart disease mortality. *Circulation* 109:2705–11.

Hedellin, M., Chang, E.T., et al., 2006. Association of frequent consumption of fatty fish with prostate cancer risk is modified by COX-2 polymorphism. *Int. J. Cancer* 120:398–405.

Heine-Bröring, R.C., Brouwer, I.A., et al., 2010. Intake of fish and marine n-3 fatty acids in relation to coronary calcification: The Rotterdam Study. *Am. J. Clin. Nutr.* 91:1317–23.

Henriksen, C., Haugholt, K., et al., 2008. Improved cognitive development among preterm infants attribuable to early supplementation of human milk with docasahexaenoic acid and arachidonic acid. *Pediatrics* 121:1137–45.

Heude, B., Ducimetière, P., et al., 2003. Cognitive decline and fatty acid composition of erythrocyte membranes—The EVA Study. *Am. J. Clin. Nutr.* 77:803–8.

Higgins, J.P., Flicker, L., 2003. Lecithin for dementia and cognitive impairment. *Cochrane Database Syst. Rev.* 3:CD001015.

Hoenselaar, R., 2012. Saturated fat and cardiovascular disease: The discrepancy between the scientific literature and dietary advice. *Nutrition* 28:118–23.

Hoffmire, C.A., Block, R.C., et al., 2012. Associations between omega-3 polyunsaturated fatty acids from fish consumption and severity of depressive symptoms: An analysis of the 2005–2008 National Health and Nutrition Examination Survey. *Prostaglandins Leukotr. Essent. Fatty acids* 86:155–160.

Hossein-nezhad, A., Spira, A., et al., 2013. Influence of vitamin D status and vitamin D3 supplementation on genome wide expression of white blood cells: A randomized double-blind clinical trial. *PLoS One* 8(3):e58725.

Hu, W., Ross, J., et al., 2011. Differential regulation of dihydroceramide desaturase by palmitate versus monounsaturated fatty acids. *J. Biol. Chem.* 286:16596–605.

Huang, T.L., Zandi, P.P., et al., 2005. Benefits of fatty fish on dementia risk are stronger for those without APOE epsilon4. *Neurology* 65:1409–14.

Iwata, N.G., Pham, M., et al., 2011. Trans fatty acids induce vascular inflammation and reduce vascular nitric oxide production in endothelial cells. *PloS ONE* 6:e29600.

Jackson, P.A., Reaya, J.L., et al., 2012. Docosahexaenoic acid-rich fish oil modulates the cerebral hemodynamic response to cognitive tasks in healthy young adults. *Biol. Psychol.* 89:183–90.

Jakobsen, M.U., O'Reilly, E.J., et al., 2009. Major types of dietary fat and risk of cornary heart disease: A pooled analysis of 11 cohort studies. *Am. J. Clin. Nutr.* 89:1425–32.

Jenkins, D., Josse, A.R., et al., 2008. Fish-oil supplementation in patients with implantable cardioverter defibrillators: A meta-analysis. *Can. Med. Assoc. J.* 178:157–64.

Johnson, E.J., 2012. A possible role for lutein and zeaxanthin in cognitive function in the elderly. *Am. J. Clin. Nutr.* 96:1161S–5S.

Karppi, J., Laukkanen, J. A., et al., 2012. Serum lycopene decreases the risk of stroke in men: A population-based follow-up study. *Neurology* 79:1540–7.

Kato-Kataoka, A., Sakai, M., et al., 2010. Soybean-derived phosphatidylserine improves memory function of the elderly Japanese subjects with memory complaints. *J. Clin. Biochem. Nutr.* 47:246–55.

Kelly, L., Grehan, B., et al., 2011. The polyunsaturated fatty acids, EPA and DPA exert a protective effect in the hippocampus of the aged rat. *Neurobiol. Aging* 32:2318.e1–15.

Kennedy, A., Martinez, K., et al., 2009. Saturated fatty acid-mediated inflammation and insulin resistance in adipose tissue: Mechanisms of action and implications. *J. Nutr.* 139:1–4.

Kennedy, A., Martinez, K., et al., 2010. Antiobesity mechanisms of action of conjugated linoleic acid. *J. Nutr. Biochem.* 21:171–9.

Keys, A., 1953. Atherosclerosis: A problem in newer public health. *J. Mt Sinai Hosp.* 20:118–39.

Kiecolt-Glaser, J.K., Epel, E.S., et al., 2012. Omega-3 fatty acids, oxidative stress, and leukocytes telomere length: A randomized controlled trials. *Brain Behav. Immun.* 28:16–24.

Koletzko, B., Lien, E., et al., 2008. The roles of long-chain polyunsaturated fatty acids in pregnancy, lactation and infancy: Review of current knowledge and consensus recommendations. *J. Perinat. Med.* 36:5–14.

Kremmyda, L.S., Vlachava, M., et al., 2011. Atopy risk in infants and children in relation to early exposure to fish, oily fish, or long-chain omega-3 fatty acids: A systematic review. *Clin. Rev. Allergy Immunol.* 41:36–66.

Kromhout, D., Bosschieter, E.B., et al., 1985. The inverse relation between fish consumption and 20-year mortality from coronary heart disease. *New Engl. J. Med.* 315:1205–9.

Lai, Y.H., Petrone, A.B., et al., 2013. Association of dietary omega-3 fatty acids with prevalence of metabolic syndrome: The National Heart, Lung, and Blood Institute Family Heart Study. *Clin. Nutr.* 32:966–9.

Lassek, W.D., Gaulin, S.J., 2013. Maternal milk DHA content predicts cognitive performance in a sample of 28 nations. *Matern. Child Nutr.* doi: 10.1111/mcn.12060.

Laurent, G., Moe, G., et al., 2008. Long chain n-3 polyunsaturated fatty acids reduce atrial vulnerability in a novel canine pacing model. *Cardiovasc. Res.* 77:89–97.

Lecerf, J.M., 2010. Acides gras et dégénérescence liée à l'âge (DMLA). *Cahiers Nutr. Diét.* 45:144–50.

Lecerf, J.M., de Lorgeril, M., 2011. Dietary cholesterol: From physiology to cardiovascular risk. *Br. J. Nutr.* 106:6–14.

Lewis, M.D., Hibbeln, J.R., et al., 2011. Suicide deaths of active duty US military and omega-3 fatty acid status: A case control comparison. *J. Clin. Psychiatry* 72:1585–90.

Li, J., Xun, P., et al., 2013. Intakes of long-chain omega-3 (n-3) PUFAs and fish in relation to incidence of asthma among American young adults: The CARDIA study. *Am. J. Clin. Nutr.* 97:173–8.

Liang, B., Wang, S., et al., 2008. Impact of postoperative omega-3 fatty acid-supplemented parenteral nutrition on clinical outcomes and immunomodulations in colorectal cancer patients. *World J. Gastroenterol.* 14:2434–9.

Lophatananon, A., Archer, J., et al., 2010. Dietary fat and early-onset prostate cancer risk. *Br. J. Nutr.* 103, 1375–80.

Ma, L., Dou, H.L., et al., 2012. Lutein and zeaxanthin intake and the risk of age-related macular degeneration: A systematic review and meta-analysis. *Br. J. Nutr.* 107:350–9.

Maddock, J., Berry, D. J., et al., 2013. Vitamin D and common mental disorders in mid-life: Cross-sectional and prospective findings. *Clin. Nutr.* 32:758–64.

Maillard, V., Bougnoux, P., et al., 2002. n-3 And n-6 fatty acids in breast adipose tissue and relative risk of breast cancer in a case-control study in Tours, France. *Int. J. Cancer* 98:78–83.

Malpuech-Brugère, C., Mouriot, J., et al., 2010. Differential impact of milk fatty acid profiles on cardiovascular risk biomarkers in healthy men and women. *Eur. J. Clin. Nutr.* 64:752–9.

Mangialasche, F., Westman, E., et al., 2013. Classification and prediction of clinical diagnosis of Alzheimer's disease based on MRI and plasma measures of α-/γ-tocotrienols and γ-tocopherol. *J. Intern. Med.* 273:602–21.

Marangoni, F., Poli, A., 2010. Phytosterols and caardiovascular health. *Pharmacol. Res.* 61:193–9.

Martinelli, N., Consoli, L., et al., 2009. A 'desaturase hypothesis' for atherosclerosis: Janus-faced enzymes in omega-6 and omega-3 polyunsaturated fatty acid metabolism. *J. Nutrigenet. Nutrigenomics* 2:129–39.

Mayo-Wilson, E., Imdad, A., et al., 2011. Vitamin A supplements for preventing mortality, illness, and blindness in children aged under 5: Systematic review and meta-analysis. *Brit. Med. J.* 25:343:d5094.

McCann, J.C., Ames, B.N., 2008. Is there convincing biological or behavioral evidence linking vitamin D deficiency to brain dysfunction? *FASEB* 22:982-1001.

McNamara, R.K., Jandacek, R., et al., 2013. Lower docosahexaenoic acid concentrations in the postmortem prefrontal cortex of adult depressed suicide victims compared with controls without cardiovascular disease. *J. Psychiat. Res.* 47:1187–91.

Merle, B., Delyfer, M.N., et al., 2011. Dietary omega-3 fatty acids and the risk for age-related maculopathy: The Alienor study. *Invest. Ophtalmol. Vis. Sci.* 52:6004–11.

Meyer, B.J., Grenyer, B. F., et al., 2013. Improvement of major depression is associated with increased erythrocyte DHA. *Lipids* 48:863–8.

Micha, R., Mozaffarian, D., 2010. Saturated fat and cardiometabolic risk factors, coronary heart disease, stroke, and diabetes: A fresh look at the evidence. *Lipids* 45:893–905.

Michalski, M. C., Genot, C., et al., 2013. Multiscale structures of lipids in foods as parameters affecting fat bioavailability and lipid metabolism. *Prog. Lipid Res.* 52:354–73.

Migrenne, S., Cruciani-Guglielmacci, C., et al., 2011. Détection centrale des acides gras et contrôle du bilan d'énergie. *Cahiers Nutr. Diét.* 46:289–95.

Milte, C.M., Parletta, N., et al., 2012. Eicosapentaenoic and docosahexaenoic acids, cognition, and behavior in children with attention- deficit/hyperactivity disorder: A randomized controlled trial. *Nutrition* 28:670–7.

Mirmiran, P., Hosseinpour-Niazi, S., et al., 2012. Association between interaction and ratio of ω-3 and ω-6 polyunsaturated fatty acid and the metabolic syndrome in adults. *Nutrition* 28:856–63.

Mitri, J., Muraru, M.D., et al., 2011. Vitamin D and type 2 diabetes: A systematic review. *Eur. J. Clin. Nutr.* 65:1005–15.

Mobley, J.A., Leav, I., et al., 2003. Branched fatty acids in dairy and beef products markedly enhance α-methylacyl-CoA racemase expression in prostate cancer cells *in vitro*. *Cancer Epidemiol. Biomarkers Prevention* 12:775–83.

Mohr, S.B., Gorham, E.D., et al., 2012. Does the evidence for an inverse relationship between serum vitamin D status and breast cancer risk satisfy the Hill criteria? *Dermatoendocrinol.* 4:152–7.

Molto-Puigmarti, C., Plat, J., et al., 2010. FADS1 FADS2 gene variants modify the association between fish intake and the docosahexaenoic acid proportions in human milk. *Am. J. Clin. Nutr.* 91:1368–76.

Montgomery, P., Burton, J.R., et al., 2013. Low blood long chain omega-3 fatty acids in UK children are associated with poor cognitive performance and behavior: A cross-sectional analysis from the DOLAB study. PLoS ONE 8(6):e66697.

Morgan, E., 1997. The aquatic ape hypothesis. Souvenir Books, London.

Morse, N.L., 2012. Benefits of docosahexaenoic acid, folic acid, vitamin D and iodine on foetal and infant brain development and function following maternal supplementation during pregnancy and lactation. *Nutrients* 4:799–840.

Moukayed, M, Grant, W.B., 2013. Molecular link between vitamin D and cancer prevention. *Nutrients* 5:3993–4021.

Mozaffarian, D., Katan, M.B., et al., 2006. Trans fatty acids and cardiovascular disease. *New Engl. J. Med.* 354:1601–13.

Mozaffarian, D., Lemaitre, R.N., et al., 2013. Plasma phospholipid long-chain ω-3 fatty acids and total and cause-specific mortality in older adults. *Ann. Intern. Med.* 158:515–25.

Mozaffarian, D., Micha, R., et al., 2010. Effects on coronary heart disease of increasing polyunsaturated fat in place of saturated fat: A systematic review and meta-analysis of randomized controlled trials. *PloS Med.* 7:e1000252.

Narandran, R., Frankle, W.G., et al., 2012. Improved working memory but no effect on striatal vesicular monoamine transporter type 2 after omega-3 polyunsaturated fatty acid supplementation. *PloS ONE* 7(10):e46832.

Park, J.S., 2010. Astaxanthin decreased oxidative stress and inflammation and enhanced immune response in humans. *Nutr. Metab.* 7:18.

Parletta, N., Cooper, P., et al., 2013. Effects of fish oil supplementation on learning and behaviour of children from Australian Indigenous remote community schools: A randomised controlled trial. *Prost. Leukotrienes Essent. Fatty acids* 89:71–9.

Patel, P.S., Sharp, S.J., et al., 2010. Fatty acids measured in plasma and erythrocyte-membrane phospholipids and derived by food-frequency questionnaire and the risk of new-onset type 2 diabetes: A pilot study in the European prospective investigation into cancer and nutrition (EPIC)-Norfolk cohort. *Am. J. Clin. Nutr.* 92:1214–22.

Pierre, A.S., Minville-Walz, M., et al., 2013. Trans-10, cis-12 conjugated linoleic acid induced cell death in human colon cancer cells through reactive oxygen species-mediated ER stress. *Biochim. Biophys. Acta* 1831:759–68.

Pot, G.K., Majsak-Newman, G., et al., 2009. Fish consumption and markers of colorectal cancer risk: A multicenter randomized controlled trial. *Am. J. Clin. Nutr.* 90: 354–61.

Pottala, J.V., Yaffe, K., et al., 2014. Higher RBC EPA + DHA corresponds with larger total brain and hippocampal volumes. *Neurology* 82:1–8.

Poudyal, H., Panchal, S.K., et al., 2011. Omega-3 fatty acids and metabolic syndrome: Effects and emerging mechanisms of action. *Prog. Lipid Res.* 50:372–87.

Presse, N., Belleville, S., 2013. Vitamin K status and cognitive function in healthy older adults. *Neurobiol. Aging* 34:2777–83.

Psaltopoulou, T., Kosti R.I., et al., 2011. Olive oil intake is inversely related to cancer prevalence: A systematic review and a meta-analysis of 13800 patients and 23340 controls in 19 observational studies. *Lipids Health Dis.* 10:127.

Puska, P., 2009. Fat and heart disease: Yes we can make a change—The case of North Karelia (Finland). *Ann. Nutr. Metabol.* 54(suppl 1):33–8.

Quinn, J.F., Raman, R., et al., 2010. Docosahexaenoic acid supplementation and cognitive decline in Alzheimer disease: A randomized trial. *JAMA* 304:1903–11.

Ramsden, C.E., Zamora, D., et al., 2013. Use of dietary linoleic acid for secondary prevention of coronary heart disease and death: Evaluation of recovered data from the Sydney Diet Heart Study and updated meta-analysis. *Brit. Med. J.* 346:e8707.

Ran-Ressler, R.R., Sim, D., et al., 2011. Branched chain fatty acid content of United States retail cow's milk and implications for dietary intake. *Lipids* 46:569–76.

Ravnskov, U., 1995. Quotation bias in reviews of the diet-heart idea. *J. Clin. Epidemiol.* 48:713–9.

Renaud, S., de Lorgeril, M., et al., 1995. Cretan Mediterranean diet for prevention of coronary heart disease. *Am. J. Clin. Nutr.* 61:1360S–7S.

Salisbury, A.C., Amin, A.P., et al., 2011. Predictors of omega-3 index in patients with acute myocardial infarction. *Mayo Clin. Proc.* 86:626–32.

Samieri, C., Féart, C., et al., 2011. Olive oil consumption, plasma oleic acid, and stroke incidence: The Three-City Study. *Neurology* 77:418–25.

Sanders, T., Filippou, A., et al., 2011. Palmitic acid in the sn-2 position of triacylglycerols acutely influences postprandial lipid metabolism. *Am. J. Clin. Nutr.* 94:1433–41.

Sarris, J., Mischoulon, D., et al., 2012. Omega-3 or bipolar disorder: Meta-analyses of use in mania and bipolar depression. *J Clin Psychiatry* 73, 81–6.

Schmidt, J., Liebscher, K., et al., 2011. Conjugated linoleic acids mediate insulin release through islet G protein-coupled receptor FFA1/GPR40. *J. Biol. Chem.* 286:11890–4.

Schulz, M., Hoffmann, K., et al., 2008. Identificcation of a dietary pattern characterized by high-fat food choices associated with increased risk of breast cancer: The European Prospective Investigation into Cancer and Nutrition (EPIC)-Postdam study. *Brit. J. Nutr.* 100:942–6.

Sekikawa, A., Miura, M., et al., 2014. Long chain n-3 polyunsaturated fatty acids and incidence rate of coronary artery calcification in Japanese men in Japan and white men in the USA: Population based prospective cohort study. *Heart* 100:569–73.

Shikany, J.M., Vaughan, L.K., et al., 2010. Is dietary fat "fattening"? A comprehensive research synthesis. *Crit. Rev. Food Sci. Nutr.* 50:699–715.

Shimokawa, T., Moriuchi, A., et al., 1988. Effect of dietary alpha-linolenate/linoleate balance on mean survival time, incidence of stroke and blood pressure of spontaneously hypertensive rats. *Life Sci.* 43:2067–75.

Simopoulos, A.A., 2010. Genetic variants in the metabolism of omega-6 and omega-3 fatty acids: Their role in the determination of nutritional requirements and chronic disease risk. *Exp. Biol. Med.* 235:785–95.

Siri-Tarino, P.W., Sun, Q., et al., 2010. Meta-analysis of prospective chohort studies evaluating the association of saturated fat with cardiovascular disease. *Am. J. Clin. Nutr.* 91:535–46.

Song, Y., You, N.C., et al., 2013. Intake of small-to-medium-chain saturated fatty acids is associated with peripheral leukocyte telomere length in postmenopausal women. *J. Nutr.* 143:907–14.

Souied, E.H., Delcourt, C., et al., 2013. Oral docosahexaenoic acid in the prevention of exudative age-related macular degeneration: The Nutritional AMD Treatment 2 study. *Ophthalmology* 120:1619–31.

Spence, J.D., Jenkins, D. J., et al., 2012. Egg yolk consumption and carotid plaque. *Atherosclerosis* 224:469–73.

Steinberg, D., 2002. Atherogenesis in perspective: Hypercholesterolemia and inflammation as partners in crime. *Nature Med.* 8:1211–7.

Stolzenberg-Solomon, R.Z., Sheffler-Collins, S., et al., 2009. Vitamin E intake, alpha-tocopherol status, and pancreatic cancer in a cohort of male smokers. *Am. J. Clin. Nutr.* 89:584–91.

St.-Onge, M.P., Bosarge, A., 2008. Weight-loss diet that includes consumption of medium-chain triacylglycerol oil leads to a greater rate of weight and fat mass loss than does olive oil. *Am. J. Clin. Nutr.* 87:621–6.

Sublette, M.E., Ellis, S.P., et al., 2011. Meta-analysis of the effects of eicosapentaenoic acid (EPA) in clinical trials in depression. *J. Clin. Psychiatry* 72:1577–84.

Sydenham, E., Dangour, A.D., et al., 2012. Omega 3 fatty acid for the prevention of cognitive decline and dementia. *Cochrane Database Syst. Rev.* 6:CD005379.

Taguchi, H., Watanabe, H., et al., 2000. Double-blind controlled study on the effects of dietary diacylglycerol on postprandial serum and chylomicron triacylglycerol responses in healthy humans. *J. Am. Coll. Nutr.* 19:789–96.

Tan, Z.S., Harris, W.S., et al., 2012. Red blood cell omega-3 fatty acid levels and markers of accelarated brain aging. *Neurology* 78:658–64.

Thomas, G.N., Hartaigh, B.O., et al., 2012. Vitamin D levels predict all-cause and cardiovascular disease mortality in subjects with the metabolic syndrome: The Ludwigshafen Risk and Cardiovascular Health (LURIC) Study. *Diabetes Care* 35:1158–64.

Titova, O.E., Sjögren, P., et al., 2013. Dietary intake of eicosapentaenoic and docosahexaenoic acids is linked to gray matter volume and cognitive function in elderly. *Age* 35:1495–505.

Wakai, K., Tamakoshi, K., et al., 2005. Dietary intakes of fat and fatty acids and risk of breast cancer: A prospective study in Japan. *Cancer Sci.* 96:590–9.

Whalley, L.J., Deary, I.J., et al., 2008. n-3 Fatty acid erythrocyte membrane content, APOE epsilon4, and cognitive variation: An observational follow-up study in late adulthood. *Am. J. Clin. Nutr.* 87:449–54.

Woyengo, T.A., Ramprasath, V.R., et al., 2009. Anticancer effects of phytosterols. *Eur. J. Clin. Nutr.* 63:813–20.

Wu, J.H., Lemaitre, R.N., et al., 2012. Association of plasma phospholipid long-chain omega-3 fatty acids with incident atrial fibrillation in older adults: The cardiovascular health study. *Circulation* 125:1084–93.

Xin, W., Wei, W., et al., 2013. Short-term effects of fish-oil supplementation on heart rate variability in humans: A meta-analysis of randomized controlled trials. *Am. J. Clin. Nutr.* 97:926–35.

Yamagishi, K., Iso, H., et al., 2010. Dietary intake of saturated fatty acids and mortality from cardiovascular disease in Japanese: The Japan collaborative cohort study for evaluation of cancer risk (JACC) study. *Am. J. Clin. Nutr.* 92:759–65.

Yang, Z., Liu, S., et al., 2000. Induction of apoptotic cell death and in vivo growth inhibition of human cancer cells by a saturated branched-chain fatty acid, 13-methyltetradecanoic acid. *Cancer Res.* 60:505–9.

Zaalberg, A., Nijman, H., et al., 2010. Effects of nutritional supplements on aggression, rule-breaking, and psychopathology among young adult prisoners. *Aggr. Behav.* 36:117–26.

Abbreviations

ADA: American Dietetic Association

ADHD: attention deficit hyperactivity disorder

AFSSA: Agence française de sécurité sanitaire des aliments (devenue ENSES)

AI: adequate intake

AICR: American Institute for Cancer Research

AJR: Apports journaliers recommandés

ALIENOR: Etude: antioxydants, lipides essentiels, nutrition et maladies oculaires

AMD: age-related macular degeneration

AMT: Apports maximaux tolérables

ANC: Apports nutritionnels conseillés

ANR: Apports nutritionnels recommandés

ANREF: Apports nutritionnels de référence

ANSES: Agence française de sécurité sanitaire de l'alimentation (anciennement AFSSA)

APOE4: Apolipoprotein E allelic form 4

AREDS: Age-Related Eye Disease Study

AS: Apports suffisants

BME: Besoins moyens estimés

BMI: body mass index

CAC: Codex alimentarius commission

CETIOM: Centre technique interprofessionnel des oléagineux métropolitains

CIQUAL: Centre d'information sur la qualité des aliments (ANSES)

CIRC: Centre international de recherche sur le cancer

CLA: conjugated linoleic acid

CM: chylomicron

CNERNA: Centre national d'études et de recommandations sur la nutrition et l'alimentation

CNIEL: Centre national interprofessionnel de l'économie laitière

CNRS: Centre national de la recherche scientifique

COGINUT: Etude cognition et nutrition

COX-2: cyclooxygenase-2

CRP: C-reactive protein

CSHPF: Conseil supérieur d'hygiène publique de France

DAG: diacylglycerol

DHA: docosahexaenoic acid (22:6 n-3)

DMLA: Dégénérescence maculaire liée à l'âge

DRI: dietary reference intake

E3N: Étude épidémiologique de femmes de la MGEN, composante française de l'EPIC

EAR: estimated average requirement

EARNEST: Early Nutrition Programming

EFSA: European Food Safety Authority

EPA: eicosapentaenoic acid (20:5 *n*-3)
EPIC: European Prospective Investigation into Cancer and Nutrition
ESPGHAN: European Society for Pediatric Gastroenterology, Hepatology and Nutrition
EURAMIC: European Community Multicenter Study on Antioxidants, Myocardial
      Infarction, and Breast Cancer
FAO: Food and Agriculture Organization
FAOSTAT: Statistics division (FAO)
FDA: Food and Drug Administration (the United States)
GMC: genetically modified crop
GNP: Gross National Product
GRAS: substances generally recognized as safe (the United States)
HDL: high-density lipoprotein
IDL: Lipoprotéines de densité intermédiaire
IFN-γ: interferon-γ
IMC: Indice de masse corporelle (poids en kg/taille^2) (= BMI)
INCa: Institut national du cancer
INRA: Institut national de la recherche agronomique
INSEE: Institut national de la statistique et des études économiques
INSERM: Institut national de la recherche médicale
INVS: Institut national de veille sanitaire
IOM: Institute of Medicine (the United States)
IRDES: Institut de recherche et documentation en économie de la santé
ISSFAL: International Society for the Study of Fatty Acids and Lipids
IUB: International Union of Biochemistry
IUPAC: International Union of Pure and Applied Chemistry
LCT: long-chain triacylglycerol
LDL: low-density lipoprotein
MCI: mild cognitive impairment
MCT: middle-chain triacylglycerol
MEDHEA: Etude française "Mediterranean diet and health"
MMSE: mini mental state evaluation
NAS: National Academy of Sciences (the United States)
NIH: National Institute of Health (the United States)
OMS: Organisation mondiale de la santé (WHO)
PAQUID: Etude de personnes agées QUID
PCB: polychlorobiphenyl
PCDD: polychlorodibenzo-*p*-dioxine
PCDF: polychlorodibenzofurane
PERILIP: Perinatal Lipid Nutrition (dietary recommendations for pregnant women)
PNNS: Programme National Nutrition-Santé
PPARγ: Récepteur γ activé par les proliférateurs de peroxysomes
PSA: prostate serum antigen
RDA: recommended dietary allowance
SACN: Scientific Advisory Committee on Nutrition (United Kingdom)
SAD: seasonal affecting disorder
SCT: short-chain triacylglycerol

SUVIMAX: Etude «supplémentation en vitamines et minéraux antioxydants»
TDAH: Troubles du déficit de l'attention avec hyperactivité
TNF-α: tumor necrosis factor-α
TRANSFAIR: Etude européenne multicentrique
UL: tolerable upper intake level
USDA: United States Department of Agriculture
VLDL: very-low-density lipoprotein
WAPM: World Association of Perinatal Medicine
WHO: World Health Organization (OMS)

Index